THE GULF: CHALLENGES OF THE FUTURE

THE GULF: CHALLENGES OF THE FUTURE

THE EMIRATES CENTER FOR STRATEGIC STUDIES AND RESEARCH

THE EMIRATES CENTER FOR STRATEGIC STUDIES AND RESEARCH

The Emirates Center for Strategic Studies and Research (ECSSR) is an independent research institution dedicated to the promotion of professional studies and educational excellence in the UAE, the Gulf and the Arab world. Since its establishment in Abu Dhabi in 1994, ECSSR has served as a focal point for scholarship on political, economic and social matters. Indeed, ECSSR is at the forefront of analysis and commentary on Arab affairs.

The Center seeks to provide a forum for the scholarly exchange of ideas by hosting conferences and symposia, organizing workshops, sponsoring a lecture series and publishing original and translated books and research papers. ECSSR also has an active fellowship and grant program for the writing of scholarly books and for the translation into Arabic of work relevant to the Center's mission. Moreover, ECSSR has a large library including rare and specialized holdings, and a state-of-the-art technology center, which has developed an award-winning website that is a unique and comprehensive source of information on the Gulf.

Through these and other activities, ECSSR aspires to engage in mutually beneficial professional endeavors with comparable institutions worldwide, and to contribute to the general educational and academic development of the UAE.

The views expressed in this book do not necessarily reflect those of the ECSSR.

Published in 2005 by
The Emirates Center for Strategic Studies and Research
PO Box 4567, Abu Dhabi, United Arab Emirates

E-mail: pubdis@ecssr.ac.ae
pubdis@ecssr.com

Website: http://www.ecssr.ac.ae
http://www.ecssr.com

Distributed by The Emirates Center for Strategic Studies and Research

ISBN 9948-00-705-0 hardback edition
ISBN 9948-00-704-2 paperback edition

Contents

Global Developments and the Gulf

Iran and Gulf Security

Iraq and Gulf Stability

List of Tables

Abbreviations and Acronyms

ABC	American Broadcasting Company
ABM	anti-ballistic missile
AHDR	Arab Human Development Report
ARAMCO	Arab American Oil Company
BBC	British Broadcasting Corporation
CBS	Columbia Broadcasting System
CEDAW	Convention on the Elimination of all Forms of Discrimination against Women
CIA	Central Intelligence Agency
CIC	Coordination International Commission
CONUS	Continental United States
CPA	Coalition Provisional Authority
CNN	Cable News Network
C-SPAN	Cable-Satellite Public Affairs Network
DoD	Department of Defense
ECSSR	The Emirates Center for Strategic Studies and Research
ECOWAS	Economic Commission of West African States
EU	European Union
GC	Governing Council
GCC	Gulf Cooperation Council
GDFE	General Directorate for Female Education
GDP	gross domestic product
GMEI	Greater Middle East Initiative
GNP	gross national product
HEU	high enriched uranium
IAEA	International Atomic Energy Agency
ICG	International Crisis Group
ICISS	International Commission on Intervention and State Sovereignty
ILO	International Labor Organization
IISS	International Institute for Strategic Studies

IMF	International Monetary Fund
IOR-ARC	Indian Ocean Rim Association for Regional Cooperation
ISAF	International Security Assistance Force
LEU	low enriched uranium
MEPI	Middle East Partnership Initiative
MLSA	Ministry of Labor and Social Affairs
NAM	Non-Aligned Movement
NBA	National Basketball Association
NBC	National Broadcasting Company
NAFTA	North American Free Trade Agreement
NATO	North Atlantic Treaty Organization
NGO	non-governmental organization
NPT	Non-Proliferation Treaty
ORHA	Office of Reconstruction and Humanitarian Assistance
QRCS	Qatar Red Crescent Society
RDJTF	Rapid Deployment Joint Task Force
SSM	surface-to-surface missile
UAE	United Arab Emirates
UN	United Nations
UNDP	United Nations Development Program
UNESCWA	United Nations Economic and Social Commission for Western Asia
UNIFEM	United Nations Development Fund for Women
UNSCOM	United Nations Special Commission for Iraq
USCENTCOM	US Central Command
USMC	United States Marine Corps
WMD	weapons of mass destruction
WTO	World Trade Organization

FOREWORD

The Arabian Gulf faces multifarious challenges arising from international, regional and internal developments. In the past, international pressures have often stemmed from the region's possession of abundant energy resources, its strategic location and special economic significance. However, the region is now feeling the deeper impact of global developments on several fronts. In economic terms, the Gulf states are contending with the sweeping effects of globalization and the paradigm shift towards knowledge-based economies. At the social level, these new developments have also introduced significant changes that are transforming long-held cultural traditions and conventions.

Meanwhile, the international spotlight has remained fixed on the Middle East since the tumultuous events of September 11, 2001, followed swiftly by a global anti-terrorism campaign and the US-led military operations in Afghanistan. In 2003, events in Iraq captured world attention, culminating in a controversial US-led military campaign that enforced regime change, setting the country on a gradual path to democracy but leaving behind a trail of bloodshed, destruction, violence and fierce resistance. All of these catastrophic developments have thrown up unprecedented political and security challenges for Iraq's neighboring states and for the Arabian Gulf region as a whole.

With a view to identifying and examining the pressing issues confronting the Gulf, an elite international gathering of decision makers, renowned experts and reputed scholars focused their attention on the subject of *The Gulf: Challenges of the Future*, at the ECSSR Ninth Annual Conference, held from January 11-13, 2004 in Abu Dhabi, UAE.

The insightful conference presentations compiled in this volume shed light on significant global issues such as shifting security parameters, redefining of government responsibilities, the wider implications of globalization, investment in knowledge-based growth, the challenges of representative governance and the powerful reach and influence of the media.

In its particular focus on Gulf security and regional developments, the book also examines closely the future impact of a potentially nuclear-capable Iran and a weakened, democratic Iraq on Gulf security and stability, and the widening scope of the GCC as a forum for effective regional cooperation and economic integration. In addition, the book explores the expanding role played by women in the Gulf as they avail of opportunities for higher education and wider participation in the workforce while moving gradually towards political empowerment.

ECSSR would like to express its appreciation to all the eminent speakers for their active participation in the conference and for sharing their perspective on important global and regional developments. A word of thanks is also due to ECSSR editor Mary Abraham for coordinating the publication of this book.

Jamal S. Al-Suwaidi Ph.D.
Director General
ECSSR

INTRODUCTION

Challenges Confronting the Gulf Region: An Overview

With the advent of the 21st century, the Arab Gulf region has undergone a transformation phase. In the aftermath of the events of September 11, 2001, it has witnessed considerable changes in its geopolitical environment and in its political realities. The most notable of these changes are the tendency to launch direct external military interventions (as demonstrated by the occupation of Iraq in 2003), foreign pressures (mainly from the United States) exerted upon the region's states to join the war against terrorism and to implement radical internal reforms and changes.

The Gulf Cooperation Council (GCC) states in particular are confronting certain circumstances and developments that pose several internal and external challenges. The first involves encouraging ruling regimes to respond positively to the aspirations of the region's peoples for political participation and for moves towards democracy; fighting terrorism and uprooting ideologies of extremism and violence; boosting the GCC system, especially in the fields of economic integration and security/defense cooperation; rectifying the imbalance in the demographic structure of the region's states; satisfying the requirements arising from escalating population growth rates and the consequent need for expanding the scope of education and employment; and promoting the process of women's empowerment.

External challenges lie in addressing and dealing with the effects and consequences of the war against terrorism, the occupation of Iraq, the collapse of the Arab-Israeli peace process, and the need to initiate a comprehensive and collective approach towards the regional security system.

Conscious of the grave challenges confronting and affecting the future of the Arab Gulf region at the beginning of the new millennium, The Emirates Center for Strategic Studies and Research conducted its Ninth Annual Conference under the theme of *The Gulf: Challenges of the Future* from January 11-13, 2004 in Abu Dhabi, UAE.

A group of highly distinguished international experts, decision-makers and academic researchers, who participated in the Conference, discussed the recent and most important regional geopolitical changes and developments, the resulting challenges and their impact upon the future of the GCC states. The participants also sought to examine and pinpoint the policies and strategies that would assist the states of the region to meet and cope with these challenges successfully and effectively.

In his keynote address, His Highness Sheikh Hamed Bin Zayed Al Nahyan, Member of the Abu Dhabi Executive Council and Chairman of the Abu Dhabi Department of Economy, notes that the Gulf states are facing grave threats, which compel them to take cognizance of the challenges and opportunities posed by the new world order so as to develop their present status and lay the foundations of their future, as the world today moves towards mutual interdependence among nations, propelled by the information and communications revolution, the growing movement of capital and the shrinking margins for political maneuvering under the unipolar international system.

The current changes in the perceptions of security and state sovereignty were viewed by the conference as one of the universal shifts that have affected the nations of the world in general and Gulf states in particular, as security criteria have changed since the beginning of the 21st century. According to Hon. Gareth Evans, international security is facing three major problems: first, the growth of international terrorist organizations with deeply frightening agendas and highly disturbing capacity; second, the waning effectiveness of the treaty regimes trying to achieve the non-proliferation of weapons of mass destruction; third, the continuing existence and emergence of too many failed, failing, fragile or simply overwhelmed states.

While the absence of a credible political life in most parts of the region is not necessarily bound to produce violent conflicts in the region, or cause the collapse of its states, it would almost certainly affect its longer-term stability. The ineffectual political representation, absence of popular participation and inadequate government responsiveness; the lack of genuine public accountability and transparency; weakened political legitimacy and economic underdevelopment often translate into inadequate mechanisms to express and channel public discontent, creating the potential for extra-institutional protests. These in turn, may assume more violent forms, especially at a time when regional developments have polarized and radicalized public opinion.

As regards sovereignty, the present international context is a globalized and even more interdependent environment, where states are less able than ever before to insulate themselves from events around them, and for most states their capacity to completely determine their own destiny is more limited than it has ever been.

The Rt. Hon. John Major predicts that the global economy will be a principal factor in the evolution of the concept of state sovereignty, and that the nature of government itself is changing. He notes that while the power of government diminishes and absolute sovereignty of states declines, the power and influence of markets and multinationals are increasingly greater. National sovereignty is undermined by multi-state institutions – the European Union for example, as well as the International Monetary Fund, World Bank and World Trade Organization – all of which make decisions and undertake functions that were once the sole prerogative of the nation state.

Nevertheless, the global economy will not be the "assassin of state sovereignty, and the death of government is not imminent." Yet, government must maintain a balance between encouraging investment and free trade, on one hand, and fulfilling its legitimate obligations to its citizens, on the other.

On the question of the GCC policy options to deal with the challenges posed by global economy, His Excellency Sheikh Hamad Bin Jassim Bin Jabr Al Thani, First Deputy Prime Minister and Minister of Foreign Affairs of the State of Qatar, is of the opinion that for these states to meet the many challenges they face in the context of the global economy they must move forward on the path of desired economic integration. This may be achieved through the establishment of the Custom Union; the adoption of a unified Gulf currency by the year 2007 and by forming the necessary institutions for this purpose; the creation of the Gulf common market by the year 2010; and by putting into effect the Unified Economic Agreement.

According to H. E. Sheikh Hamad Al Thani, the policy options available to GCC states include: the effective participation in the negotiation rounds between the GCC states and the European Union (EU) on establishing free trade between the countries of the two groups; drafting the required strategies aimed at penetrating the markets of individual countries and regional blocs; eliminating any obstacles blocking the flow of foreign direct investments; and maintaining moderate positions in the context of energy diplomacy with regard to issues such as production and price stability.

The fundamental prerequisites needed to attain the desired level of development and modernization and build up the region's economic and social edifice include the consolidation of the climate of security and stability, the adoption of dialogue and understanding as a pillar to shape the network of regional and bilateral relationships, the settlement of the area's crises and conflicts through peaceful means, and achieving good neighborliness. Efforts to accomplish these goals should not be restricted to the GCC level but rather must be expanded to include the Middle East as a whole since both areas complement each other.

In the same context, His Highness Sheikh Abdullah Bin Zayed Al Nahyan, Minister of Information and Culture of the UAE, states that the position that the Gulf region occupies in the world can be viewed through three major issues: the imbalanced economic role played by the region; the acute difference at the technological and knowledge levels between the region and the developed world; and the variation in the standards of overall development between the region and the world as a whole.

Several indications can be cited to highlight the repercussions of the restricted economic role of the Gulf region, such as: the decline of the Gulf non-oil exports to around 0.5 per cent of world's total exports; the growth rate of the region's gross national product (GNP), which does not match even that of a country like Malaysia; the one-sided nature of Gulf economies, which have failed to become an attractive location for foreign direct investments. None of the Gulf states have appeared on the list of the top ten developing countries that have led in attracting such investments.

As for the difference at the technological and knowledge levels between the region and the developed world, the states of the region are still importing ready-made technologies without taking the trouble to localize these technologies. This is also reflected in the declining level of expenditures allocated to scientific research by Gulf states to less than 0.5 per cent of gross domestic product (GDP), compared to around 4 per cent in the developed countries.

On the other hand, since most of the Gulf states did not adopt sustainable developmental strategies until the end of last century, and also face the problem of demographic imbalance and the dominance of a foreign labor force, a huge gap has emerged between the Gulf and the world in terms of the overall development standards. The average GDP per capita income in the Gulf states is estimated at US$ 9000 per annum, putting these states on the list of medium income countries.

However, there are a number of other indicators that help to determine the position the region occupies in the world, such as the increase in the number of youth to over 50 per cent of the total population indicating a tangible increase in the qualified workforce; the new and expanding part that women are playing in the economic fields as their contribution has risen to about one third of the overall economic activities in the region even if they do not yet have the full opportunity to play a greater role in the political and social fields. In addition, the Gulf states can capitalize on the intellectual assets they possess, utilizing them in the technological and information fields, especially since these states have allocated huge investments to further develop their professional workforce. Moreover, some of these states, including the UAE, enjoy enormous opportunities for further fostering the potential for creating, designing and developing computer programs and information systems. The UAE experience in adopting free market policies has resulted in diversifying its economic base and reducing dependence on oil as the major source of income.

The sincere adoption of the values of freedom in its general sense: transparency, openness and participation will pave the way for setting up the proper creative environment that will help to bridge the existing gap between the Gulf region and the developed world.

Focusing on another topic, His Excellency Dr. Muhammad Abdul Ghaffar, Minister of State for Foreign Affairs of the Kingdom of Bahrain, discusses the dynamics of change and stability in the Gulf regional system. He notes that since the Iraqi invasion of Kuwait in 1990, the Gulf region has witnessed major political, military and security developments. The effects of these developments were clearly and substantively reflected in Gulf, Arab and regional relations and also impacted on the effectiveness of the Gulf and Arab security systems.

The perils of instability and flaws in the Gulf regional security order are not limited to external threats as they had been during the last decades of the past century. Rather they may be ascribed to domestic pressures as manifested in different types of failure related to internal political structures, the immature culture of political participation, the insufficient growth of civil societies, and the aggravation of terrorist activities and domestic violence. Those pressures could even multiply and worsen through the weakness of the economic infrastructure and the impact of certain social changes, such as population growth, expansion of the education base, and the imbalanced demographic structure.

H.E. Dr. Abdul Ghaffar concludes that the prospects of security and stability in the Gulf region will be affected for many years to come by the approaches adopted in addressing vital regional issues such as the future of the foreign military presence in Iraq; the prospects for the system of government in Iraq; the settlement of the Arab-Israeli conflict; the extent of political modernization, democratic transformation and social consensus in the Gulf region.

Dr. Shamlan Al-Issa seeks to examine the political impact of globalization on the Gulf states, referring to the ambiguous policies adopted by Gulf states towards globalization and its different implications. He also underlines the differences of opinion within the Arab and Gulf elites as to the definition of globalization and general approaches towards this phenomenon. However, there is partial consensus on the fact that globalization will lead to changes not only in the social structure but also in the social and cultural values of the peoples belonging to these societies.

According to Dr. Al-Issa, the political effects of globalization on the Gulf states lie in the erosion of the principles of national sovereignty leading to a change in the role of the government in the light of market economies, international interdependence and universal culture. In the Gulf states, the model of the *rentier* state is now suffering a setback that might lead ultimately to its demise.

Among the political implications of globalization are a number of principles relating to human rights and respect for political and intellectual choices. Hence, the Gulf states are now subject to internal and external pressures to effect democratic reform and increase popular participation, thus driving some of these states – such as the Kingdom of Bahrain and the State of Qatar – to take several steps in this direction.

In Dr. Al-Issa's view, the obstacles hindering political reforms in the Gulf states include: the absence of the will to reform among the political regimes and ruling elites; the lack of organized and efficient political opposition; the fragility of civil society institutions; the weakness of institutionalism; and the absence of democratic culture and the prevalence of a culture of authoritarianism.

Within the same context, Dr. Benjamin Barber examines the challenges of bringing about democratic reforms in non-Western states, including the Gulf states. He stresses that democracy cannot and should not be confused with Americanization or Westernization. He believes that a number of myths have impeded the evolution of democracy throughout the developing world and specifically in the Middle East.

The impressions created by these myths imply that there is one democracy and not several; that it is a form of government that can be secured quickly; that peoples can be 'liberated' and liberty can be imposed from the outside; that democracy constitutes little more than elections; and that it means outcomes acceptable to outsiders rather than a process by which the will of citizens is represented.

In Dr. Barber's judgment, the very conditions specific to the Arab world that would help nurture democracy can also be obstacles that impede its chances of success. These include the tradition of tribal leadership based on one-man or one-family; the acquisition of immense oil wealth; and the reality of a deeply Islamic society and history. The features associated specifically with the Gulf region are the small-scale of its societies and the tradition of consultative rights inherent in traditional modes of decision-making.

Wealth, for instance, could be an essential factor to achieve democracy, but it remains a double-edged weapon because it can create both prosperity and inequality at the same time. Great wealth may generate huge economic gains and benefits, but it can also be very destructive as it would provide more room for moral corruption and would also invite many other vices.

Dr. Barber concludes that democratic changes in the Arab Gulf region, indeed throughout the Islamic Middle East, are possible, feasible and desirable, since Islam is less of a restraint to democratization. Democracy, however, is neither a gift from the West nor is it dependent on Western institutions, since it can be realized in many different cultural variations and forms. Democracy will belong to those who want it and strive to make it possible.

The effects of US foreign policy in the Gulf region are also discussed. In this respect, General Anthony Zinni, USMC (Ret.) presents a military perspective, stating that the objectives outlined by both the US government and the Central Command (CENTCOM) seem consistent and interrelated. Maintaining the US presence to ensure the stability of the region, securing uninterrupted access to oil resources, and safeguarding the freedom of navigation are among the goals set by the CENTCOM. Besides, for many years, furthering the security of US regional friends and allies has been a major objective. Ties with the military organizations of the Gulf states have always remained remarkably strong even during times of political tension.

General Zinni considers that the US and each of the Gulf states have differing perspectives regarding the threats encountered by the region. Therefore, further bilateral consultations are needed to arrive at a general consensus in this matter based upon mutual needs.

Since the United States is passing through a transformation phase at the start of a new decade in terms of adjusting its policies, the time is ripe for forging clear strategies regarding the US military presence and cooperation with its allies in the Arab Gulf region.

General Zinni holds the view that the US military was able to exercise greater influence on the US foreign policy towards the region. He also feels that the United States has failed to project a proper image in the Gulf region, which is one of the reasons behind anti-US sentiments in the Gulf. He also stresses the need for a collective approach towards Gulf security, the absence of which has created several problems for the United States in the region.

As Iran plays a key role in the region's security and stability, the Conference devoted an entire section to examining this matter. Expressing an Iranian perspective, His Excellency Mr. Mohammed Ali Abtahi, former Vice President for Legal and Parliamentary Affairs of the Islamic Republic of Iran, deals with matters that impact on political stability in the Gulf region. According to him the continuation of the present situation in the region could result in a number of dangerous developments. The foremost of these are: the continuing foreign intervention in the region due to the lack of a collective security model; widening gap in the social fabric, mainly between the political regimes of the region and their peoples; and the growth of extremism.

According to Mr. Abtahi, one of the main goals of the foreign policy of the Islamic Republic of Iran is to achieve cooperation and engagement with all the countries of the region to guarantee collective security. He adds that the principle of *détente* and confidence-building remains one of the fundamental foreign policy measures followed during recent years. He notes that while the Islamic Republic of Iran continues to reject the policy of militarization and hegemony by powers from beyond the region, it is also firm about rejecting extremist movements that incite people to engage in acts of violence.

Discussing the same subject, Dr. Mahmood Sariolghalam examines Iran's regional security approaches. He expresses the opinion that Iran's security policy at the regional level is the product of a political struggle between two schools of thought: the revolutionary and the internationalist. Both sides enjoy social support and possess institutional and economic backing, each with members who belong to both conservative and reformist clerical circles.

The revolutionary group believes in the preservation of the ideological order, the clerical establishment, state control of culture and considers the Western world as Iran's enemy. The internationalist group, however, does not dissociate the domestic structure from global dynamics, national economy from foreign policy and national security from economic development. The former group believes that Iran should focus on its internal agenda and maintain a calculated distance from the international community. The latter group promotes national economic development and believes that Iran should join the World Trade Organization (WTO) and become a normal member of the international community. The former category asserts that Iran's security is guaranteed when it dissociates itself from the economic and political impositions of the international capitalist system led by the United States. The latter category presumes that Iran's national security stems from its economic interdependence with the international community and that Iran should focus on producing national wealth, employing economic diplomacy and soft politics. Whereas the threat perceptions of the former are fundamentally military and existential, the latter group perceives economic, social and soft issues as constituting basic threats to the country.

The Iranian security doctrine is also motivated by the threats implied in certain positions taken by the US and Israel, as well as the pressures exercised by the Israeli lobby which is highly influential in shaping the US policy towards Iran.

Dr. Sariolghalam suggests that certain recent events – the attacks of September 11, 2001, the war on terrorism, the war in Afghanistan and the war in Iraq – have prompted Iran to modify its behavior and change its tactical moves. Hence, there has been effective cooperation between Teheran and Washington on such issues as Afghanistan and rebuilding its future in the post-war period. Certainly, it can be said that Iran is neither considering nor seeking any military confrontation with the United States.

As a result of a combination of generational shifts, socio-economic changes, nationalist-oriented public opinion and a new regional geopolitical map, the Iranian national security doctrine will be increasingly based on its economic viability rather than ideological premises. In this context, the security policies of Iran at the regional level will be increasingly based on trade, economic interdependency and cultural exchange. Such a development will be guaranteed when the Islamic Republic of Iran moves from a security phenomenon to an economic-cultural phenomenon.

Dr. Jerrold Green highlights Iran's regional policies from a western perspective, stressing the fact that despite the consensus among western powers on the desirability of strengthening the Gulf region's security order, these powers hold different and diverse attitudes and policies towards the region's key issues. Significant disagreements are witnessed between the United States and its European partners concerning all major Middle East issues, including the Arab-Israeli conflict, the future of Iraq and Iran's role at the regional and global levels.

Generally speaking, western powers have failed to reach consensus on a common policy towards Iran. While the Islamic Republic of Iran was at the center of the United States policy of Dual Containment and subsequently included on President Bush's "axis of evil" list, the Europeans seemed to prefer the "carrot" to the "stick," opting for policies of constructive engagement with Iran. In other words, it has become obvious that the European states believe they are able to exercise their influence on Iran through engagement rather than exclusion and isolation, which has been the preferred US option despite the fact that this isolation has been punctuated by periods of *de facto* and even overt collaboration with Iran, such as the Bonn Conference held in November/December 2001, to lay the foundations of the future of Afghanistan.

Dr. Green underlines the linkage between strengthening Iran's regional role and influence, on one hand, and improving its relationships with the West in general and the US in particular, on the other. He opines that two main issues – namely, Iran's nuclear program and its support for international terrorism – have played an important role in straining Iranian-Western relations. The settlement of these questions will contribute to the improvement of these relations and boost Iran's influence in the region. The issue of Iran's nuclear program has provided grounds for consensus amongst the Western states, which generally disagree on all other aspects of Iranian foreign policy. Undoubtedly, this issue, more than any other, has undermined Western relations with Iran. This includes not only the United States, but also a number of Western European powers as well.

The establishment of a more constructive and positive relationship between Iran and the West, in general, and the US, in particular, would contribute to greater Gulf security, to more effective nation-building processes in both Iraq and Afghanistan, and help to mitigate the negative effects of a major source of stability that adversely affects the whole region. Disagreements between Iran and the West can be ameliorated partly in the Western capitals themselves. If the West is able to speak with one voice on the current major Middle East issues, particularly as they relate to Iran, significant steps could be taken towards addressing some of the most debilitating problems of this volatile region.

In tackling the Iranian nuclear program as one of the highly sensitive issues at the regional and international levels, Dr. Geoffrey Kemp's view is that Iran's desire for a nuclear-weapons capability is based upon a number of established factors, including the perceived threat from the United States, the existence of other regional nuclear powers, and the momentum exercised by a nuclear establishment within Iran's civilian and military leadership.

In the long-run, although the Gulf countries have real concerns about the acquisition by Iran of nuclear weapons, their short-term fears would focus primarily on anxiety about possible US and Israeli preemptive military strikes against Iran. The American administration and Israeli leaders have made it clear that the possession by the Islamic Republic of Iran of a nuclear bomb will be regarded as an intolerable threat. Dr. Kemp assumes that the primary motive that incited Iran to consider the nuclear option is to use it as a deterrent to prevent a US-led attack that would jeopardize its national survival and thwart its political regime, and not as a tool for regional aggression.

The recent developments in Iraq were viewed by the Conference as very critical and directly impinging upon regional and international stability. In this respect, His Royal Highness Prince Turki Al-Faisal Bin Abdul Aziz, Ambassador of the Kingdom of Saudi Arabia to the UK, considers the occupation of Iraq and the possible results and ramifications (notably, the partition of Iraq, turning it into a focal point for terrorism, or remaining under the American hegemony) as the gravest challenge for the GCC states.

H.R.H. Prince Turki stresses the possibility of meeting the challenge of the Iraqi question by supporting the Iraqi people in regaining their sovereignty and independence and in practicing their freedom of expression. All this presupposes dealing with the Iraqi people as one nation—not as separate sects, creeds and races. He underlines the necessity of dealing seriously and effectively with the United States by promoting US institutions and personalities that offer support for Arab positions, and also serve to counter the Zionist influence in the United States.

Dr. Faleh A. Jabar explores the post-conflict conditions in Iraq, assuming that the new Iraq may well emerge as a moderate Islamic state—a liberal, pluralistic polity, with a vibrant civil society and a market-based economy. However, Iraq will be consumed by the tensions of change and transition. It will emerge as a very weak nation in military terms and will pose hardly any direct military threat to its neighbors in the foreseeable future.

According to Dr. Jabar, such a situation would stimulate regional security cooperation between Iraq and GCC states, and a strong Iraq-US military cooperation with a view to enhancing Iraq national defense capabilities. However, such US collaboration and military presence on Iraqi soil would enhance Iraq's strategic importance and diminish that of major players in the GCC.

Dr. Jabar points out that the main challenge that Iraq will pose to the Arab Gulf region will shift from defense and security to the politics of change and regional alliances. However, as Iraq is not a closed society, its politics and society will be affected, in turn, by what regional players, especially in the Gulf would be willing to do or offer, given that Iraq is in dire economic straits.

The fallout of this process of drastic domestic restructuring of Iraq's political system will have profound consequences for its neighbors such as Iran, Turkey and Syria. However, the strongest impact will be felt by the GCC states.

Although still in its infancy, political liberalization in Iraq has thus far freed civil society forces and institutions that will, if balanced and stabilized, have strong demonstration effects on societies in the GCC states. Such a radical change would galvanize the role of civil society in the region and heighten pressures and demands for more reforms, notably where reforms have long been overdue. Hence, Dr. Jabar concludes that the GCC states will have to take this change in political culture and realities into account and launch well-considered reforms.

Addressing the same theme, Mr. Frederick Barton notes that Iraq at present is in the midst of multiple and simultaneous transitions, and that concrete progress has to be made in four areas in order to achieve success in post-war Iraq: restoring public safety throughout the country; formulation of a clear vision for the political transition in Iraq; expanding the scope of engagement by Iraqis in the reconstruction process; and creating transparent and consistent dialogue channels for exchanging views and information among all parties concerned.

Post-conflict challenges require the vision of a new Iraq that enjoys the consensus of the Iraqi people: an open, safe and non-aggressive country that respects the basic freedoms of its people and invests its resources in their well-being and prosperity.

It seems that in post-war Iraq the anxiety over scenarios of failure is a more familiar experience for most Iraqis: chaos with inter-communal violence, a failed state with "jihadist" control; a playground for regional mischief makers; or a weak US-puppet state.

As regards the reconstruction process and the challenges it encounters, Mr. Barton points out that despite the importance of this process for post-war Iraq, there has been a total lack of preparation for post-conflict reconstruction on the part of the coalition, and therefore the process is being carried out at a slow pace. Nevertheless, the coalition is being reminded that war should not be fought if the peace cannot be won. The fundamental question raised here is: Can the reconstruction process in Iraq be successful in the absence of the United Nations and a broader alliance?

Bringing about stability in post-war Iraq is a difficult task in the absence of economic recovery. His Excellency Mr. Kamel Al-Kilani who was the Minister of Finance in the Iraqi Interim Administration, addresses the challenges faced by the Iraqi economy, including dealing with cumulative debts estimated at US$ 120 billion; reconstructing the devastated infrastructure; rehabilitating and developing the public sector; boosting the private sector and activating its participation in investment plans; rehabilitating the banking sector and improving the living standards, given that most Iraqis are now living below the poverty line. The major economic objectives envisaged by the economic policies of the new Iraqi regime are: achieving a high level of economic growth; promoting economic openness; integration with the international economic community; setting up a social security network; and creating an attractive, advantageous environment for investments.

On the future prospects for the region, Dr. Bassam Tibi maintains that if the Gulf states fail to adapt to today's changing world, they would not be able to attain their goals. He stressed the need to develop a political culture within which democracy and a participatory society can be established. The Gulf region, in particular, and the Middle East in general, should be able to achieve these objectives without having to depart from their own heritage and culture, given the fact that democracy is the cornerstone of political stability which, in turn, is associated with economic progress and prosperity.

Dr. Saleh Abdulrehman Al-Mani examines the challenges confronting the prospects for cooperation among Gulf states, taking into account the fact that the GCC states are going through a period of change. These states previously followed the centralized state model, but since the 2003 Iraq War they have moved towards a different system of governance and are now exposed to external pressures and internal demands for reforms.

During the past three years, a war-like international system has emerged, exporting chaos and wars to the Gulf region. In this system, the GCC leaders have limited choices. Hence, according to Dr. Al-Mani, they must return to the roots of their societies and work towards developing a new social contract. Moreover, the relationships between the GCC elites need to be developed, both in their respective countries and among the states themselves. The GCC leaders should discuss their political agendas at the regional level and interact with the public at the local level in order to broaden the overall impact of their policies.

The Gulf region has always been renowned for trade, and this has been a factor in fostering peaceful relations between countries in the region. Historical relationships in the Gulf also provide clear evidence of cultural interchange between different states. Long-distance trade, in contrast, has promoted international competition for resources like oil, apart from becoming a destabilizing factor undermining peaceful coexistence.

In the same context, Dr. Abdul-Rida Assiri addresses the role of the GCC in promoting the policies of regional coexistence, stressing the fact that such coexistence is the only guarantee for the future of this region, as without it, the nations in this region cannot sustain their standards of prosperity.

The events that marked the end of the seventies and the beginning of the eighties of the last century – such as the Iranian Revolution, the Iraq-Iran war, the Soviet invasion of Afghanistan – have had significant and vital effects that expedited the need for the formation of the GCC. The crisis caused by the Iraqi invasion of the State of Kuwait demonstrated the importance of having the GCC, on one hand, and also the incapability of this organization to confront the invasion and the subsequent threats, on the other. However, the War in Iraq and the toppling of the regime of the former Iraqi President Saddam Hussein in 2003 highlighted the need to establish a security system comprising neighboring states and reducing the likelihood of confrontation, which has plagued the region for more than two decades.

Dr. Assiri asserts that regional coexistence requires adopting a number of interconnected strategies which guarantee the concept of coexistence. At the national level, political systems should be built on the basis of effective public participation in decision-making and the prevailing assumption that coexistence between democratic states is easier and more enduring than coexistence between undemocratic states.

At the regional level, there are a number of strategies, the most important of which are: the establishment of arms control regimes that guarantee a balanced regional system in which all parties feel secure; the development of a set of "confidence-building measures," among the states of the region; and the adoption of the concept of "universal security" comprising cooperation in the economic and environment fields.

At the supra-regional level, coexistence includes links with trans-regional and global institutions, which sustain coexistence among the region's states, such as joining the Indian Ocean Rim Association for Regional Cooperation (IOR-ARC).

Dr. Assiri assumes that no effective role can be envisaged for the GCC in achieving regional coexistence in the absence of consensus among its members on a common strategy to achieve such coexistence—first, between themselves, and second, between them and their neighboring states. Hence, he suggests the establishment of an institutional framework to bring together the GCC states and their neighboring countries.

The Conference sought to tackle some of the shifting roles in the Gulf region, especially that of the mass media. Mr. Jamal Khashoggi, Media Advisor to the Ambassador of the Kingdom of Saudi Arabia to the UK, observes that the print media, the most influential in the political arena, had to search for a new role since the collapse of oil prices in the 1980s. The emergence of the Qatari Al-Jazeera satellite channel is one of the positive factors promoting change in the role of the regional media. As an example, he cites the role that the Saudi mass media – the print media, in particular – has played in addressing the issue of internal reforms, which has become the key topic of most opinion articles in the majority of Saudi newspapers.

Mr. Khashoggi urges the removal of all types of support provided to media institutions in the region in order to ensure their independence from any pressure that might be exercised by sponsoring parties seeking preferential treatment.

Dr. Munira Fakhro outlines the changing role of women in the Arab Gulf region, noting that the increased educational and employment opportunities provided to women in the Gulf, as a result of the discovery of oil, has played a significant role in altering their position in Gulf societies. For example, women now represent a considerable proportion of both higher education students and the labor market. However, most women in the Gulf are working in the education sector, while their participation in the private sector is extremely limited due to social constraints and dependence on an expatriate labor force.

The economic and political changes that the region has undergone since the discovery of oil have brought about many structural and substantive changes in the Gulf family and led to the evolution of the smaller "nuclear" family.

Nevertheless, the changes that the family structure has undergone have not affected the traditional roles of both men and women; and therefore, in addition to their traditional roles, working women have had to shoulder their responsibilities as members of the labor force. By undertaking these two roles without the traditional assistance from their extended families, working women are obviously encountering many additional pressures.

The new economic and social role generated by the educational and employment opportunities allowed to women has not been associated with the policies and societal institutions that should have been designed and set up to further enhance women's role and empowerment. Hence, new advanced personal status laws need to be enacted by the Gulf states. Indeed, Gulf legislation pertaining to working women is inconsistent with international standards in terms of wages and working hours.

Despite the fact that the women's associations in the Gulf are exerting considerable efforts in support of the new role of women, their influence – as well as that of other civil society organizations – on the political decision-making process remains limited. Moreover, they are still too powerless to confront religious and tribal forces existing in Gulf society.

Dr. Badria Al-Awadhi examines the challenges and opportunities that globalization poses to women in the Gulf area, referring to the concerns about the negative repercussions of globalization on the political, economic and social status of women in the Gulf region.

Gulf women face massive political and legal challenges. Politically, these are reflected in their lack of participation in the political process. Although all Gulf national constitutions provide for equity between men and women in rights and duties, the major obstacle lies in the ways and means of activating the right of women's political participation which remains a controversial political, religious and social issue in some parts of the Arab Gulf region. The 1990s witnessed the start of debates regarding such a right, and its implementation, which has begun due to initiatives by the political leaderships in Sultanate of Oman, the Kingdom of Bahrain and the State of Qatar.

The legal situation of Gulf women demonstrates unequal treatment in some fields. Although all Gulf legislations recognize women as citizens, women are denied certain citizenship rights that ought to be acquired by birth. The most evident example of such bias is seen in the personal status laws, which are not in conformity with the new developments in the status of women in Gulf societies, and the nationality laws that deny children the nationality of their Gulf mothers if they are married to foreigners. In addition, the effective legislations lack the necessary provisions that protect women from crimes perpetrated against them at home (domestic violence) and at the workplace. As for employment, Gulf countries do not always implement signed international accords, particularly those regarding work and maternity. Hence it is important for women to participate in the decision-making process and in the drafting of laws in order to safeguard and consolidate their rights.

As such, Dr. Al-Awadhi concludes that the most important challenge that women face in the Gulf region is to promote awareness of their rights, both among themselves and among the general public, so that the defense of such rights is viewed both as a religious and patriotic duty.

In conclusion, it is hoped that the holding of the Conference under this theme, and the open-minded contributions, discussions and analysis presented during its proceedings will initiate a serious, in-depth dialogue among intellectual and political elites on issues affecting the future of the Gulf region, with a view to laying the foundations for stability, prosperity and a brighter future for the peoples of this region.

Keynote Address

The Gulf: Challenges of the Future

H.H. Sheikh Hamed Bin Zayed Al Nahyan

It gives me great pleasure to welcome and thank you, on behalf of His Highness Lieutenant General Sheikh Mohammed Bin Zayed Al Nahyan, Crown Prince of Abu Dhabi, and President of the Emirates Center for Strategic Studies and Research (ECSSR), for attending the ECSSR's Ninth Annual Conference. We deeply appreciate your interest in actively participating in this Conference, which brings together such an elite group of international decision-makers, experts and specialists to enrich the discussion on important issues and topics addressed by the Conference by sharing their valuable expertise and research studies.

We are well aware that the Gulf region today is facing grave challenges and threats that are impacting upon its present realities and redrawing the map of its future. These threats compel the states of the region to take cognizance of the challenges and opportunities posed by the new world order so as to develop their present status and lay the foundations of their future.

The world today is moving towards mutual interdependence among nations, propelled by the information and communications revolution, the growing movement of capital and the shrinking margins of political maneuvering under the unipolar international system.

While debates in academic circles focus increasingly on the globalization of the international system and its impact on all aspects of life, we still believe in the uniqueness of cultures and nations, variations in their strategic environments and their different vital national interests, which affect their course towards globalization. These differences have produced a variety of changes, the effects of which are manifest in societies, nations and states all over the world.

Owing to its natural wealth and its human and intellectual resources, the Gulf region has the capacity to take advantage of the new developments of the era, and indeed contribute constructively to human progress, provided it succeeds in achieving the optimal utilization of such potential resources. Furthermore, stimulating the principle of rational dialogue between the nations and governments of the region will certainly create the momentum needed for building trust and exploring a better future.

In my view, your participation in this Conference to address and analyze the future challenges confronting the Arab Gulf region will undoubtedly initiate constructive discussions and dialogue that will not end with the conclusion of this Conference. This stems from the fact that the Conference concentrates on vital issues closely related to Gulf citizens and contributes towards crystallizing their aspirations and fulfilling their hopes for a prosperous and secure future.

In conclusion, I would like to express my sincere wishes for the success of this Conference. I hope that it will contribute effectively to an enriching and meaningful dialogue, and help to formulate practical recommendations that will achieve its ultimate objective.

Section 1

Security and State Sovereignty

1

Shifting Security Parameters in the Twenty First Century

Hon. Gareth Evans AO QC

Looking out at the world as it stands early in this new century, there is both good news and bad news on the security front.

The good news is that the actual number of armed conflicts, not only between states but also within them, is in a declining trend. The overall number of people being killed in battle has also been steadily decreasing. In addition, although it may not feel like it, even the total number of international terrorist incidents has been steadily dropping—although, as suicide bombing has emerged as the weapon of choice for terrorists, the number of casualties has not fallen.[1] Looking objectively at the total picture, despite conflict and mass violence continuing to be bad, it is significantly less so than it was a decade ago.

This has all been facilitated to a great extent by the end of the Cold War, which removed a major source of ideological and great power conflict, and stopped the flow of resources to warring parties in proxy wars in the developing world and to authoritarian regimes of left and right. Yet, the international community is also handling these situations better than before. There has been an extraordinary upsurge in conflict prevention and management activities by governments, regional organizations, World Bank, NGOs like my own International Crisis Group,[2] and other international actors.

The United Nations itself was given new freedom, with the end of the Cold War, to play an active security role, and did so of course very effectively in the context of Iraq's invasion of Kuwait. And in other areas, despite inappropriate mandates, inadequate resources, lack of political commitment by key players and many other problems, there is certainly at least a prima facie case that the much-maligned UN has made a real difference in reducing the risk of war, both between and within states.[3]

So much for the good news. The bad news is that, for all the security problems that the global community has become better at solving, the ones that remain are very big ones indeed. These are growing and the capacity to deal with them is diminishing. There are some major things going wrong, both in terms of risks on the ground and in the way in which the global community is dealing with them.

A World at Risk

In terms of risks, there are three generic ones that are causing justifiable alarm. The first is the growth of *international terrorist organisations* with deeply frightening agendas, and deeply disturbing capacity—more so as we see the development of Al-Qaeda unfolding not so much as a single organization, but as a global franchise.[4]

The second concern is the waning effectiveness of the treaty regimes trying to achieve the non-proliferation of *weapons of mass destruction*. Until now, none of these has been more important or effective than the nuclear Non-Proliferation Treaty (NPT). This confounded the predictions of the 1960s that there would be 20-30 nuclear weapons states by now but has been looking ever more fragile in the last few years, even with the apparently better news coming out of Iran, Libya and perhaps even North Korea in the last few weeks.

The third concern is the continuing existence and emergence, right across the arc of instability from West Africa to East Asia of too many *failed, failing, fragile or simply overwhelmed states*, which do not resemble the fully functional sovereign actors constituting the basic premise of international law and international relations theory. These are states in which, as a result of government action, inaction, incapacity or huge internal division, often put their own peoples at extreme risk; and their debilitation is such that they often pose a risk to the peoples of *other* countries through their export of terrorism, drugs, other crime, fleeing refugees, health pandemics or environmental catastrophe.

In terms of the way in which the global community is dealing with these various risks, there are also three big things going wrong. There is a weakening of confidence in the *rules* that are supposed to govern the use of force. There is a weakening of confidence in the *institutions* making such rules and trying to enforce them. And there is little or no consensus about the *strategies* that are needed to deal with the great risks of terrorism, weapons of mass destruction proliferation, and the threats - both internal and external - posed by fragile or overwhelmed states.

A Lopsided World

Making all these problems harder to handle is the cross-cutting problem of US power.

The world has never in its history seen a global power balance as lopsided as that which currently prevails. The United States is, quite simply – militarily, economically and culturally – the biggest dog that has ever turned up on the global block (to employ a favorite metaphor of Bill Clinton). Furthermore, its behavior – whether actual, perceived, anticipated, feared or imagined – is the catalyst for a great many reactions by other countries and peoples, some rational and some irrational, that bear upon global, regional and national security.

The military budget numbers alone make clear the immensity of US authority. Before even getting to Iraq – which continues to cost the Pentagon nearly $US1 billion a week – US defense expenditure is now running at nearly $400 billion a year, just about as much as the whole rest of the world together. In percentage terms, this is over 40 per cent of global defense expenditure, and eight times the US share of global population. In dollar terms, it is higher than the combined total of Britain, France, Germany, the entire European Union, China, Russia *and* the seven 'rogue states' identified by the

Pentagon as its most likely adversaries (including Iran and North Korea).

The possession and exercise of dominant power has always generated a variety of reactions within other countries – to play the role of devoted acolyte; to compete, by building counter-weight authority; or to be more or less impotently enraged – and we can see plenty of examples of all three reactions in the world around us. Acolyte sentiment is strong in my own country, Australia, at least under its current government. Competitive sentiment is strong in continental Europe, although it is generally acknowledged – albeit a little reluctantly in Paris – that even on the most favorable assumptions, it would take decades for the EU, or even China, to catch up economically, let alone militarily, with the US. The reality of US power, both absolute and relative, is going to be there for decades.

The more worrying phenomenon is the outright rage and hatred which US dominance is also generating, particularly concentrated in many parts of the Arab and Islamic world. Some of this sentiment can readily be dismissed as not all that different to the hostility that has always been expressed toward the top dog of the day—from the Romans to the Ottomans to the Imperial British. The United States cannot do anything to alter its size, or much to moderate its relative clout, or anything to redress the disappearance of the Soviet Union as a power balancer.

However, what is also manifestly generating a great deal of hostility is not just the fact of US power but the way in which it is being, or at least perceived as being, presently exercised. The charge sheet is familiar: a distaste for multilateral institutions and processes; a disposition to make rules for others which it will not apply to itself; a tendency to dress up *realpolitik* as high morality – "to talk like Athens, but behave like Sparta" and a tendency to be highly selective in its assessment of others' behavior, especially when it comes to its friends.

In relation to the last point, it has not escaped notice that at least as many of the world's current serious security problems are associated with US friends as with its avowed foes. Pakistan has been seen as the foremost among the global nuclear proliferators; several sources of funding, recruitment and refuge for Al Qaeda and other radical Islamists drawing inspiration from it have turned out to be located within Saudi Arabia and Pakistan; and Israel's obduracy on the Palestinian issue has been as unhelpful in resolving that issue as any misbehavior on the other side, while its possession of nuclear weapons continues to be a serious inhibition on rational global non-proliferation policy.

No one can sensibly ask the United States to abandon its own national interests—it cannot and it will not do so, any more than any other country would if it had the choice. However, what one can ask of the United States is to consider carefully what its national interests actually *are* in this highly interdependent, globalized world, and recognize that it is no longer sufficient to define them narrowly in terms of security and economic benefits. No country, however big or powerful, can independently solve all the problems that affect it, whether the issue is terrorist violence, weapons proliferation, narcotics or other organized crime, refugee flows, health pandemics or environmental spillovers. The cooperation of others is essential, and that cooperation will only be gained if there is a willingness to help others to solve their problems. In short, all countries, including the United States should think of national interest not just in terms of traditional security and economic interests, but as embracing a third element: the national interest in being, and being seen to be, a good international citizen.

The problem that many countries perceive with the United States at the moment is not any lack of engagement with the rest of the world, but a determination to conduct that engagement almost wholly on its own terms. Some in the Bush administration, Secretary Colin Powell conspicuous among them, have always understood that this is not only irritating to others but ultimately counterproductive for the United States. However, there are others who remain to be persuaded.

We are very far from seeing a fully cooperative approach prevail across the three big contemporary security problems I identified earlier—lack of confidence in rules, in institutions and in strategies. However, that is no reason for not continuing to try to improve policy in all three areas—as difficult as it is, in this lopsided world of the 21st century, to make progress on anything without US support.

The Problem of Rules

There has undoubtedly been a growth of cynicism and skepticism about the existence and binding nature of the international rules governing the use of force. Under the guise of acting to meet threats of one kind or another, some states – the United States in particular – have been seen as making up rules as they go along, going to war when they should not, and not going to war when they should.

There are three different situations here that should be disentangled: the right to take military action against another state in self-defense; the right to take such action against a state posing a threat to any other states or individuals outside its borders; and the right to intervene against a state when the only threat involved is to those within it.

Self Defense

On self defense, Article 51 of the UN Charter clearly acknowledges that there is an "inherent right of individual or collective self-defense if an armed attack occurs against a Member of the United Nations"—which can be exercised without prior UN Security Council authorization (as it was by the US, without much argument from anyone, in Afghanistan after 9/11).

It is well accepted in international practice that this right extends beyond an actual attack to a threatened one, at least where the threatened attack is 'imminent.' What is very much challenged is the US notion, asserted in relation to Iraq in 2003, that the right to react in self-defense extends without any restraint to situations where the threatened attack is neither actual nor imminent—and where the state reacting (in this case, the United States) is effectively the sole judge of whether there is a real threat at all.

The problem is not so much with the notion of pre-emption as such. Countries have never been expected to wait until an imminently threatened attack became actual, and it is perfectly possible to imagine threats, including the nightmare scenario combining rogue states, WMD and terrorists, which are very real indeed, albeit not imminent. Yet international unease has to be expected when, as Sandy Berger has put it, this Administration has "elevated pre-emption from an option every President has preserved to a defining doctrine of American strategy."[5]

Ultimately the question boils down to credible evidence, and whether – assuming there is time to consider alternatives – the military attack response is the only reasonable one in all the circumstances. This is why there is still so much focus on the issue of whether credible and compelling intelligence was indeed available to support the war in Iraq. It is difficult to argue with the proposition that, if the whole international security system is not to descend into anarchy, the less imminent a threat and the weaker the evidence of its reality – as clearly was the case in Iraq 2003 – the greater the need to win Security Council support for the proposed response.

External Threats Generally

Moving beyond self-defense cases to respond to external threats generally, Chapter VII of the UN Charter clearly empowers the Security Council to take any action "necessary to maintain or restore international peace and security." The Council can, and does at its complete discretion from time to time, authorize or endorse the use of force by blue helmets, Multinational Forces, "coalitions of the willing" or individual states, as well as endorsing (sometimes after the event) military action by regional organizations operating under Chapter VIII—such as ECOWAS in Liberia and Sierra Leone.

The Iraq situation has rung all the changes on this theme. The Security Council gave such an authorization for the attack on Iraq in 1991, after its invasion of Kuwait; but did not do so in 2003. However, it is a reasonable bet that it would have been prepared to do so if there was more time to test Iraq's apparent failure to cooperate with the international inspectors, and if obstruction had continued.

Internal Threats

For wholly internal threats, raising the issue of so-called humanitarian intervention, the UN Charter is conspicuously unhelpful. Article 2.7

expressly prohibits intervention "in matters which are essentially within the jurisdiction of any state," although this is in tension with language elsewhere acknowledging individual human rights, and a mass of law and practice over the last few decades which have set real conceptual limits to claims of untrammeled state sovereignty, not least the Genocide Convention.

The Security Council can always authorize Chapter VII military action against a state if it is prepared to declare that the situation, however apparently internal in character, does in fact amount to a "threat to international peace and security"—as it did for example in Somalia, and eventually Bosnia, in the early 1990s.

However, more often than not, even in conscience-shocking situations like Rwanda in 1994, it has declined to initiate or authorize any enforcement action at all. Most people accept that the Security Council should continue to be the first port of call in these situations; the question is whether it should be the last. This is the issue that was brought to a head by NATO's intervention in Kosovo in 1999, bypassing the Security Council. It has been brought to a head again in Iraq 2003, with the emergence of the argument – as other rationales in terms of bombs and terrorists dropped away – that it was Saddam's murderous tyrannizing of his own people that made him a suitable case for humanitarian intervention treatment.

When it is Right to Fight: Six Criteria

The most urgent need in the international security debate, from whatever point in the ideological spectrum one approaches it, is to try to re-establish consensus about what the basic rules, or principles, governing the use of force should be, and how they should be applied in practice. I suggest that in *all* cases there is a basic over-arching checklist of six principles, or criteria, that must be worked through in determining whether it is right to fight—applicable whether the threat is external or internal; or whether the threat is constituted by armies marching, by WMD acquisition, by terrorism, or by tribal machetes.

These six criteria – a threshold test of seriousness, four prudential criteria and a legal test – were essentially those agreed by the International Commission on Intervention and State

Sovereignty[6] which I co-chaired with my Algerian colleague Mohamed Sahnoun three years ago. The ICISS commission, as it has become known, published its report: *The Responsibility to Protect*, setting out these principles, in December 2001. The formulation of the criteria owes much to the traditional 'just war' theory, but these principles owe their force much more to their intuitive acceptability than to any theological doctrine. They are certainly intended to reflect universal, and not just Western, values. They are as follows:

1-Just Cause: *Is the harm being experienced or threatened sufficiently clear and serious to justify going to war?*

For external threats to others, as with self defense, everything depends on the quality of the evidence. Actual behavior is one; merely threatened behavior is something else: to establish a threat, plausible evidence of both capability and intent to cause harm is required.

For internal threats, the threshold criteria to justify coercive intervention need to be tough. Unless the bar is set very high and tight, excluding less than catastrophic forms of human rights abuse, prima facie cases for the use of military force could be made across half the world: the only rule book would be the whim of the potential enforcer, and any prospect of mobilizing consensus for international action in the cases most deserving it – for example, another Rwanda – would fly out the window.

It was these kinds of considerations that led the ICISS commission to propose that the 'just cause' for intervention in these internal cases should be narrowly confined to two kinds of situation: *large scale loss of life*, actual or apprehended, with genocidal intent or not, which is the product either of deliberate state action, or state neglect or inability to act, or a failed state situation; or *large scale 'ethnic cleansing*,' actual or apprehended, whether carried out by killing, forced expulsion, acts of terror or rape.

For Iraq 2003 this threshold test cuts both ways. It would certainly have been satisfied a decade and more ago (when the West could not have cared less), but probably not in more recent years, as tyrannical as Saddam's regime continued to be in other ways. I find it difficult personally to accept as a trigger for war a 'humanitarian intervention' ground which is only seriously advanced as other grounds for military action retrospectively evaporate. However, it is hard to denounce Saddam's murderous behavior for twenty years, as many of us have done, then object, at least on this ground, when someone finally proposes to do something about him.[7] Call honors even on this one.

2-Right Intention: *Is the primary purpose of the proposed military action to halt or avert the external or internal threat in question, even if there are some other motives in play as well?*

For Iraq 2003, in the case of the United Kingdom, the judgment of history may be that the decision to go to war was wrong-headed, but at least palpably sincere.

For the United States, the jury may well be out a good deal longer on the question of intention, given most observers' experience that the only common motivation was regime change, for reasons not having an awful lot to do with Saddam's atrocities towards his own people. There may have been a genuine fear of physical attack with WMD by Saddam or those he might assist. However, a variety of other considerations, each coming down to a regime change as the bottom line, all seemed to rank higher in the motivation table — if not a hand in Iraqi oil production, then certainly considerations like bestowing the values of American democracy on the Arab world; or asserting absolute US military authority for its demonstration effect; or just being seen to be doing *something* to keep up the momentum of response post 9/11 (This should never be underestimated as a motivation for any government).

For Australia, if there was any other motive than following the leader (and earning Frequent Fighter Points along the way) I have yet to hear it credibly argued — but having been out of the country for the last few years, I may have missed something.

3-Last Resort: *Has every non-military option for the prevention or peaceful resolution of the crisis been explored, with reasonable grounds for believing lesser measures will not succeed?*

It continues to be strongly argued by opponents of the war in Iraq – with more and more credibility in retrospect – that there was ample time for the inspection process to have been carried through, and that resorting to military action in March 2003 was at the very least premature. That response does not answer an argument based on Saddam's past treatment of his own people, but it is a pretty good answer to the other rationales for war.

4-Proportional Means: *Is the scale, duration and intensity of the planned military action the minimum necessary to secure the defined human protection objective?*

In the case of Iraq, the question has to be asked whether some 5,000 civilian deaths and 10,000 military deaths – assuming that those guesstimates are at least roughly accurate – were an appropriate trade, from an Iraqi perspective, for the end of Saddam Hussein's capacity to persecute his people.

5-Reasonable Prospects: *Is there a reasonable chance of the military action being successful in meeting the external or internal threat in question, with the consequences of action not likely to be worse than the consequences of inaction?*

Military action can only be justified if it stands a reasonable chance of success, and will not risk triggering a greater conflagration or a greater peril.

This has to be called at the outset, not with the benefit of hindsight, but it has been from the beginning, and certainly remains now, a tough one for proponents of war in Iraq. We cannot finally assess the balance of consequences until we know how long Iraq's post-war misery will last, whether it is going to become a democracy or a theocracy, whether the war has indeed concentrated the minds of other dictators, and whether Al-Qaeda and like networks will indeed find it easier to recruit. However, the outlook on most of these fronts was not very encouraging before the war, and it is getting worse, not better, during the post-war phase.

6- Right Authority: *Is the military action lawful?*

As international law now stands, if the Security Council says "no," that means "no." However, should the absence of Security Council endorsement be the end of the intervention story? Is legality the whole story or, as many have argued, are there not wider questions of legitimacy as well? What if the Security Council fails to approve military action in another Rwanda-type, utterly conscience-shocking situation that just about everyone else thinks cries out for action? A real question arises as to which of two evils is the worse: the damage to international order if the Security Council is bypassed, or in the damage to that order if human beings are slaughtered while the Security Council stands by?

The ICISS Commission's response to this dilemma was to give a clear political message: if an individual state or ad hoc coalition does step in, fully observe and respect all the necessary threshold and precautionary criteria, intervene successfully, and ends up being seen to have so acted by world public opinion, then this is likely to have damaging consequences for the stature and credibility of the UN itself. That is pretty much what happened with the US and NATO intervention in Kosovo: the UN cannot afford to drop the ball too many times on that scale.

On the other hand, in Iraq 2003, the contrary argument has been put with some force – that compliance with the six criteria was on balance so weak, particularly on the issues of last resort and reasonable prospects – that the Security Council would have lost global credibility had it supported military action.

It will be a long haul to gain general acceptance in principle of the relevance and utility of all six criteria,[8] and an even longer haul to have them systematically applied in practice in every case – and when they are applied it will not mean the end of argument about particular cases, as we have just seen. The alternative to making a serious effort to enforce the international rules we have, and to supplement them with further principled guidelines and criteria of the kind here proposed in the areas where there are gaps, is to abandon the field to those who are more comfortable with the ad hoc exercise of power – who do not really want to be limited by rules and principles, who feel constrained by international process, who see multilateral cooperation in very narrowly self-interested terms.

However, a world that appeals to people like this is not one in which most people in the world want to live.

The Problem of Institutions

The effectiveness of the international security system depends not only on the rules in place but the credibility of the institutions making and enforcing them. There are multiple problems in this respect at the global, regional and national level which need to be addressed.

The basic one is that nearly all our contemporary security architecture, from the UN to NATO to the configuration of most country's armed forces, was designed by policymakers preoccupied by the primary concern of statesmen for centuries, states waging aggressive war. Of course we cannot assume that the human habits of centuries have changed, that in our contemporary, interdependent, globalized world the rewards of war will always be much less than the risks, and that aggressive war is a thing of the past, with Iraq's invasion of Kuwait being the last instance in this respect. However, as was said at the outset, we can certainly be confident that for the foreseeable future, traditional warmongering will be less of a policy problem than how to deal with terrorism, WMD proliferation as well as failed and fragile states. It is in these areas that our institutions have not been functioning as flexibly, effectively and credibly as they should.

United Nations

Secretary General Kofi Annan[9] is just one of the many who have argued that UN institutional architecture is badly out of date and desperately needs reform. In his report of September 8, 2003, Annan highlights the fact that the institution was designed for its original membership of 51 states, not its current 191, and that the composition and powers of the Security Council reflected the power balances of 1945, not the world of 2003. The decisions of the Security Council, he said, "increasingly…lack legitimacy in the eyes of the developing world, which feels that its views and interests are insufficiently represented among the decision-takers." And he has now appointed a High Level Panel – on which I have the pleasure of sitting along with others such as Amr Moussa – to come up with answers on this and related issues.[10]

The basic question we have to confront is whether the powerful Permanent Five, whose veto can block any Charter change, will ever be willing to relinquish that power, or share it with the likes of India, Brazil and Nigeria. Will Britain and France, for a start, ever be prepared to subsume their identity in a single EU seat?

And there are other questions as well. Will, for example, a legion of developing countries in the General Assembly ever be prepared to abandon the old-thinking which has blocked any talk of a new role for the now out-of-work Trusteeship Council in managing non-colonial states in distress?

Will they ever be prepared to support even the establishment of a serious conflict assessment and analysis capability within the Secretariat?[11] And will any country, developed or developing, ever be seriously prepared to vest the necessary authority and resources to create a standing military rapid reaction force to do the Security Council's bidding when emergencies demand it?

Everybody acknowledges that the United Nations has its uses, and may even be indispensable as a source of legitimacy and a vehicle for global burden sharing. Even the United States periodically goes through learning experiences in this respect. However, continued UN credibility and the maintenance in perpetuity of the present Permanent Five privileges simply do not mix: sooner or later one will have to give way to the other.

A good start would be for the Permanent Five to reach a gentleman's agreement (actually proposed once by the French), in which they would undertake, in the absence of their own vital interests being involved, not to exercise the veto to obstruct humanitarian intervention missions for which there was otherwise majority support on the Council. However, there is no evident enthusiasm at the moment among the Five for even this limited start.

Others

United Nations reform is a dispiriting business but these are all issues on which we must persist in a spirit of optimism. We must also continue to work to strengthen the security role of regional organizations. NATO is a potentially important new recruit to the role of UN enforcer not only in Europe but out of area, and the African organizations have already shown that their role can be absolutely crucial. However, there is much more that others can do in defense cooperation, conflict prevention and crisis management – including throughout Asia and the Pacific, and here in the Middle East with the Arab League and the Gulf Cooperation Council.

So too must we continue to urge individual countries, in Europe and elsewhere, to strengthen their own defense capability, building greater self-reliance in strategic lift and other areas of conspicuous shortfall. As Washington not unreasonably comments from time to time, complaints about US military dominance would carry more weight if other countries did a little bit more. Part of the answer is by providing the soft power (development funds and the like) while the US provides the hard power, although it is not the whole solution.

The Problem of Strategies

In the conduct of international relations, on issues of war and peace as elsewhere, rules and institutions are only as good as the intelligence with which they are applied. The big problems I identified at the outset – global terrorism, weapons of mass destruction proliferation and coping with fragile, collapsed and divided states – all need sensible response strategies. How well have we been doing in developing them?

Terrorism

The problem with the global war on terror that has been waged since 9/11 is that its primary achievements so far seem to have been more war and more terror. Osama bin Laden is still alive, and Al-Qaeda is down but not out. Its affiliates and offshoots and imitators in South East Asia and elsewhere are damaged but certainly not destroyed. In Iraq, where the terrorist connection was the least plausible of all the reasons for going to war, terrorist violence has now become the most harrowing of all its consequences. In Israel, with the collapse of the roadmap process, the suicide bombers are back with a vengeance. Nobody anywhere is confident that the 'big one' cannot or will not happen—an attack bringing together the sophistication and ruthlessness of the attack on the twin towers with the use of nuclear, chemical or biological weapons.

There are at least two general lessons we can learn from what has happened so far. One is that to wrap everything up in the language of a "war on terrorism" or a "war on evil" doesn't contribute much to clear operational thinking. A war against evil is, almost by definition as many have said, unlimited and interminable. The concept does not help us much in identifying points of entry, and there is certainly no obvious exit strategy. There are big risks in ignoring those problems which are not easily subsumed under the mantle of a war against terror. Perhaps there are even bigger risks in wrapping security problems in that mantle – like those in Iraq, Iran and North Korea – which at most are only marginally connected to terrorism.

The second lesson is how little the fundamentals of conflict and mass violence have actually changed since 9/11. The great dangers come from political problems – some of them with underlying economic and social causes – that are unresolved, unaddressed, incompetently or counter-productively addressed or deliberately left to fester, until they become so acute they explode. Part of the fallout of such explosions can be terrorism, but this is not a self-driving concept or an "enemy" per se. It is not even an ideology, as anarchism was in the 19th century. Rather it is a tool or a tactic, resorted to almost invariably by the weak against the strong – by weak individuals, weak groups and weak states.

Since power relativities have changed to the point where virtually everybody is weak in comparison to the United States and since 9/11 has shown the way, there is considerably more risk today that those in serious dispute with Washington, and by extension its allies, will use terror as a tactic to compensate for that weakness.

However, the core problems go back to political issues, broadly defined. Tough internal police action could be part of the answer, and so could military force, which was legitimately used in Afghanistan for punitive, retaliatory and in effect self-defense purposes. However, whether in the hands of the United States, Israel, Saudi Arabia or anyone else force can never be an effective substitute for the traditional hard work of dealing with those core problems.

From a Western perspective, the right strategy for dealing with global terrorism involves operating at five different levels simultaneously: first, homeland defense; second, pursuit and punishment of known perpetrators; third, and most crucially, building frontline defenses in the countries of origin of the terrorists themselves, by building up the capacity and will of those countries to act both internally and cooperatively with the wider international community; fourth, addressing the political issues that generate grievance; and fifth, addressing the underlying social, economic and cultural issues that generate grievance.

The real point of addressing the so-called underlying political and economic causes of terrorism is not to try to destroy the motivation of every individual terrorist. Most of the 9/11 perpetrators were not poor, cared little about the Palestinians, and were driven by religious rather than political passion. The real point is to neutralize support for terrorists in the communities in which they live, and above all to generate the will to act against them and the capacity to act against them by the relevant governments and authorities.

And it is that job that we are not now doing very well, especially in the context of the Israeli-Palestinian conflict.

There *is* a peaceful two-state solution, which has been outlined by many and mapped in great detail eighteen months ago by the International Crisis Group (ICG),[12] and now again in the Beilin-Rabbo Geneva Accord.[13] The leaders on both sides know perfectly well what is needed to meet the basic aspirations of the Palestinians and guarantee Israel's security, and what is acceptable to the majority of ordinary Israelis and Palestinians—but left to themselves, they have proved utterly incapable of delivering.

Solutions can be reached, but only if the US-led Quartet – playing it straight down the middle – assists, monitors, militarily supervises, and above all *leads* the process every inch of the way. To abdicate the responsibility to address the real and resolvable political grievances that lie at the heart of the terrorist violence being perpetrated, to do nothing to give moderate Palestinian leaders the capacity to deal with the extremists in their midst or the Palestinian people the will to oppose them, is to condemn both Israelis and Palestinians to more killings and endless tragedy.

Weapons of Mass Destruction

International legal regimes generally, and arms control treaties in particular, play a critical stabilizing role in the international community. Here it has to be acknowledged that the present US administration has been leading by example in the worst possible way, and not just in relation to the Kyoto Climate Change Convention and the International Criminal Court. After playing an important leadership role a decade ago in securing a tough international inspection regime for chemical weapons, the most recent contributions of the United States have been to scuttle a draft protocol seeking a similar enforcement mechanism for biological weapons, withdraw unilaterally from the Anti-Ballistic Missile (ABM) treaty (which continues to have implications for long term strategic stability in North East Asia in particular), and to assert the US right to develop a new generation of nuclear weapons, including the so-called bunker-busters.

With both the nuclear Non-Proliferation Treaty and the Comprehensive Test Ban Treaty faltering, the risk is now more real than before that the few nuclear weapons states may become many, even with the destruction of Saddam's regime in Iraq and the encouraging news coming out of Libya, Iran and even North Korea.

The recent entry of India and Pakistan to the club – joining the five original declared nuclear powers and Israel – should remind us once again of the simple but powerful conclusions of the Canberra Commission on the Elimination of Nuclear Weapons: so long as any state retains nuclear weapons, others will want them; so long as any state has them, they are bound one day to be used, if not by design then by accident; and any such use will be catastrophic for humankind.

Unfortunately, the compelling force of these conclusions continues to leave the relevant policy makers unmoved, for reasons that are never very compellingly explained. In the post Cold War world, after a series of progressive reductions, why it is necessary to ultimately retain any nuclear weapons for balance of terror reasons is not at all clear.

Why nuclear, chemical or biological weapons are needed to deter rogue states – or in the case of Israel, any of its neighbors – from producing or using them when current generation conventional weapons provide all the deterrent or retaliatory muscle ever needed, is never explained. Why a nuclear or CBW armory is needed to deal with terrorist groups or individuals producing home-made weapons of this kind is simply impossible to explain.

This double standard issue is crucial. As we know from ordinary life, saying "do as I say but not as I do" cuts no ice with anyone. It certainly cuts no ice internationally for major countries with pride and dignity and aspirations of their own. If the US and its nuclear armed colleagues on the UN Security Council really want to prevent a nuclear break out – to hold the line on the progress made in Iran and Libya, to make progress toward a nuclear weapons free zone in the Middle East, to stop a nuclear arms race getting started in North East Asia, to win global cooperation for coercive measures to be applied against proliferators and those assisting them – they simply cannot go on saying that it is legitimate for them (and selected friends) to retain a nuclear armory in perpetuity, but that no one else can be trusted to get to first base.

Article 6 of the NPT places an obligation on the original declared nuclear weapons states to commit themselves to the ultimate elimination of their armories, and it is long overdue for this to be taken seriously. There is a leadership role here that Britain, for one, could play. If Prime Minister Tony Blair really wanted to rewrite his place in the record books, and reestablish his deserved place in the Labour Pantheon, a declaration of intent to reduce the UK arsenal down to zero over a given time frame might involve the sacrifice of a little testosterone, but not any discernible military advantage.

Failed, Failing, Fragile and Overwhelmed States

The international community confronts this problem first, and most starkly, in the context of post-conflict peace-building, when the task is not only to meet immediate humanitarian needs but to ensure that the cycle of conflict does not start all over again.

If there is any single lesson we imbibed during the whole long debate on conflict prevention through the 1990s, and through the experience of every peacekeeping mission, both successful and unsuccessful, of the last decade, it was the need to address in these post-conflict situations, effectively and together, not only immediate security needs, but economic and social needs, governance and participation needs, and justice and reconciliation needs as well. In these circumstances it is difficult to imagine that so little was done to prepare adequately for the post-war peace-building task in Iraq, or that the commitment to Afghanistan's reconstruction remains so incomplete and fragile.

Every situation has its own distinct combination of needs, and it is not easy to make useful generalizations about the strategies required. However, in trying to rebuild societies in crisis, we should have learned at least ten lessons from our rather extensive collective experience in the post-Cold War years. We have to better understand the overall task; recognize that there are limits to what outsiders can and should do; allocate functions appropriately; learn how to pursue multiple objectives simultaneously; coordinate the process effectively; commit the necessary resources; understand the local political dynamics; make security the first priority; make justice and the rule of law a higher priority; and know when to get out.

Of all these lessons, probably the most critical one is the need for outsiders to have a close understanding of both the cultural norms and the internal political dynamics of the society that is under reconstruction. As the ICG first warned in Bosnia in 1996, while the urge to get legitimate local leadership in place is wholly understandable, early elections can be disastrously counter-productive if they only consolidate the existing ethnic or other divisions and concentrating first on building civil society institutions can often make much more sense.

In Iraq the ICG has been warning[14] of the huge downside risks in the apparently clever arithmetical weighting of the appointed Iraqi Interim Governing Council, and the cabinet it has selected to precisely reflect Shiite and Sunni, Arab and non-Arab proportions of the population. The trouble with this system is that for the first time in the country's modern history, sectarian and ethnic identity has been elevated to the rank of the primary organizing political principle. People are now more likely to join political parties built on these lines, and secular Iraqis are feeling weakened. The irony is that the United States which for so long has feared Shiite activism in Iran and Lebanon, is now effectively promoting it in Iraq. In the context of the announced handover of sovereign powers in June, there is a scramble to broaden the base of interim government. However, this may be too late to redress the damage that has already been done.

There is a second context in which better strategies are necessary. It is not only in post-conflict situations, with states already broken by war that we need to be focusing on how to improve the quality of governance and to reduce the risk of internal disintegration, with all its potential external consequences. Building the conditions for sustainable peace, to avoid conflict or mass violence erupting in the first place, is the responsibility of all governments, supported as necessary and appropriate by the wider international community.

This is not the occasion to dwell on the particular problems of the states in this region, but it is hard to avoid the perception that there are potentially destabilizing consequences for the region unless serious corrective measures are implemented. It remains the case that not one of the 22 Arab League countries has a leader elected in a free and fair election; and as was pointed out in the Arab Human Development Report of 2002, the deficits of freedom, education and women's empowerment have left the region so far behind that the combined GDP of the 22 countries is less than that of a single middle-ranking EU country, Spain.

As the ICG has noted,[15] the absence of a credible political life in most parts of the region is not necessarily bound to produce violent conflict, but it is intimately connected to a host of questions that affect its longer-term stability:

- Ineffective political representation, popular participation and government responsiveness often translate into inadequate mechanisms to express and channel public discontent, creating the potential for extra-institutional protests. These may, in turn, take on more violent forms, especially at a time when regional developments (in the Israeli-Palestinian theatre and Iraq) have polarized and radicalized public opinion.
- In the long run, the lack of genuine public accountability and transparency hampers sound economic development. While transparency and accountability are by no means a guarantee against corruption, their absence virtually ensures it. Also, without public participation, governments are likely to be more receptive to demands for economic reform emanating from the international community than from their own citizens. As a result, policy makers risk taking insufficient account of the social and political impact of their decisions.
- Weakened political legitimacy and economic under-development undermine the ability of the Arab states to play an effective part on the regional scene at a time of crisis when their constructive and creative leadership is more necessary than ever.
- The deficit of democratic representation may be a direct source of conflict, as in the case of Algeria.

Addressing this question is the responsibility of governments but not theirs alone. Too often opposition parties and civil society have contented themselves with vacuous slogans and unrealistic proposals that do not resonate with the people and further undermine the credibility of political action.

Cooperating for Peace

The world as it stands now is a decidedly uncomfortable place. The security challenges facing policy makers are as difficult and complex as they have ever been, and at the same time they are disconcertingly different from the kind of problems – mainly related to the risk of states waging aggressive war – that generations of international policy makers have been used to confronting in the past.

The context is a globalized and ever more interdependent environment, where states are less able than ever before to insulate themselves from events around them – in security matters as with everything else – and for most states their capacity to completely determine their own destiny is more limited than it has ever been.

At the same time it is a global context in which one preeminent power has a greater position of dominance than any single country has ever had, with a greater capacity to have its own way internationally than most states have ever had, and this has either contributed to the creation of a number of contemporary security problems or made their resolution more difficult, or both.

Under these circumstances, the only productive way forward, in wrestling with the security problems of the 21st century, is for every country – from the biggest to the smallest – to recognize that, while a huge amount of responsibility continues to rest on the leaders of individual sovereign countries, no country is an island, and protection from the scourge of conflict and mass violence can only ultimately be achieved through a wholly cooperative approach.

Not everyone likes the idea of Bill Clinton as a messenger, but I recently heard him deliver a very powerful message that US policy makers in particular would do well to ponder. According to him, the United States had a choice about how to exercise its present enormous and unrivalled power: "We could use it to try to stay top dog on the global block in perpetuity. Or we could use it to create a world in which we would be comfortable living when, and that time will come, we are no longer top dog on the block."

There is not much doubt about which approach the rest of the world would prefer, nor – I would suggest – which one would most benefit the longer term interests of the United States.

2

Multinational Corporations and State Sovereignty: Redefining Government Responsibilities in a Global Economy

The Rt. Hon. John Major CH

The implications of the global economy are profound—some welcome it, some fear it, but no one doubts that it is here to stay. The task for any government is to embrace it and shape it to serve national interest. It will bring acute dilemmas but also offer great opportunities. The current discussion focuses on an issue that every country in the world must address: How can national sovereignty and the global economy co-exist?

It is controversial. On the one hand, multinational corporations and free marketeers argue for further liberalization of trade and the removal of restrictions on investment and ownership. On the other hand, anti-globalization campaigners fear free trade and the power it gives to the multinationals. They question their motives and demand restrictions upon them. At the core of this heated debate lies a fundamental issue—the future prerogatives of the nation state itself. Before directing attention to that issue, it is necessary to put this whole subject in a proper context.

Our world is in a state of flux, both politically and economically, and may remain so for the foreseeable future. Why is this? It is mainly because the structure of world power has changed. Over a decade ago the Soviet Union collapsed. Most people felt the world was safer because there was no longer the threat of a nuclear exchange between superpowers. However, few foresaw the wider changes that would follow.

If the Soviet Union had not collapsed would we have had the bitter civil wars of Bosnia and Kosovo? Probably not. Would the war on terror have been embarked upon if the Soviet Union had remained an alternate superpower? Perhaps not. Similarly, if the Soviet yoke had not been lifted, would the European Union (EU) be enlarging its membership and embracing as new entrants all the former Communist states in Central and East Europe? Certainly not. Yet this last change will create the largest and richest free trade bloc the world has ever seen. Its uniform laws and structures will further develop the global market and diminish the unfettered sovereignty of individual nation states, including *significant* powers like the United Kingdom, France, Germany, Italy and Spain.

Of course, the end of the Soviet Union is only one of the political events that have left our world in a state of flux. The conflict in Iraq – and its aftermath – has also done so. So have the tensions in North Korea and Kashmir and Iran. So has the continuing dispute over Palestine which is forever threatening to move from a low grade war to a full-fledged war.

At the same time as these political events capture our attention, equally profound structural changes in international trade and economics are underway and these will not be reversible. Consider where we are. Our world now is global, changed beyond recall by technology and communications, with greater and faster change ahead of us. How *fast* change comes – and in what guise – will depend on many factors: world economic trends, political developments and macro and micro decisions by countries and companies around the world. Suffice to say the world of tomorrow will be vastly different from the world of today.

We now live in a world without economic boundaries, which were crumbling even before technology finally shattered them. Now the prosperity of every nation and every industry is affected by events far beyond its borders, and *this* explains why the nature of government itself is changing. As globalization spreads, it has buried the centralized authority of the command economy. Even as the unfettered power of government diminishes, the power of markets increases and technology has ensured that these powerful markets never sleep.

For a quarter of a century, two events have been moving in a parallel direction—the world has moved towards becoming *one* market and national companies have grown into multinational corporations to satisfy the demands of that market. It is easy to see why: that one market is huge and so are the corporations. The 200 or so multinational corporations generate global sales of over US$5 trillion; operate in 20 or more countries; have common ownership and management; sustain 150 million jobs and have annual revenue income well above that of over half the member states of the United Nations.

Collectively, these corporations represent great power and the impact of these multinationals is undeniable: the sovereignty of the nation state is lessened and the options open to government are diminished. To an extent that is inevitable because joint ventures, inter-relationships and shared ownerships can be almost impossible to control from within a single state. International mergers blur the identity of companies. Daimler-Benz is generally considered as German and Chrysler as American but following their merger neither of these assumptions is true. The individual nation state has an uphill task keeping control over such corporate giants.

This is especially true in dealing with legal redress for malpractice. How do nation states or indeed, individuals, seek redress if actions are taken by international corporations that harm them? It is difficult since the corporations may have an ephemeral and shifting legal nature. This emphasizes the crucial importance of agreements and understandings between nation states and corporations.

National sovereignty is under attack from many quarters. It is undermined by multi-state institutions – the European Union for example, as well as the International Monetary Fund, World Bank and World Trade Organization – all of whom make decisions and undertake functions that once were the sole prerogative of the nation state.

Sometimes, as in the case of EU law and multilateral agreements, action is taken to regulate national rules in such areas as labor law, health and environmental protection. This often has an economic rationale but it can also trample on national instinct and tradition. The politicians swap an enhanced market system promising economic growth for diminished political influence.

However, we should not fall into the trap of believing that these developments are novel—they are not. Nor are they a conspiracy as some anti-globalization pressure groups would have you believe. Nation state sovereignty has been unraveling since the early days of international trade beginning in the 16th century. These days it is being carried further as trade widens and multinational ownership of assets spreads. There is nothing surprising or malevolent in all this—it is simply the way the world has gone.

The nature of the global market is that – all other things being equal – companies will invest where it is most profitable to invest, buy where it is most economical to buy, manufacture where it is most cost-effective, and locate their factories and offices where the business environment is most favorable to their interests. These are rational commercial decisions. From the perspective of the corporation and its owners, the shareholders, this approach has an inexorable logic—their view is not going to change and is likely to accelerate.

This iron-clad reality imposes a discipline on national governments to provide an attractive investment scenario in their domestic environment. Governments must do so if they wish to enjoy the tax yield and employment opportunities that result from hosting business activities within their borders. Consequently, government has less freedom, whether it is to impose exchange controls, compel domestic ownership, become too regulatory, hike taxes upon an immobile private sector or have an inefficient legal system. Too much of these controls and restrictions will cause investment to flee eventually, if not immediately.

Such restraints on freedom may be frustrating for governments but it is the world in which we now do business—transport and technology advances have given business investment the kind of mobility it has never previously known. The use of that mobility has the secondary effect of inhibiting freedom of action by domestic governments. Hence, absolute sovereignty declines.

As evidence of this trend, two current phenomena may be pinpointed: China's massive growth supported by US$50 billion of inward investment each year and the accelerating flight of investment from the industrial but high cost economies of Western Europe to the emerging economies of Central and Eastern Europe. In each case, capital has been diverted to a cost-effective market. From the businessman's point of view that is a logical decision.

Of course, in all the above illustrations, national governments have *theoretical* sovereignty to tax, regulate or control as they wish but if they do, the price that their country pays in economic terms may be severe. Their sovereignty here is conditional. It is the global market that has created this situation and these days we are all part of this market whether we wish it or not.

The Implications of Trade Agreements

The speed of growth of the global market has thrown up many complex problems. These days, the overall value of free trade is so widely recognized that – by general agreement –it has been accelerated by successive Trade Rounds aiming at tariff reductions, as well as by treaties such as the North American Free Trade Agreement (NAFTA) and bi-state trade agreements.

These agreements are not perfect—far from it. There is still too much protection, too much domestic preference and too many markets guarded against fair access from the developing world. The recent American preferences on steel are a case in point, although the international machinery has now ruled them illegal and they have been dropped. Nonetheless, they illustrate two important points: first, even the most free-trading of nations can transgress but second, if they do, there is machinery in place to put right that transgression.

However, international agreements are complex and it may be necessary to revisit some of them as clauses, once obscure and seemingly harmless, appear later in a different guise. NAFTA illustrates how they can have unintended consequences.

When NAFTA was adopted a decade ago it ensured a legal right for foreign investors to bring claims against governments if the profit-making potential of their ventures was damaged by "decisions tantamount to expropriation." Yet, what does this indefinite and woolly phrase mean in law? It was intended to prevent discrimination against non-resident companies but, in fact, it went further—it gave non-resident companies rights in the courts that were not available to their domestic competitors. Moreover, it inhibited nation state sovereignty by making justiciable actions that traditionally were perfectly within the rights of the host government.

Suppose, for example, a government legislated to protect environmental standards and, in so doing, damaged the profits of a non-resident company. The non-resident company had the right to sue the government whilst their resident company competitor could not. This was hardly fair on the resident company or – for that matter – the government. Nor was it intended.

Take another example. Suppose a government banned an additive on health grounds, because they feared it carried health risks. Once more, the domestic company would have to comply with the law uncompensated whereas the non-resident multinational might take court action for compensation. Such instances are not fanciful—similar cases have been brought and won.

Clearly, existing and future trade agreements will need to address such anomalies to ensure a level playing field for companies. The implication of this is that the impact of these agreements on state sovereignty must become a genuine issue for discussion in free trade talks.

Corporate Responsibility

It is clear that the implications of globalization for the role of government are significant: It must also be considered whether there are social obligations that should become part of the role of multinationals. This debate is only in its infancy but will grow.

However, it will be a difficult debate. Some economists, business leaders and politicians will take the classical view that the multinational corporation should be obligated *only* to maximize shareholder value. Some would add that if they go beyond it, they are at risk of censure or worse from their shareholders for not giving absolute primacy to their fiduciary responsibilities.

I understand that view but it is far too absolutist. I am a believer in the free market and the virtues of capitalism. However, I do not believe in *unrestrained* capitalism—such an approach is the law of the jungle. It is also against the spirit of the best of private enterprise that has, for nearly 200 years now, begun to accept social obligations not only to their workforce but also in a wider sense. Consider, for example, the pioneering record of companies like Cadbury as far back as Victorian times and, more recently, the huge resources that Bill Gates has directed to good causes.

Mr. Gates is not alone. Many companies make large charitable endowments, fund community schemes or provide resources for the arts or sport or education. None of this *directly* increases profit for their shareholders although it does entrench a good reputation in the minds of the public which is recognized as a very desirable asset by even the most hard-nosed businessman. Moreover, there *is* an element of altruism too—many companies do accept that their success gives them a moral obligation to the wider community. This sort of obligation applies on a macro level to international corporations and will, I suspect, become ever more important in future as economic success becomes increasingly independent of government action.

To put the point simply: no corporation can thrive without the assets of its host country – whether raw materials, intellectual property or straightforward labor – and, as their influence grows, so should their commitment to the country. Therefore, in the future, as trade negotiations open up markets, corporations should be encouraged to accept an obligation to the people of the nations within whose borders they operate. It may not be possible to formalize that obligation but it exists morally and should be honored.

This may take many forms and what is appropriate may differ. It could be social provision to their workforce; sponsorship to the wider community; investment in infrastructure or communal assets. The options are wide and should properly be the subject of discussion between government and corporation.

Other obligations exist too, notably the need for the corporation to conduct their activities in a manner that is not damaging to the interests of the host country and that, wherever possible, does offer it tangible benefits.

What will the Sovereign State still Govern?

We forget sometimes that the genuinely global economy is relatively new and how much we still need to do in order to come to terms with it.

We know that globalization has continued and magnified the long-standing trend of shared sovereignty between countries. We know that the power of the market has given the corporate world a share in that sovereignty—in practice although not by intent.

Yet we need to maintain a balance. It would be unwise to overstate the loss of sovereignty and regard the nation state as impotent and out of date, as it is neither. Nor can it be regarded as being invariably in a poor bargaining position.

States acting collectively are still dominant. They can frame economic policy, determine conditions for investment, legislate for the regulatory system, impose environmental controls, as well as utilize all the legislative parameters within which corporations must do business. These agreements are reached between sovereign governments and, although multinationals may lobby, the governments can decide collectively through negotiation. The death of the nation state should not be proclaimed too readily—it lives yet and will continue to do so. Shared international sovereignty is still a powerful weapon in the hands of government. It only falls apart if governments are too divided to agree upon a course of action.

It cannot be said that *national* sovereignty is a spent force although it is right to draw a distinction between the options open to wealthy, developed countries and those available to the undeveloped world.

We now live in a world in which half the population – a total of 3 billion people – live on less than the local currency equivalent of US$2 a day and over a billion on no more than US$1 a day. The differences are stark—one-fifth of our world has four-fifths of world income and the remaining four-fifths have only one-fifth of the income.

Moreover these differences will become starker. In the next quarter of a century the population of the world may grow by one-third and nearly all these extra people will be on the poor side of the equation. These governments are in a weaker position than more wealthy and developed countries and may need more protection from international agreements. However, the more developed states will still retain the powers that most define democracy and self government.

The more developed states will determine their own style and structure of government and the methods of electing or selecting it. In addition, they continue to have a conditional freedom of taxation upon individuals, companies, goods and services. Even where this is partly shared – as in some taxes in the EU – the position of the nation states remains firm.

Other national prerogatives that remain will include social issues such as pensions, social payments, education and development of the infrastructure. These are all bedrock responsibilities that will continue to depend upon government actions. It is not easy to see how this list will diminish unless the principle and practice of representative democracy were to fall into total disrepair—and this I do not expect.

The conclusion I draw is that state sovereignty *has* changed, *is* changing and *will* change further and the global economy will be a principal factor in its evolution. However, it will not be the assassin of state sovereignty—the death of government is not imminent.

Yet government *will* change. As it does, it must maintain a balance between encouraging investment and free trade and fulfilling its legitimate obligations to its citizens.

In truth, multinationals are neither friends nor foes but commercial and industrial realities that cannot be ignored. The challenge for government is to maximize the advantages they offer and minimize the downside. It may not always be easy but then politics never was.

Section 2

The International System and the Gulf

3

Positioning the GCC Countries in the Global Economy: Challenges and Policy Options

H.E. Sheikh Hamad Bin Jassim Bin Jabr Al Thani

The accelerated pace of global development in our era has emphasized a fundamental fact that can no longer be ignored—the impossibility of separating politics from economics, as the latter cannot be isolated from the former, and vice versa. There should be some degree of integration and harmony between the elements of political security of any state or society and the components of economic stability necessary to fulfill the priorities of progress and development that would contribute to the evolution of modern states and societies capable of interacting with the surrounding realities and circumstances.

Such a key equation, in my opinion, has governed the trends that must be adopted with a view to establishing the required pillars of progress and modernization needed by our country in this critical stage, which is fraught with changes and shifts whether at the level of our Gulf and Arab region or at international levels. Therefore, it has been generally accepted that any constructive endeavor aimed at determining the position of GCC states in the global economy, evaluating the challenges and prerequisites that we have to encounter, and planning the policies and options available to us in this confrontation should be derived from certain parameters which have to be taken into account when addressing such vital issues. These parameters must also be relied upon when shaping the goals and priorities which we should strive to accomplish.

Through such a systematic approach, we should be capable of dealing with questions of development, modernization and economic progress with greater seriousness, realism and confidence in the future.

GCC States: Political and Strategic Parameters and Priorities

In my judgment, the parameters from which we should proceed are simple and clear, and the most prominent of these are as follows:

- The Arab Gulf, in particular, and the Middle East, in general, is a region that assumes an exceptional, indeed a unique, strategic importance for the whole world; a region where political, economic and security factors are effectively interlinked in a manner unparalleled in other regions of the world. These facts, along with the developments, crises and conflicts recurring in the region, have made it the focus of the international community's attraction and concerns, and definitely pivotal to its attention and interests.
- The importance of this region stems from many fundamental factors and reasons. Some might feel that the origin of this importance can be ascribed in the first place to the abundant natural resources - such as oil and natural gas - that the region enjoys and which are extremely vital to the sustainability and revitalization of the world economy. Even though this factor is true and indisputable, it is not the only one. The area also occupies a unique geographical location at the crossroads linking the world's continents, oceans, civilizations and cultures. It constitutes a medial center controlling the land, sea and air communication routes connecting different parts of the globe, resulting in political, security, social and cultural effects, the impact of which is not only confined to the area but also reaches far beyond its own borders and vicinity.

- The Gulf region represents stark testimony that political and strategic stability cannot be separated from economic and social stability. The continuing regional crises and conflicts, both internal and bilateral, directly and chronically affect the area's prevailing economic and social conditions, and consequently obstruct the developmental and progressive efforts urgently needed in the area. During the past decades, this has driven the region in general and some of its countries in particular, into some sort of a vicious circle where security, defense and arms requirements preclude the allocation of necessary resources and energies for achieving the priorities set for development and modernization processes, a situation which, if left unresolved, would fuel the causes of tension and destabilization.
- The GCC states are well qualified to become the moderate and acceptable model that can be followed throughout the development process on the political, economic and social levels, owing to the potential, resources and positive factors contributing to the achievement of this goal.
- The GCC states shoulder multiple responsibilities that involve benefiting from, and building upon past lessons and experiences with a view to fulfilling the processes of coordination, cooperation and integration between the governments and peoples of these states. Apparently, the accomplishment of the desired goals of these processes is the ideal means by which we can consolidate political, security and strategic stability and move forward to cement economic and social prosperity and welfare.

Based on these parameters, and taking into account their implications and dimensions, we will have the capability to combat the undoubtedly complex, diverse and thorny challenges confronting us.

On the one hand, we have been facing several threats that oblige us to exercise considerable patience, persistence and gravity, and to increase our endeavor, flexibility and capability to maneuver, draw up alternative plans, and implement the required changes.

On the other hand, we live in a world of rapidly changing shifts, which force us to acquire the ability to deal and interact with, influence and take part in creating these shifts, so that the consequences will not be adverse to us or detrimental to our interests and the goals we set for ourselves.

Needless to say, the global economy is now being founded on the free market system and its continually changing variables. This no longer permits us to remain on the sidelines, relying merely on conventional data, which although relied upon in the past for stability and permanence, have now become inapplicable. Moreover, globalization, in its different economic, financial, social, cultural and political implications, has in turn become an unquestionable reality that compels us to deal with its positive and negative repercussions alike. No state, region, society or culture can afford to remain on the sidelines believing that they will be beyond the effects and influences of these accelerated global shifts. In adopting such a position we will be like the ostrich that buries its head in the sand—a stance that we cannot and should not resort to since it will lead to the loss of all that we strive to accomplish. Rather, our states and societies, proceeding from the priorities we seek to fulfill for our peoples and generations, will have to embark with courage, boldness, confidence and resolve upon effective participation in determining the destiny of our country; shaping the future of our region and peoples; controlling and benefiting from our resources and wealth to attain our goals and aspirations. This predominant challenge, in my judgment, will be the most significant for us as Arab and Gulf states in the near future.

In addition, the other daunting challenge that we will have to encounter involves the development of our societies' political and constitutional institutions in order to consolidate the pillars of political stability; the modernization of executive and administrative practices and traditions within a framework where democratic values and responsible popular participation in the decision-making process can be realized; the promotion of the principles of equity and justice before the constitution and law; and the establishment of the foundations of accountability and transparency in the governmental and administrative performance. Obviously, the accomplishment of this objective represents an integral part of the imperatives for economic stability, due to the inextricable linkage between the economic and social development process, on the one hand, and the furtherance of political development based on the rule of law and state institutions, on the other.

Through this approach, we will be able to build a modern state, develop the pillars of a solid and sustainable economy; restore a coherent society; ensure equality of opportunity based on efficiency, experience and qualifications; combat corruption, inaction, extremism, isolationism, frustration and despair; deal with the world from the position of an influential, capable and effective partner at the political, economic, social and cultural levels based upon understanding, interaction, openness and mutual respect, and within the common interests and goals linking us as Gulf, Arab and Islamic states with the outside world.

Economic, Trade and Investment Developments in Qatar

Within the aforementioned context, we have undertaken several steps and measures in the State of Qatar to bolster our country's ability to face the challenges stated above in the light of domestic, regional and international economic developments. The most important of these measures are as follows:

- Modernizing the economic establishments; encouraging private initiatives; guaranteeing market freedom, movement of internal and external investments, and equality of opportunity based upon competence, experience and efficiency.
- Enhancing the growth of gross domestic product (GDP) at an acceptable rate that raises the per capita income in Qatar to the highest level in the world. It is worth mentioning that the trade balance has produced surpluses, and inflation rate has continued to drop for the fifth successive year to reach zero. The Central Bank of Qatar's statistics indicates that during the first quarter of the current year the domestic product has risen to 5.3 per cent, and that the public budget of the fiscal year ending March 31, 2003 has realized a surplus of QR 4,120 billion, compared to QR 3,251 billion in the previous fiscal year.

- Financing the development process by means of the oil and gas sectors with a tangible increase in the role of natural gas; stepping up efforts to increase oil stockpiles; the optimal utilization of natural gas through boosting of gas reserves and raising the production of liquefied gas, gaining international markets and diversifying export methods.
- Re-evaluating the gas reserves of the northern field that has resulted in doubling the previous estimates, making it the world's largest single field and placing Qatar in the second or third rank in terms of world natural gas reserves; moving ahead in expanding the existing petrochemical and fertilizer industries and setting up new projects intended to increase the added value of present industries and diversifying sources of national income.
- Raising the private sector's contribution to the national economy by privatizing some public sector projects, resulting in the creation of new corporations. We will continue to promote the privatization process in the future whenever feasible.
- Adopting a policy of economic openness towards the international trade, financial and economic system; taking several practical steps to encourage and attract investors from all over the world and create opportunities for foreign capitals to set up gigantic projects, particularly in the gas industry.
- Enacting new laws that govern the infrastructure needed for foreign investments, such as the laws on the investment of foreign capital; trademarks; protection of intellectual property rights; trade companies and agencies; and investment funds. In addition, other laws are being enacted within the stipulations of the Doha Round for Development and under the World Trade Organization (WTO) agreements.
- The development of the stock exchange after privatizing 30 per cent of four companies under Qatar Oil Co. and Qatar Fuel Co.; implementing the partial privatization of the newly-established holding company intended to stimulate and develop the tourism sector through practical measures.
- Merging and abolishing some ministries such as those of Electricity and Water, Communication and Transport and

Information; setting up a number of public establishments that are run according to commercial rules, such as the Electricity & Water Co., the holding company which supervises the Qatar Hotel Co., Qatar Airways, the handling and aircraft services companies at Doha airport, the management of new Doha airport, and the creation of an artificial tourist island.

- Developing, improving and setting up medium and small industrial companies, including those involved in hydrocarbon industries.
- Taking scientific initiatives aimed at effecting discipline in the state's monetary and financial policies in order to reduce inflation rates, rationalize expenditures and raise per capita income to the highest level in the world. Gas exports are expected to double to reach 45 million tons in 2010, while investments in the oil, gas and petrochemical sectors would rise to US$ 28 billion to reach US$ 40 by the end of this decade.
- Passing the electronic government law and public services project with a view to restructure government ministries and organizations in charge of providing public services, believing in the great value of the informatics revolution and its role in facilitating the move towards the new economic system.
- The setting up of the Educational City; opening local branches of internationally recognized universities, such as US Cornell University's College of Medicine and Texas A&M University, in accordance with a plan designed to develop both public and high education systems; seeking assistance from advanced international institutions, organizations and research centers, like RAND Corporation and the Brookings Institution, to evaluate the progress achieved in the educational and health systems, and the possibility of operating Hamad General Hospital by prominent international medical universities and colleges.
- Implementation of cooperation agreements concluded with Arab and foreign countries, and setting up joint committees with a large number of states. Around 120 agreements of this kind have been signed by March of the current year.

As I said, we have taken these actions and measures in order to enhance our country's ability to stand up to challenges facing the social and economic development process. While we have no doubt that our brothers in the GCC states are keen to adopt the same approach, we are certain that there are several important points included in the process for achieving the desired economic integration among GCC states, in addition to some available policy options which can be followed by these states to confront current and future challenges, as I shall proceed to present.

Measures of Integration between GCC States

In my opinion, the most important points that the process of economic development should cover are the following:

- The establishment of the Customs Union in order to expedite the pace of trade activities among GCC countries.
- Speeding up the adoption of a unified Gulf currency by the year 2007 and forming the necessary institutions for this purpose, such as the GCC Central Bank, as well as other organizations needed for the creation of the Gulf common market and unified currency.
- The creation of the Gulf common market by the year 2010.
- Putting into effect the Unified Economic Agreement in the context of the economic integration strategy through endorsing executive regulations concerning the economic decisions, like, *inter alia*, the adoption of the unified (amended) standards for the acquisition and trading of stocks, opening branches for national banks; allowing GCC nationals (both natural persons and legal entities) to practice all economic professions and activities without restrictions in accordance with the regulations endorsed by the Supreme Council in its 8th Session on the performance of economic activities and professions except those mentioned in the list annexed to the decisions; organizing the ownership by GCC citizens of real estate in these states for residence and investment; and adopting the GCC Petroleum Strategy and regional contingency plan concerning the oil products of the GCC states.

Current and Future Challenges: GCC Policy Options

With regard to the policy options available to the GCC states, the following points may be noted:

- The effective participation in the negotiation rounds between the GCC states and the European Union (EU) on establishing a free trade zone between the countries of the two groups, in addition to advancing discussions between the GCC states and the United States pursuant to US President George Bush's plan regarding the free trade zone to be set up between Middle East countries and the United States and the US–Middle East Partnership Initiative (MEPI) launched by US Secretary of State Colin Powell.
- Developing practical perspectives and conceptions on the competitive potential of the services sector and drafting the required strategies aimed at penetrating the markets of individual countries and regional blocs pursuant to the calls for opening up the services sector within the economies of GCC states, as well as the positive adaptation to new issues that might be raised at the negotiations on services trade held within the WTO framework.
- Setting up a flexible, transparent mechanism to tackle the trade measures that affect the interests of GCC trade partners, and expanding fields of coordination between the different state bodies and organizations and their counterparts in other GCC states.
- Enhancing and expanding the military, strategic and security relationships between the GCC countries and influential states and groups in the international arena by strengthening economic and investment ties. Such ties could be strengthened specifically with those states and groups that possess the capability and will, and whose interests meet those of the GCC states, in order to avert regional and international threats that haunt such a vital area for the global economy, especially those that the GCC states are unable to deal with in isolation.
- Promoting the fields of trade and economic cooperation, with special emphasis on benefits derived from the effective

implementation of the free trade agreement signed with the United States in order to allow the entry of Gulf exports of oil, gas and petrochemical products.

- Attracting international investments and financial resources and gaining national and preferential treatment, within the framework of activating income-source diversification, by encouraging American and multinational corporations and financing gigantic economic and industrial projects in the GCC states, especially projects that employ creative techniques to obtain funding and advanced technology.
- Continuing to maintain moderate positions in the context of energy diplomacy with regard to production and price stability, taking into account the interests of both producers and consumers, as well as the growth of the global economy.
- Adapting to the international efforts aimed at achieving stability in the global economy and reforming the international banking system by creating a flexible mechanism to ensure the stability of oil markets, bearing in mind the continuation of the policy that oil is a strategic commodity that cannot be exploited for political considerations and purposes inconsistent with international trends and new developments.
- Further stimulating the role of the private sector and private businessmen in the process of making trade and investment decisions with a view to consolidating cooperation between the GCC states and other influential blocs in these fields.
- Overcoming any obstacles and difficulties that block the flow of foreign investments, and generating the proper financial sources to bridge the gap between the required investment rate and the savings rate, especially in the light of soaring domestic and external borrowing rates in the GCC countries.
- Boosting confidence in the GCC states' economic policies by addressing the problem of multiple bureaucratic laws, regulations and procedures eliminating the ambiguities therein and outlawing the conflict between the laws and legislations that govern the activities of different government departments, as well as promoting flexibility and transparency of the judicial system in order to speed up the settlement of trade and investment disputes.

- Preparing comprehensive plans aimed at improving the levels of development of administrative performance and solving the problems of university and school drop-outs through creating innovative channels for absorbing those willing to work.
- Activating the institutional role of the management boards of economic institutions by consolidating the financial and organizational inspection of the boards of these institutions, and setting up legal and constitutional frameworks to administer them.
- Drawing up practical plans to implement the recommendations of both the International Monetary Fund and World Bank, as well as the prerequisites for putting the WTO's agreements into effect, once the competent authorities in the GCC states start restructuring the Gulf economies in order to strengthen the means for administering these economies and to avert the negative effects resulting from the fluctuation of oil prices or the financial, monetary and banking shocks in the world economy, by strengthening the financial and banking systems; strengthening the role of the private sector in making economic and investment decisions; encouraging individual and private initiatives; re-evaluating respective laws and legislation; and creating a favorable environment for developing the renewable monetary and investment instruments.
- Setting up specialized bodies to monitor the services provided by the communication, transport and electricity networks within the framework of the strategy of integration among the GCC countries in order to expedite the executive steps towards creating the Gulf common market.
- Intensifying the flow of knowledge to all parties associated with the Gulf common market (information on markets, technology, training and qualitative education) through the establishment of modern information infrastructure to link market parties in question with the knowledge networks.
- Making use of international multilateral agreements approved by international organizations in implementing plans of internal reforms by using trade agreements, both bilateral and multilateral, as vital tools to enhance the means of reform in the political, trade and investment fields.

Conclusions

Obviously, the implementation of most of the above-mentioned options and aspirations is not an easy task, as these involve a tremendous input in terms of hard work, time and planning.

Nevertheless, the main springboard to attain the desired level of economic integration between the GCC states, involving the elements already discussed, is based primarily on the availability of the common political will. No doubt, the consolidation of the climate of security and stability and the adoption of dialogue and understanding as an indispensable pillar to shape the network of regional and bilateral relationships and alliances, are among the fundamental prerequisites needed to create a favorable economic environment capable of confronting future challenges in different fields. Naturally, this will necessitate drawing up the political, peaceful, just and ultimate settlements and solutions for the area's deep-rooted crises and conflicts away from the use of force, pressure and intimidation.

The true and actual priorities of political, economic and social development, modernization and reforms cannot be realized on the ground as long as these crises and conflicts persist, as they represent a source of tension as well as strategic and security threats, which make it incumbent upon the countries in the region to devote huge resources to confront this source of tension, and thereby make it impossible to allocate these resources to bolster the objectives of furthering development, welfare and prosperity.

At any rate, security, stability, cooperation and good neighborliness remain the key pillars on which all economic and social development and modernization efforts should be based. Both aspects must seek to complement, stimulate and enhance each other. Perhaps this is the biggest and most important forthcoming strategic challenge that will confront the region's states, societies and peoples, and the one that will leave us with no other choice but to develop the political, economic and social strategies to tackle and overcome it, whether at the GCC level or at the regional level.

4

Dialectics of Change and Continuity in the Gulf Regional System

H.E. Dr. Muhammad Abdul Ghaffar

During the last two decades of the 20th century, devastating wars have occurred that have had a tremendous impact on the stability of the Gulf region. One of these was the war in Afghanistan (1979-1989), which erupted in the context of superpower rivalry between the United States and the Soviet Union during the Cold War era. This situation has been described as a revival of the rules of the "Great Game" played in the Central Asia region between competing global powers since the 19th century.

The first of the Gulf wars, the Iraq-Iran War (1980-1988), which in terms of certain strategic objectives, almost synchronized and interacted with the war in Afghanistan, was fought in the context of regional rivalry between two major powers in the Gulf security system. The second war in the Gulf region erupted in the wake of the armed Iraqi invasion of Kuwait in 1990, resulting in overwhelming and far-reaching political, military and security consequences that are still haunting the region.

The effects of this invasion were clearly reflected not only in the shape and substance of Gulf, regional and Arab relations, but also impacted on the very political future of Iraq, which has remained vague and unstable since that time. Similar effects were felt within the Gulf and Arab security systems that were already overburdened with the unresolved complexities of the Arab-Israeli conflict, which is now more than fifty years old.

From such huge deficiencies and flaws in the Gulf regional security order and the general Arab security set-up, two urgent issues were raised forcefully – Gulf security and stability – together with the debate over the domestic situation and the sources of external threats. New dilemmas emerged in relation to the old concepts of nationalism and the principles of sovereignty and independence as a basis for theories of the nation-state as the basic unit in the contemporary international order. These dilemmas were also focused on the viability and survival of those nation-states in their interaction with a changing global environment and in the face of potential internal and external threats.

I-Aspects of Instability within the Regional Context

The recurrent pattern of armed interventions and the use of force over the two past decades – regardless of the debate over the legality or otherwise of such practices – have added to the perils of regional instability and security imbalances. Here, the danger lies in the fact that such a debate over the legality or illegality of the use of force has produced many interpretations ranging from the concepts of "preventive/preemptive war" and "humanitarian intervention" to the "exercise of the right of self-defense," in what might be tantamount to the reinterpretation of concepts and practices of international legitimacy established since World War II.

However, until the beginning of the 21st century, these sources of regional instability were not only external threats whose effects were receding considerably, but were mainly created by domestic pressures, as manifested in various failures, insufficient growth of civil societies, and immature cultures of political participation. Those pressures could even multiply and worsen through occasional ethnic or communal tensions.

Moreover, the phenomenon of being unable to maintain a sufficient degree of regional stability in the Gulf and Arab regions cannot necessarily be traced to the modest achievements of collective Arab foreign policies during the nineties or even to the fragility of domestic structures and political institutions involved in formulating decisions pertaining to international issues.

Rather, this phenomenon – the vulnerability of Arab security structures – is essentially due to deep economic and structural weaknesses, discrepancies in development levels (especially the higher levels of growth achieved by GCC states) and fluctuations in social and human development, coupled with chronically heavy dependence on foreign powers in enhancing security and military capabilities. This has been further compounded by a lack of genuine conviction in the value of collective action, even at a lowest minimum level, whether in the shape of economic integration and/or genuine political consensus that could strengthen the effectiveness of joint action and contribute to the credibility of the Arab regional system.

It is for these reasons that the Arab League, the mechanism of Arab common action, has become fully aware of the enormity and intensity of the dangers affecting the Arab regional system, and the consequent fallout in terms of eroded efficiency and credibility. These reasons have also prompted Arab leaders and heads of state to further activate collective Arab action and create a real momentum for the Arab League so as to meet new challenges and respond to new variables in international relations.

In this context, the Arab League initiated a new approach to the restructuring of existing mechanisms, and the introduction of new ones for consideration by Arab leaders at the Tunis Summit in March 2004. This new approach is inspired by greater common awareness of the delicate nature of the present phase in the evolution of the Arab world, and the need for a more positive interaction with the new developments that had a bearing on Arab national and regional security.

Within the Gulf system, these predicaments, both inherited and new – whether at the level of the GCC-states' foreign policy or the internal institutional developments in the Gulf countries – have been deeply felt by Arab ruling elites and Arab societies in the Gulf, particularly in the wake of the far-reaching consequences of the events of September 11, 2001, the collapse of the Iraqi regime in April 2003, and the military occupation of Iraq. Consequently, the ruling elites and societies have been forced to look for solutions and answers to the plights and dilemmas faced by these states.

They have been primarily active in ensuring two safeguards: first, to ensure security, societal stability and national cohesion and unity; second, to ensure the historical legitimacy of ruling regimes and strengthen their credibility through what may be described as rationalization of political legitimacy.

These two safeguards, which are vital for political continuity, have been pursued over the past few years through obvious and systematic change processes of modernization and political reforms. Those processes have been designed to produce societal openness, democratic and institutional adaptation, and cultivation of the values and culture of genuine political participation. This will establish an interrelationship between democracy and stability, as well as between democracy and socio-economic and human development.

II-Sources of Internal and External Threats to Gulf Regional Security and Stability

Gulf regional security and stability is still an issue that is an intrinsically complicated, not only due to continued external threats, but mainly because of an unprecedented escalation and proliferation of terrorism and political violence all over the Arab world as well as the Gulf region. The threat of terrorism has grown steadily during the 1990s, culminating in the gravest terrorist act of the 21st century as demonstrated by the most atrocious attacks of September 11, 2001 in New York and Washington.

Thinkers and analysts have offered a diverse set of interpretations and views on the motives behind international terrorism. Some of them acknowledge that hostile stances against US policies in the Arab and Gulf regions could be behind internal instability, especially when compounded with the repercussions of the Palestinian-Israeli conflict. Others still view the dynamics behind terrorism and instability as being the outcome of the nature and accelerated pace of the political, economic and social transformations taking place in the Gulf and Arab states.

Those who uphold such an approach still assume that the powerful moves towards consolidating the foundations of internal stability has led the GCC states to initiate plans and programs for democratic, economic reform over the past decade and to launch a systematic process of social and political modernization.

States and peoples in the Gulf region are primarily concerned with the prospects for regional security and stability in the face of several manifestations of domestic unrest or instability. This will essentially be achieved through alleviation of internal pressures, as well as through the introduction of a new regional security framework in the region, which is an uneasy task to fulfill.

In this context, some US strategists and planners have proposed the creation of an integrated regional security system that would allow other countries in the region to join the GCC security structure. This integrated security system could take the form of a regional security forum, where issues of confidence-building, de-escalation of tensions, and elimination of weapons of mass destruction could be addressed.

Compared to other states in the region, GCC states are still more stable, possibly due to social and economic development plans and the open-minded attitudes by the political systems and ruling elites – with their deep-rooted historical legitimacy – towards traditional democratic practices of social cohesion and solidarity.

External sources of threat due to the rivalries between Iraq and Iran during 1970s and 1980s, compounded by the Iraqi territorial claims in Kuwait in the early 1990s, were the most destabilizing factors in the Arab states of the Gulf. However, it is undoubtedly true that certain social transformations, such as population growth, expansion of education base, the emergence of a public role for Gulf women, the challenges of domestic development, imperatives of economic integration, and the imbalanced demographic structure, are also major factors shaping the profile of security and internal stability in most of these states.

III -The Nature of the Responses by Gulf States to Instances of Regional Instability

The degree to which internal and external forms and sources of the threats to Gulf security were profoundly affected by the events of September 11, 2001 and the aftermath of military action taken by the US- and UK-led coalition forces against Iraq in March 2003, the effects of the sources of internal tensions upon social stability and harmony, and the interrelation between this issue and that of economic reforms and political liberalization have all become quite clear in the light of developments over the past few years.

Also, the two issues of social stability and reform are related to the theme of the war on terrorism and the regional and international alliances resulting from the events of September 11, compounded by the sudden international turns and political developments that the Palestinian-Israeli conflict has witnessed since the end of 2000. Further, the arbitrary analogies between terrorist acts and manifestations of political violence – in politically or historically unrelated contexts – have had an impact on this situation.

However, notwithstanding external and regional developments (the war in Afghanistan, the global war on terrorism, and the war in Iraq) and the signs of domestic tensions due to regional or internal pressures, the requirements of the national security of the Gulf region necessitate the reactivation of the inherent capabilities and geo-strategic potential of the region as a whole.

It is likely that the introduction of such new political inputs within the framework of interrelationships could contribute to the creation of a balanced interaction between the political ruling elites, on the one hand and Gulf societies on the other. These factors are related to a broad-based process of political participation at the national level, which will open up avenues for initiatives of change, gradual political modernization towards democracy, and patterns of *shura* (consultation) in governance and politics.

The experiences of the GCC countries over the past two decades, however, reveals that inherent capabilities or the domestic fundamentals of national security alone cannot be adequate unless they are pooled through a framework of integrated regional economic and security capabilities, particularly in the wake of the drastic transformation of US foreign policy, European Union policies and also the economic integration of Asian economic entities.

Even the fundamentals of individual and collective capabilities in the Gulf might continue to have relatively little impact, unless combined with flexible, active and spontaneous responses (that is, not imposed by foreign leverages) to legitimate internal aspirations or pressures calling for political and cultural modernization and democratization.

As an early initiation of the process of political modernization and democratic progress, Bahrain represents an impressive model that has attracted not only authors and researchers, but has been the focus of attention at international forums by leaders and statesmen while presenting their own initiatives on the political future of the Middle East.

Similarly, Bahrain's experience can be viewed as a precedent and an example worthy of reflection, inasmuch as it combines flexible responses to domestic aspirations and an early interaction with the international developments that overwhelmed the world following the end of the Cold War in early 1990s.

IV-Bahrain's Experience: Dialectics of the Relationship Between Society and State

The substance of Bahrain's experience immediately raises elements of a dialectal relationship between society and state, since the process of political reform and social modernization represents a "case study" of a positive interaction between an Arab/Gulf society and the ruling political elites.

As long as the state in Bahrain has correctly stood for a balanced and delicate interaction between different social forces, its interrelationships should be stable and sound so as to consolidate its historical and political legitimacy, by enabling it to draft a formula of consensus that could resolve this dialectal relationship between state and society.

In this way, Bahrain's experience could present itself as a prelude to the resolution of broader regional dialectal relationships, either at the level of political and economic coordination and integration, or at the level of establishing defense and security structures to ensure and maintain both regional and local stability.

Within the framework of the Gulf system, such dialectics or the delicate relationship between state and society, that is, between the ruling elites and the social structures and dominant cultural values, represents one of the toughest challenges for stability and internal cohesion within the Gulf states.

It is my conviction that one of the imperatives for interaction between the trends of openness within the ruling political elites, and conservative, traditional currents in Gulf societies, is that the task of the promotion of democracy should primarily be a common responsibility, undertaken by different elites after considering the aspirations of the Gulf peoples. The consolidation of the pillars of democracy and transparency in our societies has to be achieved through a genuine endeavor emerging from within these societies, and by their own accomplishments and contributions made of their own free will.

V- Gulf Experiences and the Process of Political Reforms

It can be said that Bahrain's experience is unique within the GCC system, having been based on a number of factors which brought about an early growth in political consciousness since the early 19th century, and contributed to the birth of vibrant political and social movements embracing reform. Against this background, Bahrain's experience gained a historic momentum in 1999 with the inauguration by His Majesty King Hamad Bin Isa Al Khalifa of a new era of democratic transformations that His Majesty designated as the "Reform Project."

This Reform Project was manifested in the endorsement in February 2001 of the National Charter. Thereafter, a new Constitution was promulgated, followed by holding the October 2002 general elections, within the framework of a process of full transformation that ensured the basic elements of democratic practices, such as a broad spectrum of political views, freedom of expression, and the empowerment of Bahraini women. This could be safely described as "democracy of consensus," which has fostered a broad-based reconciliation among all the parties to this ongoing process.

A highly significant step towards good governance, based on the rule of law and democracy, was the establishment in 2003 of the Constitutional Court, to exercise the judiciary's control over the validity of laws and their adherence to the constitution of the nation.

In sum, political modernization in Bahrain is based on a clear social consensus, inspired by a rich history of education, political participation, and a maturing civil society, and enhanced by legal measures embodied in the adoption of the National Charter, promulgation of the Constitution, and ultimately supported by political factors such as the activation of public authorities and institutions in the modern Kingdom of Bahrain. All this has been achieved through an evolving, gradual and developing democratic experience. Five main pillars constitute the basis for such experience, namely, political participation, human rights, empowerment of women, freedom of the press and the role of civil society.

Regarding the development of positive relations with the traditional partners of the Gulf Arab states (the United States and the European Union) in the context of promoting political modernization, liberalization and democratization through the economic and political partnership initiatives proposed in 2003, internal moves for reform from within the Gulf political regimes themselves will always be a sound, rational and a right decision. Hence, Gulf countries will ultimately be able to decide for themselves the ways and means to achieve democratic reforms.

It is my judgment, within the very framework of active relations with traditional partners in the US and the European Union, that the role of these partners should be reflected in the pursuit of policies consistent with, and systematically supportive of, the promotion of democracy in the Gulf region and Arab world. These policies have to be framed within a clear, realistic and integrated vision that determines the plans and programs for social, economic and cultural development. Such plans and programs can help instill the values and principles of democracy in a peaceful and evolutionary way in our societies, with a view to creating a convenient climate for a prospering democracy, and to stem trends of extremism and cultures of violence. It would indeed be inconceivable to create a viable democracy in the absence of sustainable development.

Our societies and political regimes, particularly in the aftermath of September 11, 2001, were understandably preoccupied, especially with the fight against terrorism. A similar preoccupation was dominant among our US and European partners. We are all in the same boat amidst the stormy seas of violence and terrorist perils. Such a relationship will increase the responsibilities of US and European partners who stand on more solid economic and technological grounds. These added responsibilities would necessitate strong support for integrated development programs in Arab societies to help eliminate the culture of violence and rejection.

This might explain why some Arab as well as Third World intellectuals look with skepticism and concern upon Western appeals and initiatives to promote democracy, believing that the containment of terrorist threats is the prime motivation for such initiatives, and not the inherent virtues and attributes of democracy itself. Those skeptical analysts even go so far as to claim that certain Western powers have betrayed their allies and partners immediately after the downfall of communism and the end of the Cold War.

I believe that there is now a clear and full awareness among the ruling elites in the Gulf states, including Bahrain of course, of the real objectives of gradual democratization processes in Gulf societies. Nevertheless, debates between those who consider political reforms as a precondition for stability and economic growth, and others who believe that promoting democracy and political transparency should follow economic reforms (as is the case of China), are still unresolved.

Objective analysis and review of modernization and development processes in the Gulf system reveals genuine transformations in the political landscape of the region as a whole. For example, the political elite in the Kingdom of Saudi Arabia led by His Royal Highness Prince Abdullah Bin Abdul Aziz Al Saud, Crown Prince, Deputy Prime Minister and Commander-in-Chief of the National Guard, has launched several initiatives for dialogue and development of the local administration system. One of the manifestations of this process has been the decree adopted by the Cabinet in October 2003 to organize partial municipal elections.

In the State of Qatar, the new Constitution, which was put forward for approval through a plebiscite held in April 2003 is a true expression of a new period of significant reforms, since it offers Qatari citizens the rights to vote, nominate and the elect members of the legislature.

Again, the first general election held in Oman in October 2003 to form the Consultative (*Shura*) Council, was but one remarkable positive manifestation of the continuing evolutionary process in the Gulf region, especially with the empowerment of Omani women by securing their membership in the Shura Council, which has the right, for the first time, to study and comment on laws, legislation on economic, educational, cultural issues, and submit questions for the interpellation of Ministers on such issues.

Challenges that affect Gulf security and stability have acquired new dimensions with the downfall of the former regime in Iraq. It is our assessment that the overall Gulf system – as reflected in the GCC countries, together with influential regional and adjacent powers – should build upon and develop the existing framework to counter new security challenges and threats against a backdrop of a changing regional environment and major shifts in the roles of the main players, particularly with the constant dangers of the proliferation of nuclear weapons in the Gulf region in particular and the Middle East in general.

While strategic power balances have been drastically altered in the wake of the Iran-Iraq War of 1980-1988, the Gulf War of 1990-1991, and the War in Iraq of March 2003, several regional challenges and impediments remain a source of deep concern due to their direct effect on Gulf security and stability within the overall defense and strategic posture.

VI-Conclusions

The dialectics of change and continuity in the Gulf regional system no longer depend solely – as they generally did during the last decades of the past century – on containing the sources of external threats and/or on regional balances of power. Indeed, in the years to come, these dialectics will be further affected in a deeper and more profound way by the interaction between Gulf political elites and their societies in the context of democratic modernization and advances.

This interaction could be the prelude to reformulation of inter-Gulf relations towards the development and implementation of resolutions aimed at furthering economic and social integration and unification. This would also contribute to the reshaping of Gulf foreign relations in the light of restructured forms of political and economic partnerships between the Gulf Co-operation Council, the Middle Eastern, and the Euro-Atlantic systems led by the European Union and the United States.

Among the economic and strategic imperatives is the need to consolidate domestic political, economic and social structures in the Gulf region in order to be more advanced, highly developed and open, so as to interact on an equal footing with the patterns of political, economic and strategic partnerships concluded with the European Union, the United States and the Euro-Atlantic community in general.

Finally, both the dialectics of change and continuity and the prospects of security and stability in the Gulf region, will be affected for many years to come by the approaches employed in addressing vital regional issues such as the future of the foreign military presence in Iraq; the prospects for the system of government in Iraq; the settlement of the Arab-Israeli conflict; the extent of political modernization, democratic transformation and social consensus in the Gulf region; as well as the degree of the positive or negative effects that such sensitive issues – whether collectively or individually – will have on the security and regional stability of the Gulf region.

Undoubtedly, those issues, whether political, security or social, with their domestic and foreign dimensions, will represent future challenges for the Gulf region. They will force the Gulf political elites and societies to devise a suitable response at the appropriate time.

5

What is the position of the Gulf in the World?

H.H. Sheikh Abdullah Bin Zayed Al Nahyan

The question posed today and still awaiting an answer is: What position does the Gulf region occupy in the world? Here we confront three major issues that would explain the realities of the current situation. These are: the imbalance in the region's economic role; the acute difference in technological and knowledge levels; and the variation in standards of overall development. All these issues cast their shadows over the prospective role of the Gulf region and its position in the new world order.

Internationally, the economic role of the Gulf region represents a huge imbalance since the Gulf non-oil exports constitute only around 0.5 percent of the world's total exports. At the same time, the region's gross national product (GNP) stood at about 2 percent of the world GNP. The growth rate of the Gulf GNP would not match even that of a country like Malaysia. In addition, despite the efforts exerted to diversify the economic base, the Gulf economies have remained one-sided with oil representing the largest share of total exports, and the region has failed to become an important venue to attract foreign direct investment. Hence, none of its countries have appeared on the list of the first ten developed or developing countries that are leading in terms of attracting such investment.

In addition to the acute difference in technological and knowledge levels, countries in the Gulf region are importing accessible, ready-made technologies in order to implement their projects and programs without taking the trouble to acquire sufficient experience to localize and develop these technologies in certain fields like oil or food sectors. All available indications show a widening knowledge gap between the Gulf region and the developed world. Expenditures allocated to scientific research and development are estimated at less that 0.5 percent of gross domestic product (GDP) compared to more than 4 percent in the developed countries.

Moreover, these obvious differences are intensified in the course of the general development process and aggravated by the problem of demographic structure and the dominance of the foreign labor force in the Gulf states. It was not until the end of the last century that these states adopted sustainable and comprehensive developmental strategies, resulting in a big gap between the Gulf and global development rates at the political, economic, cultural, scientific and social levels.

Funds assigned for health care programs in the Gulf region are far less than the international levels. The GDP per capita income does not exceed US$ 9000 per annum, which is equal to that of medium income countries. Around 3 percent of the GDP of the Gulf states is allocated to education systems whereas this percentage could rise to 6 percent in the developed states.

Indications of the reduced role and status of the Gulf region in the world order are so many, illustrating that we have been relegated to the role of influenced consumers rather than influential producers. And having lost the power of initiative, the size of our contribution has been declining.

However, when exploring the future prospects of the Gulf region, several facts have to be taken into consideration, the most important of which are as follows:

- During the past few years, the Gulf region has witnessed significant political, economic and social changes. The number of youth has exceeded 50 percent of the total population of which 77 percent have good command of writing and reading, indicating a tangible increase in the qualified workforce in which women play a growing role. At present, the contribution of women to general economic activity in the Gulf area has risen to more that 33 percent even if they do not yet have the full opportunity to play a more effective role in the political and social fields.

- The information and communication revolutions have had a great impact on the relationship between the individual and society, giving the former the capability of expressing opinions freely and of getting acquainted with the attitudes and sufferings of others without being confined by geographical frontiers. Thus, the individuals are able to increase their contribution to the development of society owing to the open-minded position adopted by states like the UAE that have benefited from the information and communication revolutions.
- Although the Gulf states, within their Arab context, have adopted the free-market economic systems, they have failed to realize the ideal benefits of development brought about by the free-market policies employed elsewhere in the world. Consequently, such a situation has made it imperative to shape new, and more positive economic policies in the Gulf region, which has suffered from a continued recession due to the inability to keep pace with global economic developments that are re-asserting the role of the state in effecting economic recovery and social justice.
- During the past years, a substantial portion of private investments in the Gulf region has been directed to the information sector without establishing the required solid linkage between the physical structure of the modern information sector and the production and services infrastructure, which is still modest and simple in most of Gulf countries. In 2002, the GCC states, for example, spent around US$ 6 billion in the information/communications markets without realizing concrete gains either in the output of the information sector or in the field of human resource development.
- At the cultural level, in the wake of the events of September 11, 2001 the whole world is anxiously watching the Gulf region because of the trends and movements which advocate religious

and ethnic notions as their fundamental starting points. At the same time, the region cannot isolate itself from the ongoing cultural developments. Therefore, it has become quite difficult to separate the effects of cultural progress worldwide and the values and principles it generates, on the one hand, from the Gulf religious and cultural discourse and the influence of its underlying domestic conditions, on the other.

- Nevertheless, the picture is not so gloomy, as there is some light at the end of the tunnel. The intellectual assets that we possess can be capitalized on, particularly in the technological and information fields, especially since countries of the Gulf region have allocated huge investments to further develop the professional workforce. Besides, some of these states, including the UAE, enjoy enormous opportunities for further fostering the creative potential of designing and developing computer programs and information systems. Our outstanding experience in UAE in adopting the free-market policies has resulted in diversifying our economic base and reducing dependence on oil as the major source of income.

Allow me to present some perspectives that might contribute to our discussions on the future of the Gulf region under the new world order:

- Freedom in its general concept is the only way to unleash creative achievements in all fields of life. Transparency, openness and participation are the most important pillars of power and authority, since they represent the tools by which society can carry out the duty of shaping its future. Adopting these values with conviction and putting them into practice will pave the way for setting up the proper creative environment to help bridge the existing gap between the Gulf region and the developed world.
- Designing a strategic plan aimed at allowing the Gulf region to participate in, rather than merely benefit from, the knowledge revolution.
- Having a free media, which constitutes the Fourth Estate, strengthens the ability of any society to reveal and correct its

errors, and ensures the right path leading to a better future. This, of course, necessitates avoiding high-flown pretensions, false slogans and mottos.

- The development of the education system with a focus on providing the youth with the proper educational requirements is the solution for the problems of extremism and violence.
- We bear a collective Gulf responsibility to support Iraq and contribute to the reconstruction of this fraternal country and relieve the people of this ancient civilization from the suffering of three decades of tyranny and oppression.
- In the light of our expectations concerning the return of Iraq to the mainstream of Gulf development efforts, and the constructive and positive role that Baghdad will play in this multi-dimensional effort, it is also inconceivable to isolate Iran from this collective Gulf effort. The activation of Iran's role in this effort involves joint endeavor on the part of the Gulf as well as on the Iranian side.

The Gulf region is facing enormous current and future challenges, some of which have resulted from new conditions and developments that have emerged during the past few years, while the others are unresolved, outstanding challenges awaiting our response. The states of the region are called upon to shoulder their responsibilities and invest their potential, will and resolve in order to overcome these challenges to attain the desired goals.

Section 3

Global Developments and the Gulf

6

The Political Impact of Globalization on the Arab Gulf States

Shamlan Yousef Al-Issa

A number of factors make the Arab Gulf a strategic region of worldwide importance. The most important of these is its strategic central position, availability of oil (energy), and the fact that it is a pivotal region that has long attracted much attention through various historical epochs. Consequently, the region has been exposed to many wars and suffered from many crises.

The Gulf states, specifically the six states that form the Gulf Co-operation Council (GCC) are experiencing interactions that affect their social, economic and political structures. The most notable among these is rapid democratic change. The societies of the Arab Gulf states are generally young, with youth accounting for more than half the population.[1] As a result of the educational renaissance witnessed by these societies, political consciousness developed, together with an awareness of the global situation, internal socio-political reality, and the need for a qualitative shift in the process of change. Hence these societies have started to call openly for broadening the base of political participation, sustaining pubic liberties and human rights and reforming all aspects of life.

The Gulf region, in particular, is witnessing economic reforms. This stems from the fact that the crises experienced by the region have revealed the weakness of its economic structures and forced the countries in the region to abandon the concept of the *rentier* state, particularly since they suffer from growing debt, a setback in the progress of development projects and even their occasional failure, a general weakness in education and its outcomes, a disparity between academic and vocational education and the needs of the labor market, as well as heavy reliance on a foreign workforce. All these factors have adversely affected the state structure and stability in the Gulf region.

It is no longer feasible for the Arab Gulf states to continue a policy in which the main approach to problem-solving is to allow issues to get resolved gradually over time. What is required now is to face up to challenges. It is not belittling to review our perception of the concept of "state," nor is it a mistake to acknowledge that the state project has failed in the previous stage. What is belittling is to have political development and internal reforms imposed on us by external decrees. This is manifest in the American project which advocates building democracy in the Middle East, from Bahrain to Morocco. The former Director of the CIA, James Woolsey, hinted at this when he said:

> It is now time for the United States to replace all Arab regimes…absolute dictatorships, narrow-based, [ruling] family regimes…and to spread democracy in the Arab and Islamic worlds.[2]

Owing to international variables following the end of the Cold War, and the disintegration of the Soviet Union and collapse of the socialist system in Eastern Europe, in addition to the crystallization of the new world order under a unipolar leadership (of the United States), globalization emerged as an historical phenomenon undergoing periodic renewal. Though opinions might differ with respect to the history of this phenomenon, there is consensus on the fact that it represents a new stage, allowing intensification of social relations on a global level, increasing global interconnection and shrinking the world, and focusing consciousness on the world as a whole.

Globalization, with its various dimensions, has dawned as a new system affecting all spheres of life—man, society and state. In terms of perspectives and attitudes, academicians, historians, thinkers and writers differ with respect to the implications and effects of globalization on developing countries, including Arab and Arab Gulf states. However, there is a partial consensus among scholars on the fact that globalization leads, in the first place, to changes in the structures of developing societies, and changes in the system of social and cultural values of its people. Perhaps the phenomenon of globalization, which dominates the world today, is a result of the scientific revolutions witnessed by the world a long time ago. These revolutions began with the revolution in science and knowledge, followed by the revolution in communications and information and, finally, the revolution in biotechnology and genetic engineering. Naturally, these revolutions, especially the revolution in information and communications, have had effects on and implications for, the political, economic, social and cultural facets of life all over the world.

In view of the strategic position of the Gulf region, and the availability of very significant energy resources, the region has acquired a distinct and important position in the historical epoch of globalization. It is well-known that the region produces about 20 million barrels of oil daily out of the world's total daily consumption of 65 million barrels. In other words, the Gulf region satisfies 25%-30% of the world's consumption of energy and its exports of crude oil are in the region of 20% of total world exports. Moreover, it possesses 44% of the world reserves of oil.[3] Several symposiums and conferences have been held and many books and studies written on the effects of socio-economic globalization on the Arab World. However, these have not addressed in detail the economic, social and political effects of globalization on the Arab Gulf states.

This chapter aims to identify the content and dimensions of globalization, as well as its political, social and economic effects on the Arab Gulf states. The study also aims to identify the positive and negative sides of this phase in the countries and societies of the region. It will do so by addressing the dimensions of globalization, especially its effects on existing political structures in the Arab Gulf states. Focus will be on aspects relating to the diminishing internal and external roles of the state, the erosion of the concept of national sovereignty as well as the effects of globalization on national identity and prevailing social norms. Though there are differences over globalization in its totality and hesitation in deciding how to deal with it, its implications for certain aspects of life are very clear, especially in the economic domain.

The importance of this study derives from an abiding awareness of the extent to which the Gulf region has been affected by globalization and its different ramifications, and of the need to deal with it in all its different aspects. Globalization is undoubtedly coming. Therefore, any dismissal of this phenomenon, and any hesitation in dealing with it, is tantamount to swimming against the tide.

On the other hand, the difficulty of this study stems from the inability to make generalizations or reach common findings due to the lack of fixed and clear policies towards globalization among the six Arab Gulf states. Each of the Gulf states has its own idea of how to deal with globalization and its effects. Some of these states categorically reject globalization insofar as its politico-cultural dimension is concerned, but accept its economic and commercial implications. Others have pioneered the process of assimilation and fusion in the melting pot of globalization. It seems that the remaining states have not grasped its full consequences and ramifications yet. Therefore, they occupy a half-way house between the other two positions.

In short, this study attempts to answer the following questions:

- Are the Arab Gulf states still showing reluctance in dealing with globalization?
- Will it be adequate to deal only with the economic side of globalization?
- What are the effects of political globalization on the Arab Gulf states?

Globalization: Definition and Dimensions

The Definition of Globalization

Globalization is a new stage in the evolution of modernity where social relations intensify on a world scale. External and internal affairs fuse together in a manner that does not allow their splitting into separate domains. Internal affairs are linked to external ones via economic, cultural, political and human ties. The thrust of globalization is towards a shrinking world and an increased consciousness on the part of individuals and nations of this shrinking process. This aspect is described by Roland Robertson who maintains that the concept of globalization means on the one hand, compressing the world and making it smaller, and on the other, intensifying the way people conceive of the world as a whole.[4]

Some scholars believe that globalization is a spontaneous, natural development which signals the gradual merging of the world. The revolution in communications, which results from technological development, plays a major role in this process. Others believe that the concept of globalization involves a form of intent. The aim is to universalize a certain pattern of existence whereby technology prevails and the United States plays a leading role because it monopolizes two thirds of world communications.[5]

Muhammad Abid Al Jabri defines globalization as a hegemonic will, and as such a "repression and exclusion of privacy." In contrast, he believes that universalism aspires to elevate privacy to a world level. He maintains that globalization means "containing the world," whereas universalism means openness to what is universal.[6] If there is a school of thought and politics that conceives of globalization as a process that is independent of all forces whatever they are,[7] there are other scholars who think that the term globalization is the equivalent of American hegemony. This is evidenced by the serious attempts on the part of the United States to impose its authority and lifestyle on the nations of the world.[8] The concept of globalization is manifested in a number of issues, like growing international interdependence and the great revolution in the field of communications and information. Together, these signal the existence of a "world culture," or a universal culture, as an alternative to multiculturalism. There are also some emerging new challenges threatening the international community that generally transcend political borders.[9]

One can say that globalization is a process that relates to spreading information, abolishing borders between political systems and increasing aspects of similarity between social groups, societies and institutions. All this has positive or negative effects on all states, and their components. Generally speaking, globalization is a mixture of various ties and interlinked relations that surpass the borders of nation-states.

The Dimensions of Globalization

The economic dimension is one of the most important dimensions of the concept of globalization. The main pillars of globalization include: a free economic system, abolition of all traditional economic patterns and practices, and embracing an economic process leading to advancement and prosperity. A free economic system rests on the theoretical fundamentals that have been addressed by Adam Smith, David Ricardo, Jeremy Bentham and Herbert Spencer. These fundamentals comprise a market economy that is compatible with human nature. Such a market tends to enfranchise individuals, remove all shackles and enable them to be innovative. Its main prop is the principle of supply and demand and non-intervention by governments in economic affairs.

The most important feature of the new world economic order is the liberalization of world trade, cancellation of governmental control (customs and other controls) on trade, and establishing the World Trade Organization.[10] In brief, we can say that globalization is a process that involves liberalizing trade in goods and services, flow of capital and exchange of products. It also increases the typification of economic institutions and the rules of control world-wide[11] and the growing functions, roles and capabilities of multinational companies in the economic, social and political spheres.

Theoretically, economics in the Arab Gulf states is based on the free economy system, which enhances the role and efficacy of the private sector. However, these states, in practice, run counter to the theoretical fundamentals of this system as governments play a major role in administering economic activities. This role impacts on the meaning and concept of economic freedom, causing it to fall short of those theoretical fundamentals. This failure is attributable to the fact that governments in the Gulf own the oil wealth, which is the main source of their income. They control sources of income and the process of redistribution of wealth. The private sector relies almost completely on the government; this is quite evident in government-executed projects, whether in the domain of infrastructure or different industrialization projects. Moreover, governments in the Gulf states are the first and foremost legislature in matters pertinent to economic and financial policies. It transpires that the role of the private sector is marginal as far as contributing to these policies or the process of formulating them is concerned.[12]

Since the discovery of oil, particularly the rise in oil prices after 1973, the Arab Gulf states have adopted policies leading to the creation of welfare states, in which citizens rely completely on the state and the services it provides in all fields. The objective was to reinforce the degree of legitimacy of political regimes and to gain the allegiance of citizens so as to achieve political stability and continuity of governance. Owing to international variations, including the emergence of globalization, fluctuations in oil prices and increasing costs of services, the Arab Gulf states have started to suffer from budget deficits, especially because they currently face internal pressures to employ national manpower. This necessitates reviewing past economic and financial policies and implementing the following measures:[13]

- making prompt political decisions relating to privatization, and inviting the private sector to participate in setting and formulating privatization policies
- transferring the ownership of the tools and means of non-oil production to the private sector
- establishing public oil companies through which citizens can participate in the process of managing oil wealth
- getting rid of administrative bureaucracy and increasing managerial competence to guarantee speed and efficiency in production
- paying attention to regulations that correspond to the spirit of the age, which entails taking into consideration accelerating changes in all fields, particularly the economic field and at the

same time paying attention to regulations and laws establishing efficient control and regulatory mechanisms.

There are some questions to be raised regarding the capacity of Gulf regimes to deal with all the ramifications and effects of economic globalization. The most important of these questions are the following:

- Can the Arab Gulf states cope with the global economic order without compromising the nature and structures of existing political systems?
- Can the economy be liberated without liberating human beings and society?
- Can the Arab Gulf states implement the principle of the rule of law, equity, justice and transparency in traditional societies where personality cult and tribalism hold sway?[14]

In its socio-cultural dimension, globalization calls for effecting changes in the structures of developing societies and their institutions, as well as their system of social and cultural norms. This is so because the social effects of globalization, particularly on the family and the process of socialization, manifest themselves in spreading a certain pattern of global values. This pattern focuses on imposing the prevalent, hegemonic system of western values. Consequently, there is an increased concern with the implications and legislations relating to civil society, human rights, minorities' rights, women's rights, political participation and democracy in developing societies. In the domain of culture, globalization seeks to impose a single, global culture. The culture of globalization endeavors to unify the world, destroy borders and negate the efficacy of national sovereignty. In the era of cultural globalization, the dominant cultural order is embodied in an audio-visual universe and the spread of satellite television stations and major media institutions testifies clearly to this fact.

The advances in communications technologies and world information systems have effects on individuals and societies and will lead to changes in values, modes of thinking and behavioral norms. The Arab world lies in an area covered by 59 satellites specialized in communications and other activities. The area receives a large number of television satellite channels. The Gulf region is one of the areas most affected by foreign television channels (among the most important of these are CNN, CBS, ABC, NBC, and BBC). These television stations profoundly affect public opinion. They represent a factor that influences the system of values and behavior of individuals and societies. Moreover, there are many Arab satellite channels which broadcast programs internally and externally.[15]

In the light of the foregoing discussion, many questions arise with respect to the socio-cultural effects of globalization on the Arab Gulf states. Two of these questions are as follows:

- Can the traditional and tribal societies of the Gulf hold their own in the face of changes in values wrought by globalization?
- Can the culture of globalization be employed in the process of rebuilding Gulf societies?

The Political Effects of Globalization on the Arab Gulf States

Over the last three decades, the Gulf region has witnessed broad-ranging developments with respect to modernization in all domains. These developments have established a massive infrastructure that includes roads, schools, universities, hospitals and other structures that contribute to providing various services. This has been possible because of the increasing income derived from oil, though the region has been unstable in terms of politics and security. The region has witnessed a number of wars (Iran-Iraq war, the liberation of Kuwait and the war in Iraq) that have had a profound effect, as reflected in the budgetary deficits in these states.

With the start of the second of the Gulf wars (the war for the liberation of Kuwait) the United States has become the sole power dominating the world. Many thinkers believe that this war was the beginning of a new world order, leading to the era of globalization, with political effects on the Gulf region that we are going to discuss. The Arab Gulf region passed through an era of political weakness in the 1980s. This led to a state of inertness that narrowed the scope of political development, and to an increased monopoly on political power in most of the Arab Gulf states. Some scholars explain this by reference to economics since Gulf regimes have developed the concept of the *rentier* state, where citizens pay no taxes or fees. *Rentier* states take responsibility for all expenditures and provide citizens with free services, such as health and education (the Kuwaiti government, for instance, subsidizes electricity, water and the agricultural and residential sectors). Moreover, these states employ citizens on a large scale in the public sector and in government institutions. Some scholars view these privileges as a political tactic to secure allegiance and unilateral political authority.[16]

Another explanation of power monopoly and unilateral rule in the Arab Gulf states is that the tribal and clan heritage of Gulf societies does not yield a feeling of citizenship, or a sense of belonging to the state. Some maintain that allegiance is to the tribe, and these states are simply “tribes with national flags.” Allegiance to tribal sheikhs and their deputies derives from attempts to multiply benefits and gains. In this respect, Prince Saud Al-Faisal, the Saudi Minister of Foreign Affairs, maintains that the failure of political reforms in the Gulf region is inseparable from the nature of geography and sociology. He states:

> The factors that have affected Arab body politics pertain to Historical Geography, which has led to an on-going historical struggle between two values. [On one hand, we have] the nomadic values of the country, which are never stably contained in a specific geographical context, and are not subject to an established political authority. [On the other hand, we have] urban values which are embedded in urban centers that are scattered in the heart of vast deserts, exposed to conquest and subject to decline. This struggle between two modes of life has led to a disrupted, temporary and partial political experience among Arabs [and] to the existence of strong ties that transcend political commitment and merge with tribal allegiance.[17]

National Sovereignty

Globalization has caused erosion in the principles of national sovereignty. It is now evident that there is a change in the role of government—it is no longer what it used to be. What is going to transpire is not the abolition of the nation state, but a change in its role in the light of market economies and universal culture. The tendency is towards abolishing the central role of the state in the domains of politics, economics and society. Hence there is a pressing need for a strong state and government to reduce the negative effects of globalization on the citizens. The era of globalization means that a state cannot dissociate itself from universalism, or what is referred to as "international interdependence."

Hence some thinkers believe that the interests of states dictate coping with this new situation, and adapting to it rather than clashing with it. This will facilitate reorganization of the government's role in a more functional way. In this way the power and legitimacy of the state shrink and regardless of the state's political leaning, it becomes the priority of the government to preserve the confidence of the market and the global business community.[18]

What the Gulf region is witnessing today is not a reversal for the nation state nor its end, but a setback to, or an end of, the *rentier* state that has dominated the lives of individuals and societies in the Arab Gulf states for the last three decades. The end of the *rentier* state and the emergence of new liberal initiatives in some Arab Gulf states do not mean the end of the state itself because oil, which is the main source of income, remains state-controlled and the private sector does not have the option to own some of the public sector institutions. What is happening in the Gulf region today is that the financial burdens on governments have started increasing considerably because of reductions in revenue owing to fluctuations in oil prices. Moreover, there is increasing pressure on the governments to employ the national labor force in the public sector, in addition to increased arms spending, all of which have caused budgetary deficits in these states.

Hence Arab Gulf states have turned towards reforms in economic regulations and related laws and regimes. On a smaller scale, some states have turned towards formal political changes although these do not necessarily mean radical systemic changes in the political structure.

One of the most important outcomes of globalization in the contemporary world is the eclipse of political ideology as a result of the decline of authoritarian, ideology-oriented regimes. It is ironic that Arab Gulf states are witnessing a growth in ideology as represented by political Islam in spite of the intense US-led international campaign against extremist Islamic groups.

It seems paradoxical that while these Arab Gulf states accept the American presence in their countries to maintain security and stability, they reject western calls for the introduction of democracy and political modernization. The press and information institutions in some of these states criticize western countries, viewing greater political participation and support for human rights as a threat to current social and political traditions. Moreover, these states are wary of the effects of modernity, contemporaneity and globalization.[19]

Democracy and Political Participation

Among the effects of globalization in the political field, are a number of humanitarian principles and initiatives, highlighting human rights, and respecting a person's political and intellectual choices. The first among these principles is democracy, which comprises a framework within which the individual exercises his right to self-determination. Democracy here does not mean electing municipal or local councils only. It also means shifting towards a system of values based upon respect for human dignity, difference of opinion, the rule of law and belief in political pluralism. It also means abolishing the monopoly on truth, when this is practiced in the interest of one particular group and at the expense of another. Democracy is a system of values whose outcome is mature political behavior based upon the aforementioned principles.

It is well-known that the oldest democratic experiences in the Gulf region materialized in Kuwait, followed by experiences in other Gulf states such as Bahrain, Qatar and the Sultanate of Oman. However, these trends fall short of full-fledged democracy and may be more aptly described as "limited democracy." Such systems lack some of the basic elements of democracy because of limitations in public liberties, absence of political parties and organizations, and lack of political participation by women. In some Gulf states, government control still permeates most fields of life, which has the effect of narrowing public liberties and reducing the scope of public participation. Also, there is a lack of transparency and control institutions, as well as a failure by the organizations of civil society to play an effective role, which is viewed as the main instrument for effecting desired change in these societies.

The main question to be posed is as follows: Why has democracy spread in Eastern Europe, Latin America and Africa but not in the Arab World? This situation persists despite the fact that there are internal public pressures and favorable international conditions for a shift towards democracy. Also, there are some political leaders who show a readiness for adapting to change.[20] Are there internal and external circumstances that facilitate democratic changes in the Arab Gulf states? Do the political leaderships in these states wish to move towards change and democracy? In other words, do they have the political will to change?

Several public attempts have been made in the Gulf to effect political participation. These attempts differ in terms of magnitude and quality from one state to the other. For instance, Bahrain witnessed demonstrations and calls for a return to democracy and the constitution of 1973 and this set in motion a process of political reform. Thus the National Action Charter was approved in a public referendum in February 2001, followed by approval of the amended constitution in February 2002. Thereafter, municipal council elections (cancelled since 1956) were held with the participation of all political forces including women, who were given the right to vote and to be elected, in May 2002. Finally, general parliamentary elections were held in October 2002. Currently, the Kingdom of Saudi Arabia is undergoing intensive attempts to implement political reforms and public participation. As for Qatar, its political leadership has decided to speed up political participation. This is evidenced by the passing of a permanent constitution and approving it by a public referendum in April 2003, in preparation for free elections to the Consultative Assembly.[21]

Women and their Human and Political Rights

The current status and social position of women reflects the degree of progress attained by that society. Therefore, investing in the potential of women is an urgent need. The issue of Gulf women and their status is multi-faceted as it relates to the abolition of all forms of exploitation of women. It is very difficult for societies to change as long as women are disadvantaged and vulnerable. The state of affairs concerning women's position in Gulf societies exhibits some vulnerabilities and disadvantages.[22]

In spite of such disadvantages and vulnerabilities, Gulf women have made some achievements in various fields, the most important of these being education and employment. However, the percentage of women in the labor market differs from one state to the other. As for political participation as a human and political right, women still remain in the background. With the exception of Qatar, Bahrain and the Sultanate of Oman, most Arab Gulf states have not granted women their political rights to vote and to be elected. It seems that male chauvinism is still dominant in these states, and that they lack the kind of political consciousness that accords importance to the role of women, even within broad female constituencies. It has been noted also that women do not vote for female candidates, which may be attributed to male dominance in society, and to the prevalence of traditional values. Evidence of this trend may be found in the results of the legislative or municipal council elections in Bahrain, Qatar and the Sultanate of Oman.[23] (A case in point here is the municipal council election which was held in Bahrain in May 2002. Though the percentage of women participating in elections was higher than that of men, women did not win a single seat.)

As for other human rights, reality reveals shortcomings in women's rights as citizens. Women are often not entitled to social allowances, nor are they always given the right to a government residence. If a woman is married to a foreigner, her children do not automatically enjoy the right to become naturalized citizens. Women often do not occupy senior government positions even if they are better educated and more competent than men. Moreover, they are generally excluded from the judiciary and diplomatic corps.

According to family law, a woman does not have the right to marry without the approval of her guardian even if she is of age—the age of consent being 21 years in most Arab Gulf states.[24]

In the context of globalization, Gulf women will face new challenges and the process of their emancipation will be subject to pressures from fundamentalist, Islamic opposition. This is evidenced by the rejection of the political rights for women in Kuwait despite the political will supporting such rights. Other challenges that women face include the public attitude to co-education and fellowship at the workplace, spearheading activities in the organizations of civil society, and the rights associated with political participation.

Most statistics indicate that the proportion of women to men is equal in most Arab Gulf states. For instance, in Kuwait, this proportion became 50.72% in 2000.[25] In general women represent one half of society. How can full development be achieved if half of society remains idle?

Obstacles to Political Reform in the Arab Gulf States

Political reform in the Arab Gulf States faces a number of obstacles. The most important of these are mentioned below:

Absence of Political Will

One of the obstacles hindering political reform in the Arab Gulf states is the absence of the political will to reform. Some leaders do not want to bring about the necessary political transformations for fear of being accused of doing so as a result of foreign pressures.[26] Also, the ruling elites tend to reject political pluralism because it will affect the way they currently rule and control state affairs.

Lack of Organized and Efficient Political Opposition

The absence of an organized and efficient political opposition and the lack of party leaderships that enjoy the support of popular and sustainable constituencies have led instead to the appearance of sectarian, tribal and familial leaderships. This has meant the absence of effective public pressure on the political authority that will make it responsive to political reform measures.

Absence of the Organizations of Civil Society

There can be no talk of political, social and economic progress or developmental advances, without recourse to civil society with its organizations and regulations. Civil society means the existence of a group of free, voluntary organizations that represent a continuum between the family and the state, and which are designed to realize the interests of their members in accordance with the values and norms relating to individual initiative, mutual consent, tolerance, and the proper management of diversity and difference. It would be a truism to say that the efficacy of civil society is determined by the extent to which it is separate and independent from state authority.[27]

Since the beginning of state formation in the Arab Gulf region, national and social organizations appeared in Bahrain and Kuwait at the start of the last century. These were effective institutions that registered achievements in the fields of education, literary development, and public culture. However, after the discovery of oil, these institutions became subject to state control and became tools of political authority. Over the last three decades, the number of these organizations has increased in most Gulf states. However, their role has diminished and their efficiency has declined.

In the light of current economic and political circumstances in the Gulf states, which necessitate a fundamental change in political discourse of the state and a reduction in its existing burdens, the state must create a conducive atmosphere in which it hands over to society all the tasks and responsibilities that fall outside the realm of its duties. To achieve this, some bold decisions must be made, and some long-term strategies must be drawn up to encourage and establish organizations of civil society with the aim of minimizing internal tensions and achieving a general equilibrium in society.[28]

The Weakness of Institutionalism

One of the main obstacles to the process of political reform in the Arab Gulf States is the weakness of institutionalism in the political structure.

In these states personality cults and unilateral political decisions supersede the working of political structures. Ruling authorities dominate the mechanism of political decision-making. Despite the existence of parliamentary institutions, which are supposed to have the main role in passing different legislative laws, these institutions are weak, since most bills come as proposals from the government. The governments apply pressure on parliamentary groups to make them pass the desired bills. Moreover, the constitutional regimes have the ability to dissolve parliaments at will. Thus the principle of separation of powers has become a mere formality.

The Weakness of Democratic Culture

One of the most important reasons for the slow process of political reforms in the Arab Gulf states is the absence of a culture imbued with democratic values. This relates to the fact that the structure of the political and social systems is a patriarchal, nomadic, traditional one, with a propensity towards authoritarianism. In such systems the power to bind and to release belongs to the head of the family, tribe or state. For instance, in Kuwait, the state with the oldest experience of parliament and elected municipal councils in the Gulf, tribal and sectarian polarization is reflected in the structure of the representative democratic institutions (the National Assembly and the municipal councils), where elections do not involve modern political groups, but traditional coalitions. There is now a practice which has become deep-rooted—tribal by-elections to eliminate competitors and pick one nominee representing the tribe in the constituency concerned. Although illegal, this practice has become a reality. According to the results of the elections for the tenth Kuwaiti National Assembly (2003-2007), traditional tribal factions won 24 seats out of a total of 50. They were divided into 7 seats for Al ‘Awazim, 4 seats for Al Matran, 4 seats for Al ‘Ajman, 3 seats for Al Rashaida, two seats for Al ‘Otban, two seats for Al ‘Anouz and one seat for each for Shamar and Al Hawajer.[29]

The political competition between Gulf regimes and political Islam movements to win public support through traditional social structures, especially tribal, clan and sectarian allegiances, has created new trends and parties that defend factionalism, clannishness and sectarianism.

Consequently, the dominant culture in the Gulf states lacks a certain degree of freedom and choice. This in turn stifles innovation and creativity despite the fact that many decades have lapsed since independence and the foundation of modern institutions. Although education opportunities have spread, the dominant thinking is still traditional in terms of methodology and mindset.

Given the rapid changes and the emergence of the era of globalization, which contributes a great deal to opening up vistas of change, there is a pressing need in the Arab Gulf states to reformulate the modern values attached to these changes. Not only this, there is also a need to draw a clear-cut strategy with regard to social and political upbringing, and spreading the culture and the values of democracy. Without this there can be no way to meet these rapid changes. Benefiting from the forces of change in all spheres will vary from one state to the other in accordance with their current levels of development.[30]

External Initiatives for Political Reform

The Arab World, including the Arab Gulf states, is experiencing several pressures to embark on universal political reform. The most important of these are external pressures that come in the form of initiatives and recipes for desired change. The most recent is the Greater Middle East Initiative (GMEI) which is a new US initiative for the Middle East region. It is the culmination of Bush's endeavor to spread democracy and freedom in the Middle East, and Colin Powell's call for an American-Middle Eastern partnership.

The principles of the new initiative are based upon effecting change in the light of the two Arab Human Development Reports (AHDR) issued by the United Nations Development Program in 2002 and 2003. The US initiative has adopted the three main objectives mentioned in the two reports. These are:

- promoting democracy and good governance
- building a knowledge-based society
- expanding economic opportunities.

These main headings are further divided into sub-headings, with a detailed executive framework. In the field of democracy the initiative calls for supporting and promoting free elections in the countries of the region via technological assistance, training in the field of parliamentary practices, activating women's role and participation, developing independent private media, encouraging the states of the region to combat corruption and supporting the emergence of civil society. As for building knowledge-based societies, the initiative consists of a number of plans concerned with supporting primary education, eradicating illiteracy, dealing with shortages of school books, reforming educational programs, and widening the use of the Internet and other information systems. As for expanding economic opportunities, the initiative proposes raising several funds to finance what is referred to as "bridging the economic gap of the Greater Middle East," starting funds for small projects, and establishing the Greater Middle East Development Bank (GMED) to help reform-minded countries by providing development prerequisites, in addition to a partnership for a better financial system. Moreover, the project proposes facilitating the process for the Middle East states to join the World Trade Organization and establishing free trade zones to promote regional commercial exchange and joint regional projects.[31]

The United States launched a wide media campaign to promote this initiative, starting with the World Economic Forum held in Davos in January 2004. The United States aimed to provide an international and legitimate umbrella for the initiative by discussing it in three international forums held in June 2004: the G-8 summit held in the US state of Georgia, the summit of NATO member states held in Istanbul, Turkey and the joint summit meeting between the European Union and the United States, held in Ireland.[32]

As for the reactions to the US initiative in the region, the meeting between the Egyptian and Saudi leaderships, for instance, held in February 2004 rejected the Greater Middle East project. Its final statement made clear the "refusal to accept the imposition of a particular mode of reform on the Arab and Muslim states from abroad," and indicated that "the Arab states are going forth in terms of development, reform and modernization in a manner consistent with the interests and values of their people and compatible with their needs, particularities and Arab identity."[33] In reaction to this last initiative, Egypt has submitted to the Arab League a project for Arab political reform. This is considered to be a rejection of readymade foreign recipes though it does not amount to a refusal of external assistance which is consistent with this endeavor, provided that it does not conflict with the realities of the Arab World.

Conclusion

The era of globalization will open new doors to the people and states of the developing world, including the Arab Gulf states. The winds of change will blow across all fields, reaching individuals and governments unhindered by constraints or borders since they do not require permission to enter. As mentioned already, resisting this sweeping tide or ignoring it will be very difficult. This is so because doing either will relegate the Arab Gulf states to the position of backward nations lost in the darkness of underdevelopment. Diverse and difficult challenges will be encountered while the region copes with globalization. These include the challenge represented by Islamic fundamentalism and some fanatical nationalistic forces, as well as the prevailing system of traditional values.

In the light of this, the question arises: Are the Arab Gulf states prepared to face the challenges of globalization? Vahan Zanoyan's study entitled *Time for Making Historic Decisions in the Middle East*, postulates the view that the Gulf region faces painful choices and decisions today that ought to have been addressed a long time ago. He attributes this to the fact that the preferred strategy of dealing with problems in the Gulf region is "to flounder through them." In the last three decades, the Arab Gulf States have been preoccupied with preserving the status quo without giving serious consideration to addressing the problems and issues at hand through a realistic, long-term strategic vision. The prevalent policy was to buy the allegiance of different players, or to buy time in the hope that the problems would resolve themselves. The policy of floundering through crises has thus become deep-rooted. It has become the only feasible policy for the regional states in facing challenges. This is a strategy that no one dares to question because it reflects a tendency to avoid dealing with events and problems and transferring them to the next generation.[34] The plausible choices, as Zanoyan envisages them are those that tackle basic, chronic problems that have been neglected for too long. They are not those choices that present partial, superficial solutions dealing with symptoms, which have only come to the fore in recent years.[35]

In fact, the Arab Gulf states are racing against time, since coping with the dimensions and effects of globalization is in itself a formidable challenge. This study has sought to list the political obstacles confronting these states. There are a number of stumbling blocks on the way that will hinder the process of change. These include the absence of effective political will to bring about the desired change, as well as the absence of organizations of civil society. There is also the continuing strength of traditional structures and tribal, sectarian and familial allegiances, all of which may be major obstacles to change. Moreover, the following factors need to be noted: the absence of a reformist, national, effective political opposition, the fragility of democratic culture, the lack of a middle class, the low level of concern over human rights and the absence of an effective role for women.

This chapter has shown that the Arab Gulf states have reservations with regard to the political dimension of globalization. However, they show no hesitation in dealing with the economic dimension of globalization. These states are witnessing an economic openness to the world which varies from one state to the other. The Emirate of Dubai has gone a long way in the field of economic openness; Qatar, Kuwait, Bahrain and the Sultanate of Oman are trying to catch up with Dubai. In this connection, we have to highlight the fact that all the Arab Gulf states – with the current exception of Saudi Arabia – have accepted the conditions of the World Trade Organization and have obtained its membership.

7

Democracy in the Gulf? Challenges and Opportunities

Benjamin R. Barber

The focus of my presentation is on that most seductive, most enticing, most troublesome, most controversial and most problematic subject—democracy. Democracy, it may be noted, always comes at the far end of the day after economics and war have had their say, and attention has waned.

At the outset, I would sound a warning: In the ancient world, when the Athenians dominated the Mediterranean, it was often said: "Beware the Greeks, especially when they come bearing gifts." I would say to you: "Beware the Americans, especially when they come talking democracy." And I am an American, and my President has been talking a great deal about democracy.

So let me begin by suggesting what democracy is not. Most importantly, democracy (or democratization, the process of becoming democratic) cannot and should not be confused with Americanization. America is a democracy but democracy is not American. Americans do not own democracy any more than France owns literature or Persia owns rug-making.

Nor is democracy identical with modernization—something that is new, innovative and contemporary. On the contrary, its lineage is ancient and its roots in the West go back to ancient Athens, Sparta and Rome, back to medieval natural law and early Christian egalitarianism (which, like Islam, focuses on the brotherhood of man). It goes back to a great British thirteenth century charter – the Magna Carta – and the small principalities and republics and the free port cities of the Renaissance, back to the autonomous towns of colonial New England and the revolutions of 1776 and 1789. Many theorists would argue that modernity does not so much give birth to democracy as throw up obstacles in its way, undermining the conditions that facilitate its participatory character. This was certainly the argument of the great eighteenth century philosopher Jean-Jacques Rousseau.

Finally, democracy is not the same as Westernization. Although its early history certainly had Western roots, it is founded on universal human aspirations to self-governance and communal autonomy, and in recent times has appeared in many different forms in many different parts of the world. It did not start in one place and spread like a virus from one venue to another, it arose – as civilization did and as life itself did – in several different places at different times as a result of indigenous struggles and ripening conditions. Note that the recent introduction of a new constitution in Afghanistan was secured by an old and indigenous Afghan democratic institution – the Loya Jirga or national tribal council – through which the many disagreements about the new constitution were adjudicated and resolved. It was a case of old, indigenous democracy paving the way for new, "imported" democracy.

These rather general and abstract reflections on democracy are a prelude to more concrete remarks on democratization in the Gulf, and more generically, the Islamic world throughout Arab and Persian societies, because they refute a number of pernicious myths that have impeded the evolution of democratic life throughout the developing world and specifically in the Middle East. These myths give some misleading impressions: that democracy is a Western commodity; that it is a form of government that can be secured in a hurry; that a people can be "liberated" from the outside and liberty can be imposed; that democracy constitutes little more than elections; that it is a function of "good government" and top-down constitutional relations; and that it means outcomes acceptable to outsiders rather than a process by which the will of citizens is represented. Upon examination, all these claims prove to be bogus, myths by which some have attempted to impose narrow proprietary perceptions of democracy on others.

The Myth of Democracy as a Singular System

Democracy does not belong to one people, neither does it grow out of a single nation's history nor is it the property of a particular nation. Many nations – among them France and the United States – claim an "exceptionalist" story that makes them uniquely susceptible to the blessings of liberty, and such claims can be useful incentives in a local struggle for democracy. However, the blessings of liberty are a universal right, and the story of liberty turns out to be plural rather than singular. Indeed, it is wise to speak always of democracy not in the singular, but rather of democracies in the plural sense. The world's peoples aspire to democracy in different ways and follow different paths towards it. There are many stories of democracy, and hence many different democracies. Societies in the Gulf region can forge their own way, as other nations have forged theirs.

The Myth of "Overnight" Democracy

Democracy takes time. It cannot and has not been secured in a hurry. The United States took a hundred years to secure its initial independence and another eighty years under republican government to rid itself of slavery. The English have been on a democratic path for three hundred years or more and are still struggling to perfect their system. The Swiss declared their independence first in 1291, which means they have been working with democracy for more than 700 years—and women in Switzerland only received the vote in 1961. Yet in Iraq, optimists think democracy should take about six months to be secured, while pessimists give it two years. These statements are not being made to countenance endless delay or to disrespect people whose impatience pushes them to make immediate demands for their rights. Instead, these statements acknowledge that achieving free government takes time, skill, patience and a willingness to accept mistakes.

The Myth of Liberation by Force

Democracy cannot be imposed at gunpoint, even when the gun is wielded by a "good-willed liberator." War may overthrow a brutal tyrant but cannot by itself create a democracy. Liberation from the outside by force leads to occupation and anarchy rather than self-government and democracy. Furthermore, occupation and anarchy, even in a good cause, often breed strife, unrest, civil war, and renewed tyranny. The distance to be traveled from the overthrow of tyranny to the founding of democracy is evident in the chaotic condition prevailing in Iraq. It is prudent to recall that in the American War of Independence of 1776, the French did not "liberate" the Americans from the despotic hold of the British Crown. It was the Americans who seized their own liberty—to be sure, with help from the French.

The Myth of Elections

Democracy cannot be equated with elections. There are undemocratic systems that use plebiscite-type elections to ratify despotism (such was the case in the erstwhile Soviet Union) and there are free societies in which the election of representatives plays a relatively small role (as in Switzerland, where referendums engage citizens directly in legislation). It is real democracy based on citizenship that produces elections, not elections that produce democracy. In the absence of a robust civil society and a vigorous and responsible citizenry, elections will have little meaning or impact. The rush to hold elections can obstruct democracy by leapfrogging the hard work of training citizens to make elections work. Of course, it is only in a second election when a new government is tested by being subjected to public scrutiny and possible electoral defeat that democracy proves its real mettle.

The Myth that Freedom is a "Gift"

Democracy cannot be gifted by one people to another. It is produced internally by indigenous struggle and by a people who want liberty and will fight for it. Freedom cannot be a present from those who have it to those who do not. Indeed, nations willing to gift freedom (and hence a degree of their own power) to others often have self-interested agendas. Ask the courageous women in this part of the world who know that they will only be as free as their struggle permits them to become and that they will grow much older before gaining any freedom if they merely wait for others to "gift" them their dignity.

The Myth of "Good Government"

Democratic government is not always good or efficient government, though it strives to be. What democracy really means is self-government. And self-government, if it means anything at all, means the right of a nation and its people to make their own mistakes. America certainly did (and does): for eighty years, after most of Europe had abjured the abomination of slavery, the United States maintained a slave-based plantation system that was deeply at odds with its claims to liberty and rights for all. Fortunately, the British did not come and try to snatch back its sovereignty because Americans had "failed" in their republican experiment. Democracy within a nation is not something for well-wishers outside to "cancel" or "annul" when outcomes do not live up to the outsiders' expectations. When Turkey rejected the American plans to send troops through Turkish territory to establish an American front in northern Iraq last year, some Americans reacted as if that decision was "undemocratic"—because it was not in America's interest. Western powers have sometimes tried to cancel electoral results when these were not to their liking, as happened in Chile in the 1970s and Algeria in 1991 (In the latter case, when primary elections in Algeria put a moderate Islamic Party on the brink of victory, France, with American support, helped the Algerian military to "cancel" the outcome—resulting in a decade of devastating civil war in which the middle class was nearly liquidated and far more toxic forms of fundamentalism were bred than were on the horizon in 1991).

In this regard, I would like to quote a personality who has an ambiguous reputation in this part of the world—T.E. Lawrence. When confronted by British officials who were disturbed at the turn of events in colonial Arabia, Lawrence of Arabia told them:

> Better to let them do it imperfectly than do it perfectly yourself. For it is their country, their way, and your time is short.

Take note Paul Bremer. Pay attention Dick Cheney and Donald Rumsfeld.

Having examined these dangerous myths and having outlined some ensuing general guidelines, I would like to address issues of democracy here in the Gulf, not as an expert in the affairs of this great civilization – as I am neither an Arabist nor a historian of the Middle East or of Islamic society – but only as a student of democracy with a deep respect for the diversity of the world in which democracy unfolds.

At this point, let me explain that I prefer dialectical thinking to analytic pairs. Many trained in the school of Aristotle and later Kant think it must always be "either A or not-A, which is B." Either/Or – This or That. However, I would like to understand how A and B are related, how "this" helps to create "that" and how "that" is often a contrary mirror image of "this." It is in this spirit that I wrote *Jihad vs. McWorld*, which considers the forces of integration and uniformity, and the forces of disintegration and parochialism as forces that generate and sustain one another – both at the expense of democracy – rather than as polar opposites occupying different universes. Thus, when I reflect on the strengths and weaknesses of the struggle for democracy in the Gulf region, I think not of two distinct sets of characteristics and features that are isolated from each other, one "good" for democracy and the other "bad" – of dualistic "virtues" and "vices" – but rather of a single set of characteristics that both facilitate and impede democracy. They represent two sides of one coin, two moments in a single dialectic.

In simple terms, many conditions in this region that help to nurture democracy can also be obstacles that impede its chances of success.

Six features characteristic of this part of the world come to mind. Three of these are quite specific to the Gulf and the United Arab Emirates, while the other three are found both in the Gulf region and throughout the Islamic Middle East. The three features that are specific to the Gulf are the small-scale of society, the federalized or decentralized organization of the state, and the custom of consultative rights inherent in traditional modes of decision-making. The three aspects that are general to the region include a tradition of one-man or one-family tribal leadership, the accident of immense fossil fuel wealth, and the reality of a deeply Islamic society and history. A closer look at these six features will show how in each case, there are elements that nurture democracy as well as elements that are potentially problematic or obstructive to democracy.

1-Size

A small-scale diminutive society can be one of democracy's greatest strengths. After all, democracy was born in small polities like ancient Athens, the Renaissance principality and the New England townships. Its defining participatory institutions work best in small societies. Traditional political theorists such as Baron de Montesquieu, Jean Jacques Rousseau and John Dewey agree that democracy works best under conditions where a limited number of people live in close proximity and share a history, a culture and common values. They argue that large-scale societies are more suited to empire and hence monarchical government. Historically, modest democracies have shown the way to larger ones, with the small leading the great. Athens was democratic before England, Switzerland before France, Holland before Germany, Rhode Island before the United States. The UAE has traveled much further down the democratic road than some of its larger Gulf neighbors. During my recent visit to Holland and my meeting with Prime Minister Dr. Jan-Peter Balkenende, the virtues of a small democracy playing a crucial role as part of a larger Europe were much in evidence.

Yet if a diminutive size is an advantage in establishing a democracy, it can be a disadvantage in sustaining a democracy. Being a small state in a world of large states can mean weakness and dependency. Ask the people of Belgium or Poland. As a small federation, the United Arab Emirates can experiment freely with democracy, but as a small state it must be sensitive to its more powerful neighbors such as Saudi Arabia, Iraq and Iran – as well as to superpowers far beyond its borders – the European Union and the United States. The compactness that facilitates democracy exposes it to great risks. The UAE has been progressive in integrating women into work and society, but elsewhere in the Gulf, there are places where women remain severely isolated from the social and economic mainstream.[1]

2-Federalism

The United Arab Emirates fashioned its government in 1971 as a federation—a form that permits integration around economic and defense policies but affords continued autonomy of culture and politics in the seven emirates. As in the United States, Canada and India, power is divided vertically—protecting against its concentration and abuse at the central level. The devolution of power to subsidiary regions also enhances participation and a sense of citizenship in the localities. As Alexis de Tocqueville has suggested in his *Democracy in America*, "the township comes from the hand of God" and is a crucial foundation for civic and democratic freedom. Federalism and the devolution of power are, in other words, ways to limit power without privatizing or de-democratizing it. From this angle, the new government of Iraq – in danger of breaking up around fissures in ethnic, tribal and religious identity – would undoubtedly benefit from a federated governing structure. As he has done so often, Alexis de Tocqueville illuminated democracy when he suggested (in *Democracy in America*) exactly why federalism serves liberty:

> I believe that provincial institutions are useful to all nations, but nowhere do they appear to me to be more necessary than among a democratic people…A democracy without provincial institutions has no security against these evils. How can a populace unaccustomed to freedom in small concerns learn to use it temperately in great affairs?

Yet federalism can also check and limit the proper use of democratic power to achieve common ends. It can be used to frustrate the advance of rights that only central government can sustain. Tocqueville worried about the tyranny of majorities, but provincialism and federalism can secure the tyranny of minorities. In the United States, for a long time, "states' rights" was the rallying cry of those who wanted to undo the results of the Civil War and block civil rights for African-Americans. Could federal and local rights in the Emirates be used in a similar way to impede the integration of women into society or to obstruct the development of a liberal economy? Not necessarily, but surely this is an ongoing temptation and danger.

3-Consultative Rights

The third feature of political traditions in the Emirates that has an impact on democracy is a tradition of consultative rights, one that places a great value on informal kinds of consultation and participation by citizens in a governing process otherwise dominated by elites. This informal traditional process has added substance to the abstract idea of rights. And rights are, of course, a core constituent of any democratic regime. Without rights, a government can – in the name of popular sovereignty – trample upon popular liberties.

Yet the importance of rights in a democratic regime has led some to believe that the articulation and protection of rights requires nothing more than a "Declaration" or "Bill of Rights." Indeed, for some in the West, democratization of emerging societies would seem to require little more than importing a Bill of Rights and posting it in some capital city square.

However, as James Madison understood, rights are but abstract claims, and a Bill of Rights is little more than a piece of paper unless there is a body of citizens sufficiently educated in the responsibilities of liberty and competent in exercising its duties to make those rights real. Madison observed that Bills of Rights offer only "parchment parapets" from which to do battle on behalf of rights and are wholly inadequate to their defense. Rights are only as strong as the citizenry prepared to defend them. The former Soviet Union had a remarkable Bill of Rights, which guaranteed not only free speech, free assembly and the free practice of religion, but also the right to medical care, to work, to pensions, to death benefits, even to free vacations. However, this document was wholly fraudulent, since the Soviet Union had no real citizenry capable of enforcing rights of any kind, let alone this boastful list. During the period when it facilitated the slaughter of Hutus and Tutsis, the government of Rwanda was a signatory to the Genocide Convention—again, clearly a gesture without meaning.

Men and women may be "born free" in some abstract sense, but they are not born citizens, and can realize their freedom only when they learn how to be citizens. Citizens are made not born—through a process of education and life-long learning in what Tocqueville presciently called the "apprenticeship of liberty," reminding us that this is the "most arduous" of all apprenticeships. This explains why advocates of democracy – from Jean Jacques Rousseau and Johann Heinrich Pestalozzi to Thomas Jefferson and John Dewey, have placed education, particularly civic education – at the center of the democratic agenda. In the United States, both Jefferson in Virginia and John Adams in Massachusetts argued that in the absence of universal education for all who were to be citizens, the new experiment in democracy would be likely to fail. H.H. Sheikh Abdullah Bin Zayed Al Nahyan, the Minister of Information and Culture of the UAE showed great wisdom in his prudent comments on the need to put education at the center of development in the Gulf. If only authorities in Iraq had acted as wisely. Had I been in command of American troops entering Baghdad last spring, I would have put an M-1 tank or a Humvee not just in front of the energy and oil ministries, TV stations and arms depots, but also in front of every school, museum and library in the city—and thereby signalled the vital connection between education and democracy. Instead, schools and museums were left unprotected and were looted by recently released criminals or Saddam's vandals, and democracy received a devastating blow even before the democratization process could get under way. Schools are the nurseries of democracy and their care and development must be the first concern of those who believe in democracy. Saddam's palaces, used today by Paul Bremer and American occupying forces as headquarters and playgrounds for the troops, should have been made over into

Teacher Training Colleges, education facilities and childcare hospices. This would certainly have helped to foster the growth of democracy.

Whenever some one says "rights," those who understand how democracy grows need to reply "and responsibilities too." Whenever someone says "liberty," the response must be "but only citizens can be free, and we must educate them first." In short, whenever some one says, "Let's talk about democracy," the only appropriate response is "Let us talk about education, let us found universal schools."

This brings us to the second of our two sets of features that both facilitate and obstruct democracy. This second set encompasses characteristics common throughout the Middle Eastern Islamic world, and not just in the Gulf. The set includes Leadership, Wealth and Religion.

1-Leadership

The history of the UAE shows that leadership has been crucial to democratic development. The great fortune that came with strong and wise leadership by the late founding father His Highness Sheikh Zayed Bin Sultan Al Nahyan [May his soul rest in peace] and the Al Nahyan family – above all by H.H. Sheikh Zayed himself, who rose from his traditional role as Ruler of Abu Dhabi to become President of the newly founded UAE in 1971 – is evident in the maturation of its social and political institutions. Like George Washington, Thomas Jefferson, James Madison and John Adams in the new United States, H.H. Sheikh Zayed combined strong and disciplined leadership with forward-looking views that permitted the compass of power and participation to grow. Such a presence is called "founding leadership" in the literature of political theory, where the paradox is readily acknowledged that a free people may depend in their emergence on powerful and disciplined leadership. Turkey benefited enormously from the founding leadership of Mustafa Kemal Ataturk just as Germany's post-war transition to democracy was eased by the leadership of Konrad Adenauer. In a more general sense, the Middle East has come to depend on strong leaders—though by no means always wise and generous leaders.

Indeed, the vices associated with leadership are evident in how quickly it can become corrupted and serve to frustrate rather than advance the interests of democracy. Whether it is the Shah of Iran or Saddam Hussein, or other leaders in the region, we see instances of leadership that (sometimes) were benevolent at the outset, but in time became something other than that. Self-limiting strong leadership is something of an oxymoron—though not impossible. Even at its best, as the great Mexican revolutionary leader Emile Zapata once said, strong leaders can make for a weak people. As leadership extends from individuals to whole families and families become clans, and clans perpetrate their rule over generations, they cease to facilitate democratization and become obstacles instead. The great democratic "trick" is "self-limiting leaders"—men (and women) who know both how to lead and how to move aside when their job is done. Like great teachers, great leaders aspire to make themselves superfluous, to forge citizens who no longer require their leaders' wisdom to sustain democracy. They strive to imitate Cincinnatus—that nearly mythical Roman farmer who, when the Republic called him, rose to become a great General and leader; but who, when his task ended, voluntarily retired from the public eye to pursue his vocation as a farmer. Cincinnatus was George Washington's role model. Though many Americans wished him to be their hereditary king, Washington agreed only to be President for a term and then retired to his farm at Mt. Vernon. The great leader must know when his authority is needed, and when it encroaches on the emergence of a free self-governing people. Knowing when and how to lead is hard enough but knowing when and how to let go is still harder.

Finally, the two most prominent features of society in this part of the world are its extraordinary fossil fuel wealth, and its deeply Islamic religious character. Wealth is thought to be almost exclusively an advantage for democracy. Therefore, in the spirit of dialectic, I will focus on its disadvantages, which are not as frequently discussed. Islam is often (in the West) perceived as a restraining factor in the growth of democracy. Thus, in the same spirit of dialectic, I will focus on its advantages.

2-Wealth

Many people associate wealth with one of democracy's indispensable conditions, pointing to the United States as an example of liberty's connection to nature's bounty. The Middle East, and in particular countries like the UAE that are blessed with natural fossil fuel deposits, are thus thought to be particularly well suited to democratization—and, by the same token, are sometimes maligned for having made so little progress towards democracy, given their oil and natural gas wealth. However, it is useful to recall that democratic theorists have worried about the deleterious impact of wealth on liberty as much as they have celebrated wealth's democratic utility. While a sound and productive economy is certainly a useful premise for democratic development, as the economist Amartya Sen has noted, democracy rather than wealth is more likely to produce a sound economy. Although its fertile soil helped France on its road to liberty, nations like England and Switzerland with few natural resources and little fertile land also became democratic. Wealth can be used to establish schools and universities, to promote literacy and drive science and engineering—all of which are useful to democracy. It is hard to write a strong constitution on weak and empty stomachs. Nevertheless, wealth remains a two-edged sword with respect to democratization.

It is partly a matter of moderation. Moderate wealth created by hard work over generations makes a contribution to moderation and democracy (a point that Aristotle liked to emphasize). Great wealth, particularly when conferred by good fortune rather than hard work, may have other consequences. For although wealth creates prosperity, it cannot guarantee equal distribution, and great wealth is often turned into a weapon of inequality leading to a politics of resentment and resulting in anger, instability, violence and revolution.

Great wealth may also create dependency rather than independence of spirit in the people blessed by its bounty. Oil has liberated many Middle Eastern nations from dependency on foreign economies, but it has made them dependent on those same nations as consumers of their oil and natural gas, and has made them targets of those who need the oil and want to control it themselves.

Moreover, great wealth can create a dependent people—citizens who are indolent, complacent, disdain hard work and are themselves dependent on foreign workers to undertake the small tasks needed to run an economy. Their very wealth renders them vulnerable and in time they become increasingly dependent upon those workers from the outside world whom they treat as "guest laborers." The percentage of foreign workers to native citizens in the United Arab Emirates is not one of its democratic assets.

Great wealth in its very immoderateness can also incite capitalism's less benign side-effects: competition becomes predatory rapaciousness, self-interest transmogrifies into hedonism and narcissism, productivity becomes over-productivity leading to rampant consumerism, the search for profit trumps the quest for jobs, and the inclination to secure a rent on every asset becomes profit-mongering and usury. What are otherwise capitalism's virtues, useful to the development of productivity, become vices that, in their cultural and social consequences, distort the character of society and its mores. Thus does the theory of capital become a doctrine of wealth, which undermines the religious foundations of society (a theme explored in the next section).

For these reasons, democratic theorists such as Rousseau and the Puritan founders of America have argued that austerity is a surer foundation for democracy than wealth, and that moderation will secure society where riches and hedonism destroy it. Thus, in the case of the United Arab Emirates and its neighbors any moves towards democracy are more likely to be despite oil wealth rather than because of it.

3-Religion

If wealth has been deemed by too many superficial advocates of democracy to be a facilitator of liberty, religion – especially Islam – has been singled out as one of democracy's more formidable impediments, above all in the Middle East. Scholars of Islam such as Bernard Lewis have expressed concern that Islam has turned its back on modernity and positioned itself as an enemy of liberty. Less studied critics such as Samuel Huntington have proclaimed a "clash of civilizations" in which a war between "the West and the Rest" with Islam as the West's chief antagonist, is ever more inevitable. Although President Bush has been careful to distinguish fundamentalist zealots and terrorists from Islam's mainstream practitioners, Muslims will wonder whether America is targeting them rather than extremists in its Middle East adventures.

The view that there is a war between Islam and the West, and between Islam and democracy, is reinforced by extremist groups like Al Qaeda and their followers, who are eager to embrace and mirror the Manichean view that critics of Islam advocate. Osama bin Laden becomes Samuel Huntington's malevolent twin, while impassioned Presidential rhetoric invoking "us and them," and labeling nations that have no obvious connection to 9/11 as part of an "axis of evil" that America has a natural right to destroy through "preventive war" adds fuel to the fire.[2]

Certainly there is a necessary and healthy tension between religion and worldly government, between the spiritual and profane domains that define the ecclesiastic and the secular, which is as old as St. Augustine's "City of Man" and "City of God." The requirements of the body and the requirements of the soul push in different directions, even in a theocracy. Religion posits values that may not always be in accordance with the worldly mores of economy and society. Yet religion remains a foundation of common values, common practices and common will on which a polity often depends for its unity and integrity. Indeed, democratic regimes may depend on value consensus of the kind imparted by religion even more than other forms of government, precisely because their liberty pulls them apart and demands integration on another level. As Tocqueville understands:

> There is no true power among men except in the free union of their will; and patriotism and religion are the only two motives in the world that can long urge all the people towards the same end.

In a word, democracy without faith can issue in anarchy. Faith without democracy can issue in oligarchy, even theocracy. Faith and democracy forge a stable and free society, bound together sufficiently to be able to tolerate social conflict, ideological division and political dissent.

Of course some argue that the problem is not religion as such, but a particular religion, Islam. They contend either that Islam is more susceptible to fundamentalism or that it is in itself less tolerant of democracy and pluralism. It seems odd that the Christian West should make such claims about Islam, since there is no civilization that in its time has been as bloody, inquisitional, intolerant and as given to extremism and wary of the secular as Christianity. Indeed, even today there are Christian fundamentalists who despise democracy in America and elsewhere and regard themselves as being at war with modern secular society. (More than a million American Protestant fundamentalists school their children at home to keep them from the evils of a secularized, "corrupt" society.)

In fact, extremism and fundamentalism are found in every faith, often a consequence of religion under siege. Judaism has a form of ultra-orthodox fundamentalist belief that denies the very legitimacy of a Jewish state, while Hinduism has taken a virulent political form in India today, which inspired a successful and less than tolerant political movement that drove the traditional Congress Party from power for some years. Christianity continues to express itself in temperate and intemperate forms: there were Christian proselytizers ready to follow the American army into Iraq.

The claim that Islam has a special inclination to intolerance or theocracy would seem to stem either from ignorance, bigotry or some poisonous combination of the two. On the contrary, Islam has probably shown more tolerance to those living in its midst from other faiths, particularly monotheists such as Christians and Jews, than most other religions. Certainly Jews have fared worse under the pogroms and occasional genocides of Christian nations than under Muslim suzerainty (even if many Muslim nations oppose Zionism and do not recognize Israel as a state).

Islam has no monopoly on extremism – which turns out to be a tendency of all religions that are under threat. Extreme religion is religion under siege, under duress, religion in extremis. Like other religions under pressure, Islam has seen some forms of extremism – sects that seem more militant than Sunnis or Shiites and are more easily turned to rage against the West, against Christians and Jews. However, as I suggested in my *Jihad vs. McWorld*, there is a dialectic at work between aggressive Western secular commercialism and aggressively reactive Islamic extremism. Though democracy is an ideology of pluralism and moderation, it is often advanced under the cover of a monolithic and immoderate materialist monoculturalism—global markets identified with global liberty even as they undermine local liberties. Too often, what America exports under the banner of democracy are brand name pop cultural artifacts. The USA is more than the NBA, and CNN is not the UN. Disneyland is not the "land of the free" anymore than Hollywood is the "land of the brave." Yet it is sometimes hard to make this distinction on the basis of America's self-presentation.

Thus, Muslims, Jews, Hindus and others confronting America often have the impression that what America yearns for is not their liberation but their commercial colonization—not their public freedom to govern themselves but their private freedom to buy everything that America has to sell. In the joke about the Iraqi shouting "Yankee go home! (And take me with you!"), there is an ambivalence regarding America that is surely at the heart of Islam's anxiety about the West.

None of this indicates that there is anything in Islam that is inherently hostile to democracy, other than the natural tension mentioned at the outset of this discussion. Experience has shown, not just in Turkey and Morocco and countries with a more secularized version of Islam, but also in Bangladesh, Indonesia, Pakistan and India (the second largest Muslim nation on earth) and of course in the Gulf states of Kuwait, Qatar, Bahrain and the United Arab Emirates that when an appropriate balance is found between cultural values and liberty, between community and difference, and between patriotism and pluralism, Islam can be as comfortable with such democratic values as tolerance, pluralism and multiculturalism as any other religion. Ample empirical evidence may be found to support this view. However, this does not necessarily suggest that Islam will be altogether at ease or unconditionally tolerant (which serious religion is?), but it will be sufficiently so to make democracy possible.

Conclusion

In my view, democratization in the Gulf, indeed throughout the Islamic Middle East, is possible, feasible and desirable. Indeed, as the spirited dialogue generated by this conference shows, it is happening before our eyes. Islam is less of a restraint to democratization than oil wealth, while small scale society, federalism and consultative rights can aid and abet democracy (even if they are also capable of compromising it). However, democracy is neither a gift from the West nor is it dependent on Western institutions. It belongs to those who want it and take it and it can be realized in many different cultural varieties and forms. Inasmuch as you want it and will it, it is yours, whether others wish you to have it or not.

There is an old American spiritual song written when African-Americans were still awaiting deliverance from the legacy of slavery. To those waiting for outside liberators to come and free them from servitude, the lyric's opening line proclaimed: "We are the ones we've been waiting for."

Listening to the wisdom offered by those in this room, learning from the experience of this society waiting to be tapped, hearing the voices of leaders, newly empowered women, critical journalists, thoughtful scholars, as well as members of the Al Nahyan family, it seems very clear to me that YOU are the ones you've been waiting for.

So there is no need to wait any longer.

8

The Impact of US Policy on Gulf Security: The Military Viewpoint

General Anthony C. Zinni USMC (Ret.)

The first significant recognition of vital United States security interests in the Gulf region came over half a century ago when President Franklin Roosevelt made the initial contacts with King Abdul Aziz of Saudi Arabia in 1945 to discuss commercial and security cooperation. Over the next twenty-five years, the United States focused on expanding cooperation in the development of the Gulf oil industry with a modest presence of US naval forces in the region. The British retained the primary security role as it had for over one hundred and fifty years. By the 1970s, the United States had replaced the British as the primary force determining the structure of Gulf security. By that time, the Cold War rivalry was driving US policy for the region until the collapse of the Soviet Union in 1989. During this period, it became evident that the complexities of the region made a simple bipolar calculation for developing a strategy for regional stability and security far more difficult than for other parts of the world. Also during this time a number of destabilizing events occurred and caused significant changes to the cooperative security arrangements. By 1990 it became clear that these complexities and events would further complicate the development of a post Cold War strategy and security policy. The last decade of the twentieth century and the beginning of the twenty first century have witnessed the continuing trend of unforeseen events that have jolted stability in the region and the continuation of complicated and strained relationships that have challenged security cooperation.

No formal security structure, such as an alliance, was ever created between the United States and the Gulf allies. Instead a loose cooperative and informal relationship evolved over time, often referred to as "The Gulf Coalition," especially after Operation Desert Storm in 1991. This coalition was based on a series of bilateral military relations established between the United States and each of its allies in the region. This arrangement facilitated the basing of US forces, combined exercises, and other cooperative military activities. It also allowed for the conduct of sanctions enforcement and military strike operations in the Gulf area during the 1990s.

The security interests of the United States and regional allies in the Gulf have been articulated over the years as being centered on issues such as ensuring the free flow of energy, enhancing regional stability, deterring WMD proliferation, ensuring the freedom of navigation and commercial trade, preventing dominance by regional hegemons, and countering transnational threats such as terrorism and drug trafficking.

The events of September 11 and Operations Enduring Freedom and Iraqi Freedom have once again shifted the security focus and the nature of security cooperation in the region. It is important at this point, not only to examine the current impact of US policy on Gulf security but to look ahead at the implications of recent events on the future of the cooperative security structure that has been collectively maintained since the middle of the last century.

Several issues should be considered in addressing the future security posture between the United States and Gulf allies and these will form the essence of this chapter. These issues may be identified as follows:

- The role and image of US military forces in the region.
- The future presence and composition of US forces in the Gulf.
- The collective security structure needed in the region to counter future threats and protect mutual interests.
- The effect of US policy positions on future security cooperation.

The History of US Involvement in Gulf Security

The history of US involvement in Gulf security has grown and expanded considerably over the last sixty years. Understanding this evolution and the events during this period that drove US policy is critical to grasping how the United States reached the current security posture and how it should best proceed in developing a future security arrangement. Prior to the 1945 historic meeting between President Franklin D. Roosevelt and Saudi King Abdul Aziz on the Great Bitter Lake in Egypt, US interest in the region was centered on the oil industry. US oil companies had begun working joint ventures in the 1930s and these continued to grow in the decades following World War II. The creation of the Arab American Oil Company (ARAMCO) in 1944 capped this growing energy partnership throughout the Gulf region. With the exception of a small naval presence and training missions established in the 1950s, the security foundation for the Gulf remained primarily a British function as it had for a century and a half. In 1971, however, that changed abruptly when the British pulled out of the region.

The United States immediately moved to fill the void left by the British departure. The energy resources of the region were a significant prize in the zero-sum game of the Cold War that made every region a potential arena of competition for influence or dominance by the superpowers. The US strategy during the 1970s was built around the two pillars of Saudi Arabia and Iran, its most important allies in the region. The oil crisis of 1973 further highlighted the growing importance of the energy capacity in the Gulf and its value to the competing sides in the Cold War. In 1979, the US strategy was jolted by the overthrow of the Shah. His regime was replaced by a radical theocracy that now presented a new threat—one from within the region itself.

These events required the United States to develop a new security strategy for the Gulf for the 1980s. What emerged was the Carter Doctrine. It committed the US to the defense of the region against Soviet invasion or encroachment. In 1983, President Carter created the Rapid Deployment Joint Task Force (RDJTF), a war-fighting staff with the mission of planning for and executing military operations in the Gulf. This staff would shortly thereafter become the US Central Command (CENTCOM), a unified command with the same mission. During this decade, threats from Iran led the US and its allies to re-flag and protect tankers in the Gulf and to support Iraq in its war against Iran. CENTCOM was facing the requirement to defend the region from outside attack and from the intra-regional threat coming from Iran as that country attempted to achieve its hegemonic and revolutionary objectives in the Gulf. Another significant event in this decade was the founding of the Gulf Cooperation Council (GCC) in 1981. This was the first regional security organization to be formed.

The end of the 1980s and the start of the 1990s witnessed the collapse of the Soviet Union. The euphoric reactions to this end of the Cold War were pronouncements of a "new world order" and a "peace dividend." There was even speculation that CENTCOM might be disbanded or reduced to a lesser organization now that the Soviet threat was no longer present and Iran seemed to be subdued after the war with Iraq. These notions, however, were soon dispelled when Iraq opened the decade of the 1990s with an invasion of Kuwait.

The US managed to pause in its draw-down of Cold War forces to build up and deploy a large international force to the region to drive Saddam Hussein from Kuwait. In order to preserve the fragile and diverse coalition that was put together for Operation Desert Storm and to avoid a messy mission in nation building, the then US President George Bush, settled for a set of United Nations sanctions and inspections imposed upon Iraq. The aftermath of the war and the enforcement of these UN conditions prompted the United States to alter its regional strategy again. This time the strategy was termed "Dual Containment." This policy called for the enforcement of sanctions on Iraq and Iran. Iran had begun a clever post Iran-Iraq War recovery that involved the acquisition and development of what was considered "asymmetric" military capabilities. These were systems and capabilities that the United States and allied forces could have much difficulty in countering. These included missile systems, sea mining capability, fast and difficult-to-detect patrol boats, submarines, and weapons of mass destruction programs.

This new US regional strategy capitalized on the military relationships created during the Gulf War. The United States, through a set of bilateral arrangements with the GCC countries, forward-deployed a greater number of US forces, conducted more and larger combined exercises in the region, and increased military cooperation activities such as security assistance programs. In addition to the GCC nations, arrangements with Egypt and Jordan also contributed to the ability to execute the new strategy. A number of crises during the 1990s required US and coalition forces to conduct military strikes against Iraq and the rising threat of terrorists operating in the region. By the end of the decade it appeared that containment was working but the costs of maintaining it and the sporadic military operations required to enforce it were putting political strains on the arrangement from all sides. Adding to these strains was the emerging global threat of terrorism. Attacks by terrorist groups were becoming bolder, more sophisticated, and more threatening. By the beginning of the new millennium, this threat was clearly the major one faced in the Gulf region and beyond. It could no longer be viewed as just a regional problem.

The terrorist attacks grew steadily in the late 1990s and climaxed with the attacks of September 11, 2001 in the United States. Once again the United States and the region found itself with a new set of circumstances that drove a change in policy and strategy. The United States announced a more aggressive policy of prevention as it conducted Operation Enduring Freedom in Afghanistan and other operations as part of the Global War on Terrorism. In 2003, the United States led a controversial invasion of Iraq (Operation Iraqi Freedom) in line with the so-called "Bush Doctrine" that promoted a more preemptive approach to potential threats.

In examining this short history of US security approaches in the Gulf, there are three main questions that emerge: What can be gained that would help in understanding how the United States arrived at this current posture? What has been consistent or inconsistent about US policy and strategy for the region? What can it tell us about how the future security arrangement should be structured?

The Development of US Gulf Security Policy

To say the least, the Gulf has been subject to remarkable change and instability over the history of US involvement in its regional security, as described above. This has caused the US strategy to be reactive and altered significantly every decade or so. The factors that cause these strategy and policy swings will be addressed later in this chapter. There are several threads that are consistent in US policy for the Gulf throughout the years despite the changing strategies and the dynamic events and conditions that characterize the region.

The current CENTCOM strategy lays out five objectives that are generally the same as those that CENTCOM commanders have described in the past. These are as follows:

- Promote regional stability.
- Ensure uninterrupted access to resources and markets.
- Maintain freedom of navigation.
- Protect US citizens and property.
- Promote security of regional friends and allies.

From these consistently stated objectives we can see the emphasis on regional stability, the importance of commercial access (particularly to energy resources), the recognition of the critical nature of the air, land and sea routes that traverse the region, the degree of US personnel presence and investment in the region, and the need for reliable regional allies.

Internal regional stability has been a major security issue and challenge ever since the United States became involved in the area and even well before such involvement. Border disputes as well as ethnic and religious rivalries have been destabilizing and many regional nations have felt that the United States could play a constructive role in mitigating many of these existing intra-regional frictions. The US involvement has been seen by some in the Gulf region as a presence that keeps many of these frictions from becoming more serious or even exploding. In recent years, many of these long standing issues have been resolved through agreements, international arbitration and internal reform. This has been encouraging for those who would like to see the regional security structure become more formal and substantial. Some of these destabilizing issues have distracted the Gulf nations, strained relations among the regional states and made the degree of regional security cooperation more difficult.

Commercial development in the Gulf region has grown and the potential future development seems limited only by the ability to resolve the security and stability issues. The growing number of world-class ports, distribution centers, transshipment points, and commercial and financial centers make the region a critical area in the system of international trade. The advantages of geography and energy resources add to the potential to expand economic development and foreign investment. Obviously, the oil and natural gas resources of the region make it crucial to international security and stability and many argue that this interest in energy is the only thing that drives the United States or any other international interest in the region's security.

Throughout history, many of the world's major trade routes have passed through this region. The natural sea and land choke points have dictated this passage. This is still the case today with the flow of energy resources from the region adding to the importance of keeping these navigational routes open. In addition to the commercial importance of these routes, their value to moving military forces to respond to threats and crises is critical to regional and international security. This is especially valuable to the United States in moving forces from the Atlantic/Mediterranean to the Indian Ocean/Pacific.

The United States has been careful not to establish permanent bases in the region or to assign dedicated forces there. However, it has established significant military equipment pre-positioning sites, joint use bases, military headquarters and the continuing presence of rotational forces. Both US commercial interests and investment have grown over the years. This has resulted in an expanded US security requirement as a result of the growing number of US citizens and interests in the region potentially in need of protection during times of crisis.

The US approach to security for the Gulf has always been centered on one or two key allies. Saudi Arabia and Iran have been the keys in the past. In recent years, the US has broadened its approach and become more involved in the smaller Gulf states and those on the edges of the Gulf region. This expanded approach has been driven by the breakdown or straining of some relationships and by the perceived need to add more balance and flexibility to regional stability by including more states in the security relationships.

It is doubtful that these objectives will change in future US security policy for the Gulf region. Regional stability, access, freedom of navigation, protection of US interests, and the protection of allies and friends will remain consistent objectives and continue to be the principal goals of US strategy for the region.

In the past, the United States has used a "minimalist" or "economy of force" approach to security in this part of the world. The Cold War demands and threats made other regions' demands more critical, such as Europe and the Far East. In addition, the Cold War conflicts (Korea and Vietnam) drew off forces and attention. Despite the strategic significance of the Gulf region, it has always been viewed as a secondary theater. Regional sensitivities to US force presence have also been a driving consideration in determining US military posture for the Gulf. Early US military presence was primarily naval and "over the horizon." CENTCOM had never assigned forces as other Unified Commands had done. Instead, CENTCOM temporarily borrowed forces from other commands to conduct operations, enforce sanctions, conduct regional military-to-military activities, etc. This consistent approach to keep US force presence to a minimum and as "invisible" as possible, is one issue that may be questioned as events in the region unfold and future security arrangements are determined. At this point, however, there does not seem to be a move toward changing this policy for the future by establishing permanent forces in the region.

Threats to Gulf Regional Security

Although there has been consistency in security objectives for the US in the Gulf region, the perceived threats have changed dramatically throughout the years of US involvement. Originally the threat was seen as coming from outside the region from global powers that sought domination of this vital area. To respond to this threat, the United States envisioned rapidly building up forces in the region to deter or engage outside forces threatening the Gulf. At the end of the Cold War, the threat developed into regional hegemony coming from a radicalized Iran and a tyrannical Iraq. This drove the US policy from the Carter Doctrine to the policy of Dual Containment. This also led to a greater US military presence that aggressively enforced UN sanctions, such as no-drive and no-fly zones and maritime intercept operations against gas and oil smuggling.

Other secondary threats emerged over time, which have now replaced these two as primary threats. These threats are extremism – manifesting itself in acts of terrorism and destabilizing actions that threaten regional states – and the proliferation of weapons of mass destruction. These have generated the controversial new US policy, the Bush Doctrine, which guides the Global War on Terrorism and the actions taken in Iraq.

It is important to note that the Gulf states did not always view the threats in the same way as the United States. In fact, there were often differences within their ranks as to which threats presented the greatest degree of danger. For example, during the time of the Dual Containment policy in the 1990s, some states believed that Iraq was the primary threat, while to others it was Iran that posed a greater danger.

Military Cooperation in the Region

Building military-to-military relationships in the Gulf between US and local forces has always been difficult, sensitive and controversial. Since the Gulf War, however, the relationships grew closer. The effect of the recent Iraq War on the relationship remains to be seen. Since no formal military alliance has existed between the US and its Gulf allies, problems remain with regard to military standardization, interoperability, common tactics, equipment compatibility and other issues that are normally resolved through alliance agreements, such as NATO. Efforts to work more regional cooperative approaches to solve these problems have moved slowly. In the late 1990s, a Cooperative Defense Initiative was initiated to improve air and missile defense capability and some cooperative work on environmental security was begun. Reluctance in the region, however, to go beyond the bilateral relationships as the foundation of US-Gulf states security cooperation remains, and stems from many factors. These include the lack of a long historical military association, significant cultural differences, competition for influence from other militaries outside the region, suspicion of promoting military sales of expensive equipment, perceived influence of Israel on sharing of military technology, and the lack of local popular support for a closer military relationship. The United States, too, has been reluctant to go as far as some in the region would expect in a relationship, in order to overcome these factors and be assured of US commitment to the region's defense. With the exception of NATO, the US has generally resisted formal military alliances, preferring instead to work through informal coalitions.

Throughout the 1990s, in the aftermath of the Gulf War and Cold War, the military relationship did progress. Security assistance programs grew significantly; the world's largest military exercises were conducted by the United States, regional allies and other international participants; regional forces joined the US in operations that included those conducted in Somalia, the Balkans, and during the Gulf War; and more sophisticated military technology and intelligence was shared with regional forces. Most importantly during this period, the personal military relationships among senior leaders grew stronger. For many in the region, CENTCOM became a trusted military partner and the increased attendance at US military schools added to the confidence building and trust.

The key question at this point is, of course, the effect of the War in Iraq on the military relationships between the United States and the Gulf states. The lack of regional military participation in the coalition, popular disagreement in the Gulf region with the war, and the severely strained relationships resulting from the war in Iraq have all affected US-Gulf ties. No doubt a new military arrangement and posture will now be put in place by the United States in the Gulf and surrounding region, and the security focus will shift from the Dual Containment policy. What exactly this would mean is yet to be determined.

The Bush Doctrine

The President of the United States is required by law to submit a National Security strategy one hundred and fifty days after taking office. The strategy is usually laid out and tested for reaction ahead of time, in a series of speeches. The deadline has not always been adhered to and events may also delay submission. President Bush issued his administration's strategy in September 2002. This was significantly past the required deadline but was delayed due to the terrorist attacks of September 11, 2001. The strategy articulated was previewed in a number of speeches the President delivered in the intervening year and by speeches given by other senior members of the administration. The strategy was a significant departure from President Bush's political campaign statements that advocated a less "arrogant" and more multilateral approach to foreign policy. The September 11 attacks, however, changed the situation and the Bush strategy describes a nation at war. The threat, as described in the strategy, is a combination or convergence of radicalism and technology. It has been described as an aggressive and preemptive strategy. It seeks to remove existing and potential threats and not trust, as much as in the past, to diplomacy, containment and deterrence to prevent attacks by potential enemies. It was controversial on its publication and became very contentious when it was used to justify the invasion of Iraq, especially since the administration did not wait for the UN process and the inspectors' efforts to reach fruition.

The two previous US administrations consistently sought international legitimacy and coalition involvement in their interventions. This seemed to be an accepted prerequisite to action and they often went to great lengths to gain United Nations resolutions authorizing the use of force and to solicit international participation for any intervention. In the eyes of its critics, the Bush Doctrine, particularly as it applied to the Iraq operation, seemed to change this accepted approach. There was international criticism of this aggressive doctrine and fear that it might be used by others as a justification to take preemptive action. While the vast majority of nations supported the US in its Global War on Terrorism, this support did not carry over to Operation Iraqi Freedom.

Impact of the Current Policy on the US Military

Prior to the September 11 attacks, the Bush administration was set on a course of military transformation. Both presidential candidates, George W. Bush and Al Gore, ran on a platform that included recognition that the US military was badly in need of change. The Cold War structure had not been adjusted to account for changes in the potential threats, promises of technology and strategic realities of the new millennium. Speculation spread that the transformation would be radical. Some were advocating a "strategic pause" that would have the United States assume an era of relative peace and devoid of major security challenges that could give the needed time and ability to shift resources in order to develop a revolutionary transformation. There were advocates for withdrawal from certain regions of the world where US military forces were traditionally deployed or based to better support the transformation process. This included the Gulf region where US forces had mixed reviews on quality of life facilities, training potential and cooperation. Radical proposals such as the disbanding of large numbers of ground units, particularly heavy armor units, were controversial and seen by many as risky but the new Defense Department seemed committed to a bold course of action. The National Military Strategy, which follows the National Security Strategy, promised to be a significant document outlining an historic change in military structure, operating concepts, technology development and internal procedures. The Pentagon, like the White House, however, had its strategy altered by the September 11 attacks.

Instead of reduced forward deployments and changes in strategic positioning, the military found itself adding to its current commitments. Force protection and security requirements, plus operations in the war on terrorism, placed great demands on US forces worldwide. The 2003 War in Iraq and its aftermath piled pressure on an already overstretched military and further strained active and reserve units. The lack of international support placed the burden almost totally on US forces with minimal burden sharing in the form of forces willing to participate in the combat operations and nations willing to share the costs of military operations and reconstruction.

Gulf allies would not join in the operations in Iraq as they had done in the Gulf War without a UN resolution although basing, over-flight rights and other operational support needs were provided.

Lessons Learned

After over a half century of involvement in Gulf security, what can the United States learn from its experience that should help guide future policy? In addition, what can the Gulf states learn from the past relationship? The pattern of reactive strategies, changing every decade or so, has been disruptive and confusing. Although it has helped during short term crises, it has been an up and down path that has not provided for long term stability and enduring, strong security relationships in the region. What follows are some lessons that should be considered in establishing the new relationships and policies that will necessarily be required of both sides in the aftermath of The War on Terrorism and the War in Iraq.

1. As mentioned previously, the objectives or goals of US strategy seem consistent throughout the years with the strategic changes. Generally, they have been in line with those of the states in the Gulf region. Despite this, both sides have failed to state common goals and in past cases, even when there was common agreement on future actions, each side has publicly articulated a different rationale.
2. The security structure in the past has been based on a loose set of bilateral relationships between the United States and each Gulf state. Whenever cooperative action was required, it fell to the United States to try to piece together a unified approach. Gulf state participation has thus been uneven and put greater pressure on certain states that felt more exposed than their Gulf partners, when they made significant concessions or commitments to support US efforts during crises. These and other problems result from a loose bilateral "coalition" that is advertised to the public and the media as being more formal than it is in reality.
3. Gulf security goes beyond just the relationships between the US and the Gulf states. Other states on the edge of the Gulf region, such as Egypt and Jordan, and other international powers such as the United Kingdom, France and Russia have played roles in regional security issues. The role of these extra-regional states has been confusing in the scheme of US-Gulf states security relationships in many respects.
4. Military interoperability has always been a problem in the region. Standardization of equipment, common doctrinal approaches, joint training, integration of regional systems (such as air and missile defense), solid relationships between militaries and compatible security assistance programs, have all been lacking but are necessary if there is to be a stronger, more cohesive regional military capability.
5. The lack of an established consultative process has been a major issue in the past. The biggest complaint from regional states has been the lack of consultations when security issues and crises

arise. The complaint from the US has been the difficulty in getting decisions and commitments, particularly during crises.

6. The presence of US forces has always been a controversial issue. The size, type, location and visibility of US forces have been continuously debated over the years.
7. Security issues that extend outside the Gulf area or are primarily outside the area, impact on the relationship. The Middle East Peace Process, transnational threats such as terrorism and proliferation of WMD are examples of such issues. These affect decisions, cooperation and commitments.
8. There has not been common agreement on the threats between the US and its regional allies. Even within the Gulf region, states have viewed the threats differently—whether these threats are from Iran, Iraq, extremism or WMD.
9. There has not been agreement as to which countries should comprise a regional cooperative arrangement with the US. Is it the US plus the GCC countries? Should Iraq, Iran, Yemen or others be considered for the future, if conditions and relations change? Is Gulf security too limiting a construct given the issues in lessons three and seven mentioned above? These questions have never been adequately addressed in the past.
10. Probably the biggest lesson learned is that the United States needs regional allies and the regional allies need the United States.

Where Do We Go From Here?

Clearly, each of the issues above in the lessons learned will have to be addressed in building a future relationship and strategy for the Gulf. The title of this chapter could be reversed to discuss the impact of Gulf states' policy on Gulf security. The point is that US policy cannot be examined in isolation as it relates to Gulf security. It has always been tied to regional allies and their cooperation. It is too easy to dump all the problems and responsibilities of Gulf security on the United States. The regional states also have to stand up to their responsibilities. A new structure has to be defined and codified. This new structure will have to include a more formally defined relationship; a common strategy that includes regional goals, objectives, threats and resource commitments; a broader membership; stronger military cooperation; an ongoing consultative process that is operative, not just when crises loom; and a regional agreement on US military force presence.

For too long, this region has taken instability for granted. All those involved in the vitally important security of this globally critical area, have to work seriously to change this perception. The internal and external threats must be dealt with. The necessary reforms must be made—be they political, economic, social or security-related. The security aspect cannot be regarded in isolation, without considering these other factors since they are inextricably intertwined. Nor can the military aspect of security be viewed in isolation from security policy and structure. For too long, the US military has had to deal with Gulf security without having these other components addressed, and the problems of such an approach have been clearly evident.

Section 4

Iran and Gulf Security

9

Iran and International Relations: Impact on Political Stability in the Gulf

H.E Mr. Mohammed Ali Abtahi

As a citizen of Iran, and in my capacity as Vice-President for Legal and Parliamentary Affairs in The Islamic Republic of Iran,* I am pleased to begin by conveying my best compliments and gratitude to my UAE brothers and sisters at The Emirates Center for Strategic Studies and Research, and in particular to H.E. Dr. Jamal S. Al-Suwaidi, the Director General of the ECSSR, for hosting this important conference. My presentation will focus on the issue of the challenges to the future of the status quo in the Gulf and the ways and the options ahead. I shall avail of this opportunity to try to explain the prospects and possible future developments of the prevailing status quo. Also, I shall explain the position and importance of the Gulf region within the framework of regional transformations and then analyze the risks and challenges that may be encountered in the future of this region. In conclusion, I shall propose a system and a set of possible solutions to meet those challenges.

I-The Present World Situation

The contemporary world and the present international order are still going through a phase of ongoing experimentation since the end of

* H.E. Mr. Mohammed Ali Abtahi delivered this address prior to his resignation in October 2004.

the Cold War, which took place under the bipolar system, and the unilateral challenges posed by the United States of America thereafter. We can consider the present world order as "an irregular polygon." The disproportionate arms of this order are represented by the US and its more permanent allies, the European Union, the Commonwealth of Independent States, China and its area of influence in South East Asia, in addition to other relatively stable regional groupings such as the Arab League and the African Union, the Caribbean and Latin American countries and others. The most salient feature of this international order is the lack of a stable hegemonic power and the possibility that such regional groupings may influence the course and pattern of this order despite the forceful attempts by the United States to bring about a unipolar world order.

II-The Importance of the Gulf Region

The Gulf region has unique importance in the current multi-faceted world order due to four reasons: its geopolitics, strategic geographical position, its economic geography and cultural geography. In the field of geopolitics, it could be said that the outcome of the Arab-Israeli conflict is the one determining the fate of the geopolitics of the whole world. Moreover, the movement toward regional political groupings in the Middle East, with the Gulf as its central pole, is one of the most salient advantages for controlling the course of political globalization. The scope of regional political groupings in the mainly Arab Middle East and the Gulf is much larger than the potentially negative impact of the challenges posed by the designs of externally imposed regional groupings.

In relation to the strategic geography of the region, it could be said that, a close look at the world scale for weighing strategic and security importance, whether in terms of proximity to the center of the globe, or its fringes or both, reveals that there is no spot in the world that holds such massive advantages as the Gulf region.

In the field of economic geography, the Gulf region is capable of providing a total of about two-thirds of the world's oil production. The Gulf region exports around 25% of the total world consumption of oil, and the largest facilities for exporting natural gas are found in this region. Any change, whether it is a sudden or gradual decline, in the oil production networks in the Gulf can be dangerous for the world economy. Moreover, if we take into account the existing production facilities in their entirety, from the north, south, east and west, and the transportation and navigation traffic in transit, then we can say that the region provides for a market of nearly seven hundred million persons living in the neighboring regions of the Gulf.

In relation to cultural geography, the prevalence of a common religion and shared moral values that promote Islamic unity in this region can equally create unity of opinions and common vision in the social, political, cultural and economic realms. It can prepare the ground for achieving religious-cultural integration in this vital region. All of these elements constitute sources of strength for the area.

A Glance at the Security Context of the Region

The Gulf region, which is important in several ways, has experienced various security and political developments during its recent history. A number of American scholars think that the US intervention and direct military presence in this region have added further complications to what was already a highly volatile region. They assert that by tightening its grip on the Gulf region, the United States is actually intending to exert more influence on other vital areas of the world.

III -Threats and Challenges Confronting the Region

In general, the threats and challenges confronting the Gulf region may be divided into three main sections:

- At the national level (challenges that include national elements but with effects beyond the national territory).
- Within the regional level.
- Beyond the regional level.

The correlation and balance between the challenges at the national level and those emanating from beyond determine the outcome of regional challenges. The core factor that determines the outcome of regional challenges is also embodied in the issue of Palestine.

The Zionist entity is planning the strategy of widening its immediate political and security geosphere by exerting intense pressures and restrictions on the occupied territories. This in itself could endanger the national interests and security of the Arab countries, especially those bordering the Gulf region, and render them vulnerable to serious threats. The root causes of these threats emanate from an ever widening gap between the social fabric of Arab citizenry (pivotal security) on one hand and the ruling political regimes in their respective countries on the other. This is in addition to intervention by other powers from outside the regional sphere, which could result in insecurity, similar to what is happening in Iraq. The phenomenon of religious extremism and the movements associated with it represent the most important elements and impetus behind the challenges confronting this region. In fact, "religious extremism and bigotry" itself is a by-product of interventions by powers from beyond the regional setting. The trend of religious extremism is a direct reaction to the way America conducts itself in this region. However, it has caused direct and severe damage to Muslims in this part of the world. During the last years of the Cold War era, and within the framework of creating routes for intervention, the US created pockets for nurturing and developing fundamentalist *Salafi* trends, using domestic elements within the region.

In so doing, Washington sought two courses of action on two fronts:

- Inflicting maximum losses and exerting security pressure on the Red Army in Afghanistan.
- Creation of a controlled wave of religious extremism (mainly Islamic) in order to set the stage for suppressing certain Islamic movements that advocate religious enlightenment in the Islamic world.

On one hand, the outcome of this approach has been the fact that the US policy in relation to Palestine continued to subject Islamic countries in general, and Arab nations in particular, to intense moral pressures. On the other hand, the present situation of terrorism, which is the result of American intervention in the region, had the impact of creating internal instability in the Gulf region, which became vulnerable to some sort of political, economic and social stagnation. To substantiate this, I would quote a famous American scholar, Kenneth M. Pollack, who expressed similar views. He said:

> Too many (Arabs) feel powerless and humiliated….and too many feel both threatened and stifled within a society that cannot come to grips with modernity,…and these are the driving forces that heighten anti-American sentiments in the region.[1]

He goes on to say, and I quote:

> The presence of American troops fuels the terrorists' propaganda claims that the United States seeks to prop up the hated local tyrants and control the Middle East.[2]

So, if the present circumstances persist, one would be compelled to expect the following developments:

- The absence of a collective security model emanating from the will of the countries of the region would lead to the continuation of external intervention from beyond the region.
- The persistence of external intervention from beyond the region in the absence of regional integration, that is, the lack of some sort of "regional unity," will add to the challenges faced on both the national and regional levels, and create a wide gap in the social fabric, mainly between the political regimes of the region and their peoples.
- This gap will fuel the growth of religious extremism, which would facilitate the recruitment of radical fanatics and fundamentalist *Salafi* groups by providing a fertile ground for the emergence of hard-line *Salafi* trends. The rise of religious extremism within the societies of the littoral countries of the

Gulf can lead to the perpetuation of the dependency of the political and governmental structures of these nations on powers from beyond the region (mainly on the United States), a process which would also encourage more US intervention in the region.

As a researcher in theology, my concerns are not directed towards the political effects of the escalation of various foreign interventions in the region. Rather, they focus on the lingering sentiment that the religion I believe in, and which I think is going to save mankind, the very religion that I regard as a unifying element for the nations of this region, might be misinterpreted and distorted due to the possible scenario that I envisaged earlier. Extremist and fanatical modes of behavior and inclinations do not have the authority and the credentials to carry out the formidable task of "religious representation." In this context, the only winning group among the believers of any religion will be those marginal factions who strive to present themselves as religious exhorts. The chief among them are the following:

- Osama Bin Laden's group in the radical Islamic movement
- Ariel Sharon's group in the radical Jewish movement
- George Bush's group in the radical Christian movement.

I think that the basic problem in the unstable world of today, and through the last decade, has been caused by the presence of zealots in different religions. However, religion as such has never been a source of problems for the world. There is a philosophical justification of this aspect. Religion has been revealed to mankind by Allah the Almighty, who created all mankind. Allah never urged any group of persons to shed the blood of other human beings. All religions seek to spread peace and security and their main objective is the provision of happiness, security and peace for humanity. Religion has come from Allah but those who sought to exploit it have employed it as a vehicle to propagate their evil ambitions within different societies. This is evidenced by the killing of innocent people in the name of religion, despite the fact that these human beings were created by Allah the Almighty.

If we look at the realities of the world as it stands today from an optimistic point of view, we will find that it is the extremists in different religions who are fuelling the bitter sentiments of hatred and rivalry between the believers. Militants and extremists, who have no authority to represent any religion, still drag the believers and religions to suffer the consequences of their evil designs. A person like Osama bin Laden has mobilized the whole world against Islam. When did he become a representative of Islam to the extent of justifying the killing of innocent people and equating it with Islam in the eyes of the whole world? When did the Zionist entity and Ariel Sharon, who is so embroiled in acts of violence, aggression, killing and bloodshed, become the representatives of Judaism? How have the Jews chosen Sharon to represent them? How did George W. Bush, who belongs to a conservative faction of Christianity, earn the right to represent Christianity as a religion? I believe that the majority of Muslims, Jews and Christians shun the views of Bin Laden, Sharon and Bush, but what are they supposed to do? Radicals wage their wars in the name of religion, which pays a high price as a result. The basic requirement for today's world is the kind of religiosity that is imbued with tolerance and aversion to extremism, as been preached in the teachings of the Prophets Mohammed, Moses and Jesus (May Peace be Upon Them).

Thus, the only way to save religion, stop the ongoing destructive foreign interventions in the region and preserve the domestic interests of its countries, is by following the course of religious moderation, thus saving the region the troubles caused by raging conflicts motivated by religious bigotry and by paving the way for the creation of appropriate models for stable regional participation and integration within the framework of securing collective interests and reaping the related advantages.

IV- Iran and the Region

The Islamic Republic of Iran still pays attention to the issue of peace and stability in the Gulf which it considers a matter of great concern. It believes that the whole world should pay due attention to peace and stability in this sensitive area to a greater extent than before. The long-term interests of the countries of the region can best be guaranteed by making sound decisions involving all the concerned parties. One of the objectives of the foreign policy of the Islamic Republic of Iran is to achieve fruitful cooperation and engagement with all the countries of the region to guarantee collective security. The concepts of *détente* and confidence-building remain the fundamental means in foreign policy initiatives during recent years, as we have witnessed its clear and fruitful outcomes.

The Islamic Republic of Iran, as it continues to reject the policy of militarization and authoritarianism by powers from beyond the region on one hand, also insists on rejecting extremist movements that incite people to engage in acts of violence on the other hand. Iran will continue to pursue an independent path that seeks liberalization, justice, development and overall progress. That is why the whole world welcomed its call for the initiation of a constructive dialogue between civilizations and the creation of a worldwide coalition for peace.

Realizing long-term cooperation between the countries of the region, the Islamic World and the West requires greater understanding of each other, the gradual correction of false impressions and selective, narrow-minded or parochial views. It entails pursuing universal aspirations, and exerting an effort that is based on an unshakeable will, imbued with good ethics and mutual appreciation of culture.

10

Iran's Emerging Regional Security Doctrine: Domestic Sources and the Role of International Constraints

Mahmood Sariolghalam

In this chapter, Iran's national security doctrine is analyzed by employing a sociopolitical analytical framework. In doing so, emphasis is placed on the historical processes that have led to the current political and security behavior by the Iranian revolutionary elites. The Iranian Islamic revolution is considered as the outcome of a long struggle against foreign intervention in Iranian affairs and despotic rulers. Moreover, this chapter embraces the axiom that the foreign and security policy of a state is the extension of its domestic politics and structural imperatives. The hypothesis of this chapter advances the idea that Iran's security policy at the regional level is a political struggle between two schools of thought: the revolutionary and the internationalist. Both advocates enjoy social support and possess institutional and economic backing. The outcome of this confrontation will be less settled by debates, political discourse and statistical analysis of national and international trends than by critical events. In this respect, changes in the geo-strategic map of the Middle East, economic hardships, generational shifts and political exigencies to maintain the status quo will be among variables in shaping a new and more accommodating course in Iranian national security doctrine towards the Middle East region.

Domestic Sources of Security Doctrine

Historical Roots of Current Behavior

Conditions that led to the Islamic Revolution of Iran have their roots in a deep historical setting and experience. The Revolution was the outcome of a long struggle by Iranians to win independence from despotic and foreign rule. Islamists and nationalists were the usual dissenters to despotism and foreign influence. Some variations of Marxist groups also joined the popular struggle to make Iran free of despotic rule in the 1930s. Actually, some four centuries ago, the religious tradition in Iran was a calculated distance from national politics. Religious authorities normally provided advice to the political leadership and considered themselves above daily politics. Beginning in the 1800s, Islam became increasingly politicized when rulers failed to deliver efficiency or they subordinated national interest to foreign demands. Islam gradually took on a role in Iranian national "political identity." Since it was the Western countries that attempted to reap Iran's resources and control its geography, a correlation began to emerge between people's Islamic identity and the defense of their country from foreign domination. Therefore, for Iranians, following the gradual integration of Islam into their identity framework, the issue of the "West" became their most crucial cultural and political challenge in the modern age.[1]

Historically, only one person, namely the King, ruled on all issues. Whereas in Europe, groups and classes not only existed but were also independent of the state, in Iran all groups belonged to the state and the State meant the King.[2] Therefore, Iranian political structures could not create mechanisms for self-criticism or occasional self-appraisal.[3] Almost all details of statecraft had to be cleared by the court. Consequently, in this political structure, wealth, power, policy-making and influence all concentrated in the person of the King. There existed no room for creativity, non-state association, free thinking or organization beyond state control. The weakness of the polity and its inability to produce wealth and security opened the door to European and later American influence. There is an interesting contrast in the way authoritarianism emerged in Asia and Iran. In

Asia, hierarchy and functions were maintained even in an authoritarian structure and individuals were regarded as respected subjects. However, in Iran, hierarchies were considered contrary to the sacred role of the King and consequently, the individual was devalued and all attempts at organization were regarded as contrary to the sacredness of the King (or the Center of the Universe).

The individual in Iran was somehow dissolved in the community and the apparatus of the state and was not delegated a functional role to be creative. Therefore, non-state political organization as opposition to monarchical despotism and foreign intervention in Iran evolved around political ideologies. Meanwhile, Russian and British competition over Iranian geopolitics and resources led to vulnerable and highly dependent monarchies. Iran's development was undermined and delayed by both weak states and weak societies. Some one hundred and fifty years of widespread foreign influence in Iran also cultivated a political psychology of conspiracy theories leading to a culture of xenophobia. Most people blamed foreigners for Iran's underdevelopment. The idea of being "independent" and creating a republic to oppose foreign rule became the cornerstone of political thinking. These trends led to a rainbow of religious, nationalist and Marxist opposition in Iran. The Pahlavi dynasty made many attempts to modernize Iran, but in the eyes of the opposition, the Pahlavi monarchs were serving foreign interests and maintained their past dictatorial tradition. The legitimacy of Reza Shah and his son Mohammad Reza Shah was founded on British and American military and intelligence support. Therefore, opposition to foreign rule and the Pahlavi dynasty continued throughout much of the twentieth century. The idea of "indigenous" thinking, elites and source of governance stood out among the intelligentsia, political activists and the common man.

The Islamic Revolution of 1979 was born under such historical psychological and political conditions. Some years prior to this, in June 1963, the American Chargé d'Affaires in Tehran approached Iran's head of budget and planning organization. The American diplomat put a suggestion to the Iranian official: "Are you prepared to gradually proceed to become Iran's Prime Minister?" The astounded and irate Iranian official replied: "On whose behalf are you offering me this job?" The American official responded confidently that he had the instructions from Washington.[4] This incident and hundreds alike speak volumes as to the rationale of the Iranian elites in maintaining distance from the outside world in the post-revolutionary period. The Islamic state set out to be a republic, seeking self-sufficiency, Third Worldism, alignment with developing countries, the altering of the regional political matrix and focusing on ethics and justice.

In the early years of the Revolution, the clerical community was able to outperform the nationalist and leftist claims to power, forcing many of them into exile. Once the clerical community was able to consolidate its hold on power, it had the opportunity to implement its revolutionary ideals not only in Iran but also throughout the Muslim world. The Islamic nature of the Constitution was actually shaped in the second to third year of the Revolution with the clerics using their political credentials, grassroots support and national organizational and financial network. No other political or social grouping in Iran enjoys such organizational and oratory skills for effective political control and communication. Although religious, nationalist and leftist organizations in Iran all participated in the overthrow of the Shah's regime, they had no common foundation to replace it. For their part, however, the clerics benefited from a populist constituency and were able to mobilize the general public against the leftist and nationalist groups in the early years of the Revolution. Clearly, the clerics had long envisioned an Islamic state and, with the completion of the Constitution, their ideas and aspirations found a legal basis.[5] For the clerics, it was a historic opportunity to set up an Islamic state in Iran and project a model of cultural purity, sovereignty, progressiveness and economic development.

The Islamic Republic of Iran launched its offensive to change the *status quo* both within and beyond Iranian borders. While Islamic and some opposition groups were inspired by the idea to rid themselves of authoritarian governments in the Middle East, these Iranian ideas and aspirations were opposed vigorously by almost all governments in the region. The United States also opposed Iranian activities, since they all interfered with the status quo and mobilized anti-West opposition in the Middle East. As time went on, like most other revolutions, the elites of the Islamic Republic of Iran proved more interested in foreign affairs than institutionalizing an efficient and accountable republic in Iran. In reaction to Iran's attempts to change the status quo, Iraq acquired a consensus to invade Iran, the GCC was established, Islamist groups in the Middle East suffered a crackdown and further Americanization of the region took place.[6] While islands of support for the Iranian revolution among Islamists emerged, an anti-Iranian aura dominated the Arab and Muslim worlds. Relations between Iran and particularly the Arab world deteriorated sharply and security issues dominated the relations between the two sides. The symbol of opposition came from the Iraqi Ba'ath party that felt an urgency to obstruct the fervor of the Islamic revolution. Many governments provided financial and political support to the Iraqi project to impede Iranian advancements. Iraq also had the tacit support of the United States and the Soviet Union to obstruct the Iranian revolution. It should be pointed out that the analytical construct of the Iranian revolutionaries went far beyond Iran. Their domain of purpose and activity reached to the outer perimeter of the Islamic world. The universal outlook of the revolutionaries intended to rearrange the power structure of the Muslim world, bringing Islamists to power.

Consequently, the foreign policy of the Islamic Republic of Iran produced antagonism on three levels—at the Middle Eastern level, particularly in the Arab world, in the West among Iran's monarchist, nationalist and leftist opposition and at the global level leading to an American and Israeli coalition to oppose Iran. Of course, Iran believed that its intentions were just and that it was acting on its Islamic obligations. Resources to change the region were secondary; relying on the mental support of the masses was sufficient. Iranian efforts were opposed from all sides.[7] One can even claim that there was global opposition to the activities of Iran. The conceptual premise that can be substantiated here is that confrontation at the regional and international levels turned the Islamic Republic of Iran into a security phenomenon. Immediately after its early consolidation, the Iranian revolution faced existential threats and therefore focused on its survival. Managing the range of conflicts and threats emanating from its opposition demanded considerable time and energy from the Iranian clerics. It also devastated Iran's national resources and global prestige. Iran did not have the opportunity to focus on its national economic development and social reconstruction. Now, after a quarter of a century, Iran has not freed itself from potential security or military threats. Much of the attention of the leadership is devoted to maintaining and managing security threats. After the fall of the Soviet Union, Iran, North Korea and Cuba are the only countries that use the word "enemy" in their foreign policy vocabulary towards other countries. China, Russia and Eastern Europe now focus on economic diplomacy and peaceful coexistence.

Another significant conceptual premise is in order here: there is a relationship between Iran's confrontational foreign policy on the one hand and legitimating requirements of the cleric order on the other. The cleric community in Iran has an Islamic outlook. Their educational credentials are rooted in philosophy, Islamic jurisprudence, history and morals. Their industry is not economic, financial or scientific development. In a power structure, their interests lie in culture, philosophy and politics. A religious person who has strong ideological convictions may find the geography of a nation state too limited. This is perhaps a solid base to conclude that religiosity and nationalism cannot converge. The abstract mind of the Iranian clerics corresponds to their egalitarian pursuits in the Muslim world. Even Mr. Khatami in his attempts to bring about civil society, rule of law and democratic traditions could not move beyond abstractions, rhetoric and demagoguery. Iran's deep cultural interest in ambiguities, rooted in authoritarian tradition, and excessive reliance on poetry and fate as explanations of social and political events have also contributed to an abstract and philosophical frame of mind with little attention to applications and quantification.

Moreover, Iranian pursuit of political and cultural objectives beyond its borders provided its leadership with a base for maintaining a revolutionary order and power structure. If Iran were to divert its attention from the Israeli issue and the idea of egalitarianism in the Muslim world to domestic soft politics, a gradual increase of per capita income, employment schemes, tourism and biotechnology, then the leadership role and capacity of the professional groups in the country would naturally advance. Even in capitalist countries, there is a close association between the nature of foreign and security policy and the economic and industrial preoccupations of elites.[8] A consistent political dilemma in Iranian history is that adherents of nationalist, leftist and religious ideologies have had difficulty in shaping a common platform to form an all-inclusive social and political construct and to define Iran's national interests. This is why political groups have defined their survival and advancement at the expense of excluding others. A maximalist outlook founded on zero-sum game has molded Iranian political behavior. Therefore, Iranian post-revolutionary security and foreign policy behavior is a clear reflection of the legitimating needs of its elites.

Constitutional Bases of Iranian Foreign and Security Policies

The phrase "confronting foreign hegemony" is extensively used in the Constitution of the Islamic Republic of Iran. Although this phrase is a reflection of a profoundly rejected historical experience, at the time of the revolution it referred specifically to the Soviet Union and the United States. As mentioned before, Iranian suspicion of great power intervention is the theoretical underpinning of post-revolutionary behavior.[9] The political psyche of the average Iranian usually seeks an external explanation of internal problems dubbed as the "hands of foreigners," commonly referring to the British and the Americans. Perhaps no other theme dominates the Iranian Constitution more than the ideas of independence, national sovereignty and struggling against colonialism and imperialism. Article 152 of the Iranian Constitution explicitly forbids Iran from developing close and strategic relations with great powers. Therefore, much of Iranian foreign policy and security behavior stems from its bitter experience with foreign intervention, on the one hand, and the rise of Shi'a clerics to power, on the other. The latter development brought Islamic thinking into the arena of statecraft and defined Iran's national interests in terms of the Islamic world, not limiting itself to Iran's geographic boundaries.

Political independence and confronting American domination of the Middle East were the prime objectives of the Iranian revolutionaries. Therefore, it is imperative to view the Iranian revolution from a historical and evolutionary perspective. In this context, the economic development of Iran was not the prime objective of the Iranian revolutionary state. In fact, the Iranian revolution occurred at a time when much of the economic momentum and privatization took off in the Asian and Latin American regions. Beginning in the early 1990s, Malaysia, Singapore, South Korea, Taiwan, Brazil and Turkey took a great leap forward to internationalize their economies and, in the process, substantially increased their national wealth, per capita income and technologically-oriented industrial capacity.[10] In Iran however, internal cultural and external political change dominated the agenda for statecraft. Conflict management as well as crisis management had inadvertently become top priority issues in much of the post-revolutionary period. In the process of attempting to realize their regional objectives and still energized by the quick success of the revolution, Iranians underestimated the resistance and vigor of their adversaries.

Even though Iranians had long been revolutionary in spirit, Iranian society had never before experienced the political openness that has become prevalent in more recent times. Debates, critical dialogue among political groups especially in the post Iran-Iraq war became widespread. One can even claim that in no developing country as in Iran since 1989, has there been so much sophisticated, subtle and extensive criticism of the state.[11] Yet, the state through constitutional, institutional and economic mechanisms has been able to overrule society at large. This is why Iran is often referred to as a system with a complex state-society matrix.

Furthermore, in Article 153 of the Constitution, it states that it is forbidden for any hegemonic power to dominate Iran's natural resources, economy, army and other pillars of the state. These ideas were institutionalized in Iran exactly at the same time that countries such as Malaysia, South Korea, Singapore and China began to attract foreign capital and investment, and started a process of sharing wealth, management and ownership with foreigners in general and Westerners in particular. Again, historical experience is manifest in the economic realm of the Constitution. Due to colonial intervention in their affairs, Iranians developed a "sovereignty complex." While ideas of sovereignty and independence were the slogans of all developing countries in the 1950s and the 1960s, Iran during this period suffered from the Shah's dictatorship. In a sense, Iran's direct struggle with colonial ideas and intervention was delayed. Occurring in 1979, the Islamic revolution echoed all of the slogans of the African and Asian countries of the post-World War II era. In the 1970s, the concept of interdependence replaced the colonial jargon of the 1950s. Third World states realized that they need to integrate with the rest of the world, to share their resources and profits and to develop an efficient infrastructure to attract foreign investment. There was no longer the idea that a country could develop internally by relying solely on its domestic resources. There were no longer local solutions to national problems. The global system followed no geography in promoting economic interests. Moreover, development is ultimately an idea to be imitated. Memories of empire-building in the past as well as the observance of Islamic dignity prohibited Iranians from imitating the Western model of development. In the end, these cultural and historical characteristics proved to be much more powerful determinants of behavior than rational processing of the logic of the global system within which we all live.[12]

Moreover, Article 154 of the Constitution asserts that Iran "will support the rightful struggle of the weak against the strong in every part of the world." Based on this article, Iran had to maintain a distance from great powers. As a demonstration of its solidarity with the have-nots, Iran gradually reduced its extensive global operations to select Middle East and developing countries. As Moscow and Beijing in the 1950s and the 1960s, now in the 1980s, Tehran became a focal point for liberation movements around the world. Among them in particular were the Islamic movements of the Muslim world. Almost no one paid attention to the plausibility of changing the political systems in the Muslim world through ideological movements. Conceptually, the standoff was between excessive subjectivity and irrational exuberance, on the one hand, and the rational processing of the ways and means of pursuing an objective, on the other.

After the Iran–Iraq war, Tehran resumed its traditional economic relations with the major European countries and Japan. As budgets tightened, liberation movements closer to home were given priority. The Palestinian liberation movement received the greatest attention. The Palestinian issue was not defined as a conflict between the Israelis and the Arabs but as a conflict between Muslims and Zionists. Iran's support for liberation movements is partly a reflection of its revolutionary behavior but more importantly, it stems from an Islamic premise that the suffering of a single Muslim is the suffering of all Muslims. Politically, liberation movements furnished Iranian revolutionary elites with an opportunity to maintain the relevant domestic organizational arrangements. Hundreds of administrators benefited greatly and while serving as go-betweens, upheld the idea of support for liberation movements, guaranteeing themselves continued employment. Nonetheless, it should not be dismissed that in the beginning of the revolution, support for liberation movements displayed a genuine ideological zeal. However, as revolutionaries entered their middle age years, liberation movements served as instruments of political control. In the 1990s, Iran grew increasingly selective in its support of liberation movements but the idea was preserved in revolutionary circles.

As time went on and the choice between being a revolutionary state and a normal functioning state occasionally became tense, Iran separated political behavior from its economic transactions. Iranian mindset undertook varying contradictory objectives simultaneously. For example, Iran was always ready to enter into economic deals with the United States but it refused to have diplomatic relations. At the height of the Iranian–American conflict in the mid-1990s, it was not Tehran but rather Washington that terminated economic dealings and imposed sanctions on Iran. Even today, while suspicious of American political intentions, Tehran encourages economic interaction with the United States. In this respect, it is argued that Iran's political sovereignty is not jeopardized. Iranians have difficulty sharing decision-making with others. Islamic integrity and the powerful concept of sovereignty keep others at bay. What may appear as isolation by Western observers is regarded as national pride by Iranians. Therefore, constitutional and institutional foundations explain the endurance of a revolutionary and cleric order.[13]

The aforementioned conflict-ridden and confrontational atmosphere between Iran and the rest of the world provided no space for economic development. Iran's ideological foreign policy contradicted its willingness to engage economically with other countries. Perceptions of insecurity have delayed Iran's decision to open its economy. Privatization is possible when its political compatibility with the rest of the world is not seen to threaten the apparatus of the political order. The Islamic Republic of Iran has not been sufficiently at peace with the international community to deregulate its economy. Ideological foreign policy has utilized much of Iran's national resources, further consolidating state control of the economy. The continuing paradox of the Islamic Republic is that, politically, it is anti-West but if it wants to develop economically, it needs the West. The income from petroleum has helped the state to maintain itself and pursue a strategy of survival. A strategy of growth for Iran requires cooperation with the West and the abandoning of anti-American policies. However, in order to maintain their Islamic identity and political legitimacy, the Iranian leadership of Iran has refrained from entering into a coalition with any other country.

Due to its peculiar structure, perhaps it can be concluded that Iran is incapable of entering into alliances with other countries. Iran is the only country that does not accept the two-state solution in the Palestinian–Israeli conflict. However, in the economic realm, as long as trade and economic investment does not interfere with the direction, logic and nature of the political system and its political culture, it would not face any opposition. In reality, the central paradox of the Islamic Republic lies in the manner it defines itself and the international system. These definitions fundamentally serve the interests of a clerical order whose legitimacy is supported by the Constitution. The revolutionary legacy of the Iranian elites and their belief structure do not allow them to share power with the Western world. On the one hand, Iran desires to be a normal state especially when it comes to trade and economic issues. On the other hand, it wants to remain as a revolutionary state rejecting the status quo both at the regional and the international level. Nonetheless, it should be pointed out that, over the last decade and particularly during the Khatami presidency, Iran's rhetoric has been far more vigorous than its practice.

In sum, much of Iranian foreign policy and security behavior is supported by the Constitution. In fact, the Constitution places far greater emphasis on sovereignty and independence than on economic and social development. To be precise, one can conclude that the concept of political isolation is constitutionally legalized irrespective of the dynamics of interconnectedness at the dawn of the twenty-first century. History, then, has a far greater impact on behavior than objective analysis of contemporary trends and statistics.

Schools of Thought in Security Doctrine: Policy Consequences

After the Iran-Iraq war, two factions began to emerge in Iran's political apparatus—the revolutionaries and the internationalists. It is notable that similar parallels can be drawn from the Chinese and Soviet experience, when they reached a peak in the contradictions of their international conduct in the 1970s. As a result, the Soviet Union disintegrated and China acquiesced. Contradictions in Iran remain. The Rafsanjani presidency gave rise to the internationalist school of thought and provided it with organization. Akbar Rafsanjani, the Speaker of Parliament and second-in-command in the war against Iraq, realized in the midst of the war, in the mid-1980s, that the war could not proceed further, but he was conscious of the fact that it was not possible to break the revolutionary deadlock. Economists, diplomats and academics all agreed. To maintain his influence, Rafsanjani wavered between the revolutionary and internationalist schools of thought. However, for long he remained the unofficial spokesman of the latter school. Those who looked out for

Iran's global standing and the well being of the average citizen clearly found themselves in the internationalist school. For them, economic development, modernization, national wealth, access to information technology, health, education, efficiency and globalization should be the concerns of the state. The revolutionary school considered history, the plight of the Islamic world, struggle against imperialism and adherence to a code of ethical conduct as pressing issues. The revolutionary school of thought considered Islam as an all-encompassing ideology. There was no need to extend a helping hand to the foreigners. In the end, Iran had achieved no result by cooperating with the British, Russians and Americans. The international landscape is based on oppression, conspiracy and anti-Islam sentiment; it is all controlled and directed by Zionists. All others are the extension of this vast economic and political conglomerate.

Furthermore, the revolutionary group believes in the preservation of the ideological order, the clerical establishment, state control of culture and considers the Western world as Iran's enemy. The internationalist group, however, does not dissociate the domestic structure from the global dynamics, national economy from foreign policy and national security from economic development. The former group believes that Iran should focus on its internal agenda and maintain a calculated distance from the international community. The latter group promotes national economic development and believes that Iran should join the World Trade Organization (WTO) and become a normal member of the international community. The former category asserts that Iran's security is guaranteed when it dissociates itself from the economic and political impositions of the international capitalist system led by the United States. The latter category, however, presumes that Iran's national security stems from its economic interdependence with the international community and that Iran should focus on producing national wealth, economic diplomacy and soft politics. Whereas the threat perceptions of the former are fundamentally military and existential, the latter group perceives economic, social and soft issues as basic threats to the country.

From a sociopolitical perspective, the two schools of thought reflect the divergence of opinion within Iranian society. Both advocates have a strong political, economic and institutional presence in society. These ideological and therefore political divisions can be overcome by debates. The difference in the definition of national security by the two groups mirrors the divergence in their definition of psychological security. The revolutionary group feels insecure in dealing with foreigners. The internationalist has a stronger self-esteem and is willing to compete, to influence and to be influenced. The revolutionary group aims at dealing with the masses and as a consequence is an advocate of populism. The internationalists believe in social organization, political parties and are convinced that efficiency results from professionalism. Masses become meaningful when they are organized in professional groups and political parties. The confrontation between these schools of thought has had consequences for Iranian security policy and conduct. It has also damaged Iran's image beyond its borders.

The Islamic Republic of Iran has consistently confronted two types of threats: military threats and existential threats. Iraq under the Ba'ath party threatened Iran militarily. Israel and the United States have also threatened Iran with military attacks or have implied their preference for a change of regime. Such security preoccupation that continues to this day has left little opportunity for Iran to focus on non-security issues. In fact, internal cultural change, external political change and confrontation with American regional hegemony have dominated the Iranian national agenda. These pursuits have fallen within the ideological and conceptual framework of the Iranian clerical community. From a theoretical perspective then, it can be concluded that Iran's security doctrine in the post-revolutionary period is congruent with the class base and the class needs of the Iranian clerical community. Economic development requires a nationalist base that has been absent in the ideological outlook of this community. These political ideals led to greater control of national economic resources, providing extensive services for national security and foreign policy. It is a fact that increased centralization of the Iranian economy has taken place since 1979. Perceptions of insecurity have also contributed to the obstruction of the administrative and political processes for opening up the economy. Thus, Iran's security and economic doctrines reflect the political and legitimating needs of the clerical community.

Nevertheless, one should not overlook the fact that the pursuit of the aforementioned security policies enjoys the legal support of the Constitution as was spelled out before. The Iranian Constitution was written under conditions where Iranians had fought the Shah's regime and were seeking independence from foreign rule. At the same time, because of the Islamic nature of the Iranian revolution, the clerical establishment sought to change the political systems of Muslim countries. As regional realities gradually surfaced for the Iranian elites and *Realpolitik* gained credibility, it became apparent that altering the status quo was not possible. However, the dilemmas did remain in the Iranian revolutionary mindset.

The legal framework and the revolutionary inertia have led to a number of paradoxes in Iranian foreign policy behavior. Three paradoxes stand out among others. First, Iran desires to be both a normal state carrying out its ordinary state functions at the international level, while it strives to be a revolutionary state with a strong rhetoric to change the status quo. Second, Iran questions the legal and the egalitarian basis of the international system administered by major powers and industrial countries, while it seeks to become an active member of the same international system by repeatedly pursuing membership at the World Trade Organization. Third, Iran is clearly attempting to develop its economy, while it fails to accept the role of the multinational corporations or Western governments in facilitating its entry into the international technology, capital and commodity markets. There is also an interesting related paradox here. While Iran is willing to conditionally work with powerful international players to develop its economy, in the area of security, culture and politics, serious reservations are spelled out in sharing power, authority and sovereignty with the same powerful and industrial powers. The logic of the contemporary global system, however, requires that states conduct their economic, political and security affairs in an interconnected complex.

Countries, especially middle range developing countries with many vulnerabilities cannot choose between various dimensions of cooperation with the international system if they want to achieve comprehensive development. The political meaning of the American economic sanctions on Iran was to coerce Iran to view its economic and political relations with the United States as a package and to disallow Tehran to select as it wishes. From an internal perspective, this is the dilemma that Mr. Khatami has faced in his presidency. If he truly pursues the objectives he has pronounced, then he would ultimately contradict the internal logic of the Iranian Constitution and the political system to which he belongs. In a simple dichotomy, Iran is caught between accepting the norms of the international system and adhering to its domestic norms, ideological foundations and historical complexes. The outcome of these oscillatory processes has been ambiguous policies, lack of international commitment to Iran, lack of a strong premise to build alliances with other countries and an increase of vulnerabilities. Nonetheless, ambiguity, duality and indecisiveness have collectively provided the maneuvering room for the revolutionary elites to maintain the status quo.

International Constraints and Iranian Foreign Policy

An underlying threat perception dominates Iranian security doctrine, namely, American opposition to the Islamic Republic of Iran. US administrations in the last quarter century have consistently opposed the major trends of the Iranian foreign policy. Among them, Iran's support for militant Palestinian groups, Tehran's nuclear program and the Iranian approach towards the Israeli-Palestinian conflict.[14] Strong anti-Iranian stances by American officials towards Tehran have in turn engendered a confrontational approach on the part of the Iranian state. Moreover, religious symbols have exemplified antagonism between the two sides. While Iran has referred to the United States as "Satan," the United States has reciprocated by labeling the Islamic Republic of Iran a member of the "axis of evil." Nevertheless, opposing Israel's existence and the United States presents the Iranian political system with a philosophical and political *raison d'être*. As a result of Iranian opposition to the

United States and Israel and the global anti-Iranian activities and lobbying of these two countries, Iran faces a more or less complete blockage of technology transfer. If Iran wants to develop economically, it has to change the course of its foreign policy. In this context, two alterations in foreign policy are called for: adaptation to the emerging regional order in the Middle East and an attitudinal change toward the peace process. Over the last two decades, the United States has been able to increase its military and security presence in the Middle East and around Iran's neighboring countries, in considerable part due to the foreign policies of Iran and especially Iraq. Security policies of these states produced security threats to the vulnerable countries of the region, leading them to develop strategic ties with the United States. Therefore, the major object in Iran's strategic thinking in developing its security doctrine is the United States.

The quick American victories in Afghanistan and Iraq led to a moderation of Iran's foreign policy style and language. Iran began to pursue a complex mix of diplomacy and constructive dialogue with the United States. Despite the negotiations in Geneva and New York on functional issues such as border security in Afghanistan and Iraq, the fundamental differences remain. It would seem that no matter how the United States and Iran cooperate on functional issues related to their common interests at the Middle Eastern level and around the neighboring countries of Iran, in the absence of an attitudinal change on the part of Tehran toward the Palestinian-Israeli conflict, there will be no breakthrough in relations between the two countries. Israel has grown very instrumental in shaping American policy toward Iran in the post-revolutionary period. Due to effective Israeli and American lobbying around the world, high-tech exports to Iran are basically banned.

Over the last two decades, Israel has proved that its territorial influence does not lie between the Mediterranean and the Dead Sea. Israel has been actively pursuing anti-Iranian diplomacy in Europe, Russia, China, Latin America and in international organizations. The Israeli goal has been to obstruct Iranian development in any way possible. Perhaps, one can mention Israeli obstructionism in the areas of foreign economic investment, defense and nuclear programs, dual use technology and in what can be referred to as software devastation of the Iranian image at the global level. Iran is unable to purchase new commercial aircraft from the American and European producers due to sanctions imposed by the US government in part because of the Israeli argument that new commercial aircrafts might be used for military purposes. Even the Iranian nuclear program is of concern because of the ramifications for Israeli security and the macro Middle Eastern strategic interests of the United States. One can conclude that European–Iranian relations can advance only up to the point where there is no conflict with American and Israeli interests. Certainly these relations cannot be expected to reach a strategic status without tacit American support.

Since 9/11 and the removal of the Afghani and Iraqi regimes, Iranian behavior has changed. As a general premise, it would appear that "events" are more determining in the Middle East in general and Iran in particular than debates, critical thinking, futuristic discussions or philosophical pronouncements. In this respect, the quick removal of the Taliban and Ba'ath regimes was far more important than statistical discussions about the scope and effectiveness of American military power. Much of Iran's rational policies at times of crisis are a reflection of its perception of existential threats. An example in this regard is Iran's agreement to sign the additional protocols of the NPT treaty. Once it became clear that Iran's non-conformity would take the country to the UN Security Council with the prospects of turning Iran into another "1991 Iraq," the elites signed immediately.

In reaction to the emerging geo-strategic map of the Middle East, the basic threat perceptions of Iran are even more intensified. Iranian tactical adaptation, which demonstrates some flexibility, and its prolonging of existing conflicts reflect the power of events more than rational discussions. However, it should be pointed out that Iranian use of ambiguous language and its effective use of the Lebanese Hizbollah, of liberation movements as well as Islamic communities provide reliable bargaining tools vis-à-vis the United States within the larger Middle Eastern strategic matrix. The complex nature of Iran's political system and the duality of its decision-making apparatus also provide opportunities to project ambiguity and to buy time. Domestic trends demonstrate that Iran is moving into a status quo power and therefore, temptations for military confrontation or even political provocations are highly reduced. For domestic audiences, the issue of national political sovereignty is increasingly emphasized. Moreover, especially over the last five years, Iran has begun to diversify its foreign policy by opening up to Egypt, seeking observer status at the Arab League and displaying more flexibility to the European Union. Iran was the first regional country to accept the legitimacy of the Iraqi Governing Council set up by the United States.

Whereas there was a time when Iran viewed its linkages in Lebanon and Syria as ideological manifestation of its foreign policy, it appears that today these linkages are utilized as national security bargaining chips. These instruments are looked upon as significant sources of credibility and bargaining in a probable process of negotiation in a unipolar world. Nonetheless, Iranian adjustment to the post-September 11 international system is indicative of the fact that Tehran is finding it more difficult to remain a pure revolutionary state in a unipolar global order. Iran's acquiescence to the articles of the International Atomic Energy Agency (IAEA) is also another indication of its flexibility. Part of Iran's flexibility in the last few years can be explained in terms of increasing domestic vulnerability as well. Some statistics may reveal the significance of the domestic landscape. One third of the college graduates end up unemployed and live on family support. Iran ranks 146 in terms of economic openness and 90 in terms of human resource development. Since the beginning of the 1990s, some 630,000 professionals have left the country. Some 64% of the population is below the age of 33 and 25% of it is below the age of 10, which means that there are strong pressures on the economy to produce employment for the young population.[15] It appears that in the coming years unlike the earlier years of the revolution, Iran's foreign and security behavior will be intertwined with the political, economic and social exigencies of the domestic structure of the country.

Conclusion

If American unipolarity persists and a successful Iraqi model evolves, serious security consequences for Iran will emerge. With Shi'a fundamentalism in decline and Sunni fundamentalism on the rise, Iran will increasingly emerge as a status quo power. Although Iran suffers from numerous problems at the national level, seen over a period of time, Iranian national development is on the right path, despite several hardships. Iran's economic security will be far more significant than its ideological security. As a result of a combination of generational shifts, socio-economic changes, nationalist oriented public opinion and a new regional geopolitical map, Iranian national security doctrine will be increasingly based on its economic viability rather than ideological premises. Iran has for the most part retreated from its ideological pursuits to the defense of its national interests, even in the projection of its revolutionary power and energy. The Palestinian issue has remained as the only ideological and revolutionary issue in the foreign policy of the Islamic Republic of Iran. Within the government apparatus, Iran's ministry of foreign affairs and all ministries related to economic functions have played an important role in moderating Iran's foreign behavior. These institutions reflect reliable bases for the institutionalization of the internationalist school of thought. At the societal level, the academic, intellectual and "religious intellectual" communities have also been vocal in providing criticism, reappraisal and dialogue. Perhaps more than any other society in the Middle East, the Iranian society is experiencing intensive political and social change.

There is a considerable gap between the current elites of Iran and the young population in terms of life expectations, political attitudes, foreign policy outlook and definitions of Iran ten years from now. The current elites had one fundamental objective in mind when they were in their 20s and 30s: Iran's independence and national sovereignty. That has been achieved. Given the experience of the developing world, the most solid basis of maintaining national sovereignty is economic wealth. Military equipment and even nuclear weapons are no longer guarantees of national security and territorial integrity. The fact that the second generation of elites is in its early

40s and the younger generations are in their 20s and 30s, it can be expected that Iran will move in the direction of economic development, further engagement in the globalization process and political stability. In this context, security policies of Iran at the regional level will be increasingly based on trade, economic interdependency and cultural exchange. The guarantee for such a development will be when the Islamic Republic of Iran moves from a security phenomenon to an economic-cultural phenomenon.

11

Iran's Regional Policies: Western Perspectives

Jerrold D. Green

In the wake of the military defeat of Saddam Hussein and the subsequent political reconstruction of Iraq, one area of great promise and uncertainty relates to how a new security order is to be designed for the Gulf region. Gulf security has traditionally been the product of political interactions among three groups of actors and their external partners—Iraq, the Islamic Republic of Iran, and the constituent states of the Gulf Cooperation Council (Bahrain, Kuwait, Oman, Qatar, Saudi Arabia and the United Arab Emirates). Given the sometimes fractious relations amongst the three, as well as the fact that this uniquely strategic region contains the world's largest petroleum reserves, a variety of external actors have also chosen to involve themselves in the design and maintenance of a Gulf security order, some at the behest of assorted regional actors. In more recent times, the predominant external actor has been the United States.

If one were to review the history of United States involvement in the Gulf, one would find that Washington has traditionally aligned itself with one or at best two of these three sets of actors. For many years, the premier partnership was between the United States and the last Shah of Iran, Mohammed Reza Pahlavi, who functioned on behalf of Washington as a *de facto* regional policeman. After the

Iranian Revolution and the ensuing fall of the Shah, the United States briefly put its support behind the Iraqi President Saddam Hussein. This misguided and short-lived adventure came to a crashing halt with the Iraqi invasion of Kuwait and the subsequent Gulf War. Most recently, the United States has allied itself with Saudi Arabia and a shifting group of GCC member states. Although the political orientations of each of these states, on a macro analytic level, can be characterized by unanimity and consensus, microanalysis quite appropriately demonstrates significant differences in the political orientation and relations among and between all these states, each of them unique and quite properly having its own national interests at heart. In any case, with the defeat of Saddam Hussein, and the subsequent occupation of Iraq by the United States, abetted by a modest number of coalition partners, the issue of Gulf security has resurfaced. Indeed, it now falls to all regional actors, as well as to their external partners, not only the United States but also the Europeans, to determine the shape of Gulf security. Iran remains central to this enterprise although, at this point, the question as to the nature of Iranian involvement remains unanswered.

Western and Eastern Perspectives

I have been asked to address Western perspectives on Gulf security in general, most particularly as they relate to the regional role of the Islamic Republic of Iran. However, it must be noted that it is just as difficult to discuss a uniform Western perspective on anything, as it is to define a Middle Eastern or an Arab perspective. Unanimity or even general consensus is tremendously appealing as an ideal. The political reality, however, is infinitely more complex, and among the Western powers there are differing views of what they would like to see accomplished in the Middle East, what their individual national interests are in this tempestuous but important region, and how their own national goals can be achieved in the most efficacious way. The Western states on many levels enjoy cultural amity and affinity. On the other hand, the differences that divide them are no less significant than the commonalities that appear to unite them. These differences assert themselves regularly, as this collection of states jockeys for power, influence, and resources in an alluring but always inflammable Middle East. Of course Western perspectives can be discussed in a general sense. Yet, the unfortunate reality is that dissensus seems to prevail despite the obvious attractions of collective and collaborative political action. Moreover, the act of simply being "Western" is of itself insufficient justification for political, economic, or even strategic cooperation.

The most recent illustration of cracks in the Western "common front" lay in the run-up to the war in Iraq and its aftermath. The Western world evinced little unity as the war, initiated and led by the US ran its course. The United Kingdom and the United States were in agreement, or at least their governments were. However, Germany, France, Turkey and several others held a range of dramatically different views about the desirable end state not only of Iraq, but also of what constitutes Gulf and regional security. The West has traditionally been unable to agree on much of anything in the Middle East. The fault lines that divide one side of the Atlantic from the other are significant and include the Arab-Israel dispute, where the United States is generally thought to have a bias for Israel, matched by the Euro-bias for the Arab world. Attitudes and policies toward Iran is another area in which the West has been seriously divided with the Islamic Republic at the center of the United States policy of Dual Containment and subsequent inclusion on President Bush's "axis of evil" list. The Europeans on the other hand, seemed to prefer the carrot to the stick, opting for policies of constructive engagement with Iran. In other words, the Europeans, including the United Kingdom, all seemed to believe that influence could be exercised through engagement rather than by exclusion and isolation, which has been the US preference. Finally, as noted above, significant differences were witnessed between the West and the United States on the Iraq war. Many Western states opposed the war and the United States has more recently responded by denying access to Iraq to many of these countries that are now seeking opportunities to secure lucrative commercial contracts related to Iraqi national reconstruction. In short, to talk of "the West" in the singular is probably far less precise than to talk of "the West(s)" in the plural sense.

Western disunity is matched only by that of all other groupings of states. The members of the Arab League seem unable to agree on anything other than the desirability of having an Arab League. The countries of the Organization of the Islamic Conference (OIC) meet regularly, although areas of disagreement far outweigh those of unity amongst the member states. The Gulf Cooperation Council only comprises those states that are generally in agreement, but does not include relevant regional actors such as Iran and Iraq. In short, if one were to enumerate virtually all significant international organizations, including the United Nations, what one would find is dissensus coexisting comfortably alongside a fervent desire for consensus.

Consensus in a World of Dissensus

Although consensus appears to be more an ideal than a political reality, it is important to realize that states, when they deem it in their interest, are able to collaborate with those deemed even their most bitter enemies. Thus, for example, the United States and the Islamic Republic of Iran cooperated quite effectively at the Bonn Conference held from November 27 to December 5, 2001, at which the future of Afghanistan was meant to be determined. American Ambassador James Dobbins and Iranian Ambassador Javad Zarif would breakfast together almost on a daily basis, comparing notes and coordinating their strategies to help Afghanistan to reach a meaningful political solution to its myriad problems. The point being emphasized is that despite all the logic that seeks to position particular nations in convenient categories and factions, the exigencies of the moment, as well as political interests, ultimately determine where these states will finally be positioned. Moreover, these positions are final until they suddenly change due to new political alliances and challenges. The most recent example of this is the current United States-Libyan amity resulting from decisions on both sides that the time for rapprochement has finally arrived.

Eastern and/or Western states can be bound together in common action with other states that agree with them, irrespective of geography or ideology if they believe that such collective action is in their interest. Israel has long sought, for example, to exploit the non-Arab qualities of regional states like Turkey and Iran, so as to create an entente among such non–Arab political actors in the Middle East. This succeeded for a time with Pahlavi Iran, while more currently, Israel has forged a durable link with Turkey. Close strategic links between Turkey and Israel have even survived the periodic ascent of assorted Islamic-oriented political leaders and parties in Turkey, with pragmatism far outweighing ideology in helping to define this important relationship. It must be underscored that the notion of having one set of policies and political actions which the West would deem appropriate for Iran, counter posed against another set of values and political postures being adumbrated by the states of the Middle East, is overly simplistic, unrealistic and too dependent upon linear thinking. All states will evaluate the political behavior, both real and ideal, of other states such as Iran, based on their own perceptions of their own national interests. Thus, a peculiarly "Western," geographically determined standard for Iranian political behavior is unlikely to be found. Rather, all states seeking to interact with Iran will have an expectation of Iranian political behavior, resulting not from Iran's character or interests, but rather from the perceived interests of these states themselves.

Understanding Iran

What further complicates external expectations of Iran is, of course, the behavior and expectations of the Islamic Republic itself. Foreign policy is always a consequence of internal politics. Iran is an exceedingly diverse and fractious nation with a very unsettled domestic political life and character.[1] It is multi-ethnic, multi-lingual, and even multi-religious despite its appellation as an "Islamic Republic." The Iranian political system is extremely complex. Although outside experts favor interpretations based on a somewhat oversimplified bi-modal, two-camp model, with moderates on one side, represented by President Muhammad Khatami, pitted against extremists or radicals on the other, embodied by the Supreme Leader, Grand Ayatollah Ali Khamenei, the reality is far more complex.[2] Indeed, there are significant and frequently bitter differences amongst Iranian leaders. Over the years, President Khatami himself has repeatedly mentioned with dissatisfaction and frustration how powerless he is. Yet, it is also clear that these tactical intra-elite differences coexist alongside a general, across-the-board commitment to the prosperity and future of the Islamic Republic of Iran as an ongoing enterprise. Although both strategic and tactical differences do exist, all political systems have significant and sometimes quite rancorous differences of opinion at the top. Indeed, outside interpreters of Iran frequently read into the country's political conditions their own needs, hopes, and expectations rather than accurately analyzing and interpreting the conditions on the ground.

As noted above, the Iranian political system is extraordinarily complex. It militates against risk-taking and makes it exceedingly difficult for any individual or group to make decisions that may compromise the stability of the system, challenge or jeopardize carefully constructed alliances, or alienate other actors. This is not to say that the system is not frequently characterized by deep fissures and significant bitterness, which it is. Nevertheless, much of the true business of Iranian political life takes place away from prying eyes and behind closed doors. The exceedingly personalized quality of Iranian politics is not new, and existed during the Pahlavi era and before. Although the revolution may have injected Islam into Iran, the country has always been characterized by a highly complex and even stylized social and political structure. Thus, the very design of the Iranian political system makes significant political initiatives or course corrections extremely difficult if not impossible.

Revolution Fatigue

The gridlock of the Iranian political system is accompanied by a significant and indeed growing sense of political isolation.[3] This isolation is not solely the product of a revolution that has grown stale and shopworn over the years. Iranians feel beaten down by an economy which refuses to improve, a system which is incapable of making difficult decisions, and an inhospitable external world which simultaneously covets Iranian resources while, in the view of many Iranians, explicitly opposing and taking a condescending stand to the experiment called the Islamic Republic.[4] Iran in the first decade of the twenty-first century finds itself surrounded by neighbors from whom it is alienated in varying degrees, almost across the board.

Afghanistan highlights such regional tensions. A particularly critical time for these tensions emerged during the Taliban era when Iranian-Afghan relations were exceedingly poor. This is because Islam in Afghanistan, as interpreted and practiced by the state under the Taliban, was both anti-Shi'a and anti-Iranian, while the ethos of the Taliban was strange and alien to the Iranians. The Taliban's main sponsor, Pakistan, itself characterized by a significant degree of anti-Shi'ism, is still regarded in Tehran as anti-Iranian.[5] To the West, Iran's neighbor Iraq, under Saddam Hussein, engaged Iran in a brutal eight-year bloodletting. This war affected all aspects of Iranian society, and Iran today still remains scarred by this brutal and ultimately meaningless conflict.[6] To the northwest is Turkey, with its legacy of Ataturk, serving almost as the antithesis of the Islamic Republic of Iran. Although there are certainly Islamist tendencies in Turkey, these are decidedly of a Turkish nature, and Turkey with its generals has little common cause with Iran and its Ayatollahs. Finally, to the north, the states of Central Asia are lurching in their own directions, eager to elicit external support while at the same time suspicious of all neighbors due to the particular interests and insecurities of this collection of fledgling regimes.

Recent events have enhanced rather than attenuated Iran's isolation. Tehran perceives the United States as a renewed and growing threat to Iranian security on virtually all sides.[7] The United States is deeply entrenched in Afghanistan, and although its control does not seem to go appreciably beyond Kabul and certain other urban centers, the political future and fortunes of this hapless state are believed, by Tehran at least, to be largely in the hands of the United States. In the wake of September 11, Pakistan's President Musharraf found new relevance in the eyes of the United States.

Washington and Pakistan now find themselves in deep collaboration with one another, despite the fact that there are significant elements in Pakistan who yearn for the days of the Taliban and are themselves deeply anti-American, not to mention anti-Iranian, in their orientation. Furthermore, the Government of Pakistan is clearly in less than full control of its country and territory. In short, Islamabad's close ties to the United States serve to accentuate already significant Iranian concern about Pakistan.

Iran is deeply troubled about the security situation on its eastern borders. Westerners in their discussions with Iranians always favor discussions of the Arab world and the Gulf region to the West, reflecting their own political interests. In reality, Iranians themselves are far more worried about the situation to the east. Iran's eastern borders are far from secure. The inability of Tehran to stem the flow of drugs into Iran is a major problem. Somewhat ironically, the American defeat of the Taliban has led to an increase in opium production in Afghanistan rather than its diminution. Pirates, gangsters, warlords, and others prey on Iran's porous eastern border and even engage in a lucrative kidnapping-for-ransom business. Scores of Iranian paramilitary police officers have been killed fighting running battles with drug smugglers and kidnappers. To compound this, Iran has always been deeply concerned about Pakistan's nuclear weapons capability. Indeed, not long after Pakistan's initial test of a nuclear device, Iranian Foreign Minister Kamal Kharrazi quickly made his way to Islamabad to talk to the Pakistanis about their nuclear program.[8]

Iran is no more comforted by the situation in the west, which from its perspective is not much better. Although the scourge of Saddam Hussein is gone, Iran confronts a new challenge in an American-occupied and influenced Iraq. To date, Iran has been reasonably tolerant of the United States occupation and has regarded it in a pragmatic and generally rational fashion.[9] Nonetheless, there are significant forces in Washington that are exceedingly suspicious of Iran and they certainly have their counterparts in Tehran. There are many Iranians who believe that in the light of the conquest of Iraq and the apprehension of Saddam Hussein, that the Islamic Republic of Iran may be the next object of military attention by the United States. As the future of Iraq is currently being written, with numerous political factions struggling for dominance, Iran is vigilant largely because it seeks, and if necessary will demand, a political formula sensitive to its regional and national concerns. Whether or not Iran actually expects Iraq to be ruled by an Iranian-style religious government is subject to debate and interpretation. However, what is known with certainty is that Iran will not accept a political order that excludes the Shi'a factions and not simply any Shi'a faction, but rather those who share the general worldview of their co-religionists in Iran. Thus, the significant and central role of the United States in Iraq has enhanced Iranian feelings of isolation.

Iran has no genuinely close political ties with any of its neighbors. The Arab Gulf states have been pragmatic in their views on Iran, and Tehran in return has been no less business-like. Debate has diminished somewhat over the islands that Iran has taken from the United Arab Emirates, but this contentious issue has not and will not ever disappear. Moreover, because the problems of the Greater and Lesser Tunbs and Abu Musa remain, they serve as a constant reminder for some Gulf Arabs about the way in which they are regarded by Iran. In general, however, the Arab Gulf states have been extremely realistic in their engagements with Iran which are for the most part constructive, positive and free of ideology. Nonetheless, the significant American role in the Gulf, as well as growing political instability in places like Saudi Arabia which has been the target of multiple terrorist attacks, elevate tensions throughout the region as a whole and thus Iran remains exceedingly sensitive and vulnerable to the vicissitudes of these regional political developments, crises and fears.

Iran From Within

In addition to its regional isolation, Iran also has a sense of profound and disturbing internal isolation. What is meant by this is that the vast majority of Iranians appear to be alienated from their own government and their own political system. As indicated above, twenty five years after its triumph over the Shah, the Islamic Revolution has grown tarnished. Around 60% of the population of

Iran is under the age of 25 and does not even remember the Revolution. For them, inadequate opportunities for employment and housing, an extremely competitive system for admission to the universities, and an intolerant, sluggish, and remote political system heavily influenced by the religious sector create a sense of apathy and helplessness. For some years, this dissatisfied group had pinned its hopes on President Khatami. However, given the nature of the Iranian political system, President Khatami has had to function "as a knife without a blade," to borrow a phrase used by one of his predecessors, Mehdi Bazargan. Although youngsters as well as middle class Iranians still respect President Khatami, his inability to stimulate any significant political change has contributed to a growing and debilitating sense of popular hopelessness about the drift which afflicts the Islamic Republic and which makes the future ominous for many Iranians. A particularly interesting example of this internal isolation was the response to the awarding of the Nobel Peace Prize to an Iranian woman attorney, Shirin Ebadi, for her efforts to promote democracy and human rights. Appearing with her hair uncovered to accept her prize in Oslo and socializing openly with men and shaking their hands, the behavior of Ms. Ebadi was deemed an affront by some Iranians who regarded her comportment as an implicit rejection of the values of the Islamic Republic. Most countries would be thrilled that one of their own nationals had won such a distinguished award as the Nobel Peace Prize. Yet, Iran is as conflicted over this victory as it is over so many other things. The fact that the winner is a female professional and a somewhat Western-oriented one at that, suggests to some Iranians that by conferring the prize on her rather than on a more typical and officially sanctioned representative of the Islamic Republic, the world was at once embracing Iran, while rejecting the Islamic Republic. This sense of duality characterizes many aspects of modern life in Iran.

External Views of Iran

The outside world seems to regard Iran as almost an eccentric country that is at once a nuisance and a treasure house. In part, this is because Iran has chosen a sociopolitical and cultural formula that has not been replicated anyplace else. Although many countries were founded on revolutions, including the United States, none used Islam as their ideology. Certainly the Islamic world is replete with those who have chosen to base their political actions on Islam and purport to speak on behalf of Muslims everywhere. Yet none have triumphed, as did Ayatollah Khomeini and his supporters. No other Islamic protest movement has spawned a successful revolution and a subsequent national political order although the world has seen countless incidents of Islamic-based political action, some violent and some not. Nevertheless, the most notable experiment of this kind came to fruition in the Islamic Republic of Iran and is still thought by some, in particular the United States, to offer a significant challenge to the outside world.

Iran and the United States remain deeply suspicious of one another.[10] This has its origins not only in Washington's attachment to the Shah, whom the Iranians regarded as a despot, but also a history of other American-sponsored activities in Iran which Iranians can tirelessly enumerate, paramount among them being the Mossadeq affair.[11] Although Iranians genuinely believe that the United States Government is ill-disposed towards their country, they are deeply attracted to the United States as a country and a culture. The significant Iranian population now living in the United States, as well as the fact that many Iranians currently resident in Iran were educated in the United States abets this. In the mid-1970s, for example, Iran provided more foreign students to American institutions of higher learning than any other country in the world. Thus, put simply, Iranians like the United States but detest its politics. Despite the absence of the pressure and the tensions that characterize American-Iranian relations, it is fair to say that European-Iranian relations are not dramatically better. Although Europe and Iran have reasonably functional economic and political relations, Iranians are well aware of the debates in France about whether Muslim girls should be allowed to wear headscarves in school. They are also familiar with the status of Turks in Germany as well as the negative attention paid to Muslims throughout Europe in the wake of the September 11 attacks. Thus, tolerable political relations, as well as at times robust economic ties, are in no way synonymous with deep affinity or strong, close political or cultural relations. Iran yearns to be accepted by an outside world of which it is exceedingly suspicious and whose perceptions of it are equally ambivalent. Iranians believe in the superiority of their own system, certainly in comparison to their less democratic neighbors. For despite what may be said, President Khatami was elected in two democratic elections which, by regional standards at least, were comparatively free, open and honest. Iran's sense of its own persecution alongside its feeling of superiority due to its revolution and its adherence to Islamic tenets almost guarantee an absence of close partners. Indeed, Iranian isolation, both internal and external, is a significant factor in explaining Iranian political behavior as well as an important consideration in determining what its regional role should be both in the Gulf and elsewhere.

Impediments to a Broader Iranian Role in Gulf Security

It would be clear to most observers that without Iran, anything approaching ideal Gulf security cannot be achieved. And to the detriment of all, there are a number of areas in which the tensions that characterize Iran's relationship with the West in general and the United States in particular, actively limit the role that Iran could be expected to play. Perhaps the greatest roadblock to a broader Iranian involvement, as well as to closer ties between Iran and the West, lies in the current controversy surrounding Iran's nuclear program. Recent interactions with the IAEA suggest that Iran's honesty about the nature, scope, and direction of its nuclear program has been questionable. Furthermore, the issue of Iran's nuclear program has provided somewhat unusual grounds for agreement amongst the Western states that generally disagree on virtually all other aspects of Iranian foreign policy. Indeed even Russia, which has been one of Iran's primary supporters in the nuclear realm, has quietly admitted regret at the degree to which it has supported Iranian nuclear ambitions.[12] Although detailed analysis of the problem of Iran's nuclear program is being dealt with elsewhere in this volume, it is important to acknowledge that this issue, more than any other, undermines Western relations with

Iran. Importantly, this includes not only the United States, but also a number of powers in Western Europe. Once an understanding on Iran's nuclear program is reached, Iran will be far better positioned to seek a broader regional role. Although Iran clearly merits such an enhanced regional role, given its own status as a significant force in the Gulf region, this role will be more easily achieved with the forbearance of the Western powers and particularly the United States. Iran's political currency will most certainly rise once the nuclear issue is resolved.[13]

The other significant impediment to enhanced Iranian influence is uncertainty about Tehran's role in promoting and supporting international terrorism. Although there is a vigorous debate about Iran's culpability, this is primarily an American concern rather than a more general Western one. Nonetheless, given the global scope of the war on terrorism, any hint of Iranian non-support for this campaign could have a significant impact on its relationship with the Europeans as well as the Americans. In a general sense, there are roughly two sets of views about Iran. One view is that Iran itself is and has been a victim of international terrorism. Evidence cited here are the predations of the *Mujahedin-e-Khalq* that has pursued a vigorous and generally ineffective but nonetheless costly campaign of terrorism against the Islamic Republic of Iran. There have been countless instances of Iranian Government officials and others being killed in terrorist incidents, including innocent civilians. Although the successful invasion and occupation of Iraq may have neutralized the *Mujahedin*, Iranians see themselves as much victims of international terrorism, as the United States and other states in the post-September 11 era. Iran also feels that it is a victim of terrorism on its eastern border, as discussed above, as it is difficult to disentangle violent terrorism from more traditional criminal activity. Thus, the Iranians believe that the challenges to the security on its eastern border emanating from Afghanistan and perhaps Pakistan are an admixture of terrorist concerns as well as more conventional criminal activities. Europeans seem, for the most part, to accept this Iranian view of itself, although there have been certain exceptions. For example, there is the recent case of the former Iranian diplomat who was apprehended in the United

Kingdom for being linked to the attack on the Jewish community center in Buenos Aires. For reasons that are not fully clear, the United Kingdom ultimately decided to release this individual from custody. Thus, there is some uncertainty in the West about the exact proportions of Iranian victimization by and/or responsibility for international terrorism.

The American view is more straightforward and less ambiguous. The United States believes that Iran has played an active and significant role in supporting international terrorism. Indeed, it is generally thought that the two factors that most directly persuaded President Bush to include Iran on his "axis of evil" list alongside North Korea and Iraq were Iran's nuclear program and Tehran's support for international terrorism. A key indication of the latter was the shipment of arms intercepted by the Israeli Defense Forces on the Karine-A, on January 3, 2002. The captain of the ship claimed that the weapons originated in Iran and were apparently destined for the Palestinian Authority. Certainly the American view of Iran's role in international terrorism has been influenced by the Israeli view. However, there are divisions here as well, with some Israelis, including many senior government officials, arguing that Iran is the single greatest threat to regional stability and thus to Israeli national security. Others in Israel do not share this view. In the United States, Israel and elsewhere, the intentions issue, which is best reflected by Iran's presumed support for international terrorism, mixed with the capabilities issue, in which Iran's nuclear program is cited, are linked, thus fueling perceptions of Iran as a significant threat.

It can reasonably be argued that there is some validity to both the American and the European positions on Iran's record in the realm of terrorism. That is, given the complexity of the Iranian system, it is not inconceivable that Iran is no longer involved in international terrorism, although in the past such terrorism was part and parcel of Iranian foreign policy. This could be the case now or in the future. Finally, it is possible that both of these strains coexist simultaneously, representing different tactical views of different segments of the Iranian elite, as characteristics of a system in which there are competing worldviews and competing tactics used to justify them.

Iran bitterly resents being portrayed as a terrorist state rather than as a victim. For example, it regards its involvement with Islamic Jihad and Hizbollah as support for legitimate national liberation movements and not terrorist groups. Obviously, the United States and Israel strongly disagree and in the post-Saddam era, this has led to a particularly thorny new controversy pitting Iran and the United States against one another once again. Iran has reportedly rounded up a significant number of Al Qaeda activists, including some senior members of the Al Qaeda leadership. There is some uncertainty about the status of those apprehended, as in some instances Iran has returned them to their countries of origin, presumably to be dealt with harshly by their own governments, which are unsympathetic to Al Qaeda. There are others whom Iran is still holding, and it is these that the United States is particularly concerned about. Washington would very much like to have access to these detainees. Failing this, it would like Iran to turn them over to their home governments, but Iran appears reluctant to do so.

From an Iranian perspective, there is a comparable issue. Tehran finds the American position vis-à-vis the Mujahedin-e-Khalq no less frustrating than Washington finds Tehran's treatment of its Al Qaeda captives. After the successful invasion of Iraq, the United States took over a number of Mujahedin facilities. For a brief period of time, the United States encircled the Mujahedin, but inexplicably allowed them to retain their weapons. Iran would like the senior leadership returned to Iran to be tried for its crimes. Tehran has also publicly announced that all Mujahedin members are welcome to return to Iran where they will be free to take up their lives again without penalty provided they renounce their anti-Islamic Republic activities. What would seem attractive here for both countries would be some sort of a *de facto* if not explicit trade. That is, the Al Qaeda activists would either be made accessible to the United States or returned to their home countries whose governments are generally collaborating with Washington on the war on terror. At the same time, Washington would turn the Mujahedin-e-Khalq activists over to Iran. Neither side favors a direct trade, and indeed for good reason. It would be unseemly for the two countries to crudely exchange suspects in terrorist activity. What would make more sense would be for each side to do “the right thing.” That is, considering that the Mujahedin-e-Khalq features prominently on the United States Department of State’s list of terrorist organizations, Washington should take decisive action against those under its control to make certain that they never again engage in terrorist activity. Conversely, Iran should take similar action against the Al Qaeda leaders and activists that it has apprehended. Although there are undoubtedly those on both sides who believe that this is a reasonable and proper course of action, there are others in Washington and in Tehran who do not share this view. They seem to regard both the Mujahedin and Al Qaeda members as potential “capital” to be held in reserve and possibly used against their adversaries. There have been periodic rumors of an understanding being reached between Washington and Tehran on the fate of some of the detainees but whether or not this is the case is impossible to ascertain at this time. Nonetheless, it is important to acknowledge that a major impediment to closer ties between the United States and Iran is in the realm of the war on terrorism. Each side needs to recognize the legitimacy of the other side’s campaign as each has been horrifically affected by terrorism as practiced by both the Mujahedin-e-Khalq and Al Qaeda.

There is a striking irony that on both sides there are those who favor collaborative action in the war on terror, while others oppose it. Yet it is important to acknowledge that to date, the issue of terrorism serves as a major impediment to improved relations between both states.[14] The issue of moral equivalence is not being argued here. It is not self-evident that the United States and Iran are somehow equal because both have been victims of terrorism. The Iranian system by design as well as by action lacks transparency and accountability. There are weighty elements in the United States, as well as elsewhere who still believe that Iran has been and remains a willing participant and sponsor of international terrorism. It is incumbent upon Iran to clarify the record. The problem is that Iran is a large and proud nation, unwilling to be judged by outsiders and certainly not by the United States. Furthermore, it is by no means certain that Iran has abandoned all support for terrorism or, to put it differently and somewhat more charitably, it is possible that certain segments of Iranian leadership may well be supportive of terrorist activity, to the detriment of the Islamic Republic as a whole. What is noteworthy here is the recent behavior of Libya, which may be setting a new international standard for dramatically and quickly improving relations with the West including the United States. Libya's apparent willingness to eliminate its WMD capabilities while permitting complete and open international inspections is unlikely to appeal to Iran, whose situation is dramatically different from that of Libya. Nonetheless, it is apparent that many in the United States and elsewhere will now point to Libya as an exemplar of what is expected from Iran. It is doubtful that Iran would open itself up to this degree of intrusion by the outside world, but the reality is that Iranian isolation could be significantly diminished but at a cost that many Iranians would consider unacceptably high.

A New Role for Iran in Gulf Security: Positive Elements

Despite the significant impediments to closer United States-Iranian relations that are presented above in a particularly stark and rather discouraging way, there are certain factors which make it evident to all concerned parties that an enhanced role for Iran in defining and promoting Gulf security is in the collective interest of all states, both intra- and extra-regional. That is, in a situation of genuine Gulf security, there are winners and no losers. In other words, if Gulf security permits free and unimpeded access to major petroleum reserves and the ability of petroleum producers and purchasers to trade in a safe and secure fashion, there are no negatives. Given that Iraq is no longer a regional predator and is now likely to support a new Gulf security regime, it is apparent that a role must be found for Iran in this order as well. The reason for the inclusion of Iran is not merely to embark on an idealistic quest for equity for all regional actors, but rather pragmatism and efficacy. True and enduring Gulf security can only exist in a world where the GCC member states, Iraq and Iran all believe that their interests are being addressed and protected and that they all have roughly equal weight at the table. Iran ought to be part of a regional security order simply because there cannot be a comprehensive regional security arrangement without Iran. Recent history has demonstrated periods of seeming stability, punctuated by bursts of intense instability which have proven extremely costly not only for the regional powers themselves, but also for interested external parties, including the West, Japan, and others. Thus, it is in the world's collective interest to seek a means to achieve such stability and it is incumbent upon all regional actors as well as their external allies, partners and sponsors to support it.

The conundrum as shown above is the significant impasse between the United States and Iran. Although some of these issues also have an adverse and debilitating impact on Iranian-European relations, the reality is that they are far less acute and rancorous. Iran continues to have diplomatic ties with all major European states although it has no open, direct diplomatic relationship with the United States. Even the United Kingdom, which is closest in its position to the United States, has diplomatic ties with Iran. Although there have been periodic problems in this relationship, it has endured for some time. Thus, the United States is alone in its attempt to isolate Iran, despite the fact that this isolation has been punctuated by periods of *de facto* and even overt collaboration such as the Bonn conference on Afghanistan cited above. Other examples of *de facto* cooperation between Washington and Iran even include the American instigated war in Iraq. For although the Iranians are not at all happy about the United States acting so close to its own borders, from the Iranian perspective, a region without Saddam Hussein is certainly preferable to one with him. And although Iran would undoubtedly have preferred that a force other than the United States had removed Saddam, his removal was reason for some cheer in Iran, which had been a victim of Iraqi aggression just as many others throughout the region. Thus, Iran and the United States can occasionally move in the same direction, although regrettably for both, this has been accomplished in a very short-term and limited fashion. Can the United States and Iran take the necessary steps to collaborate in the design and execution of a new security order for the Gulf along with other concerned actors?

It is clear that Europe and Iran could reach such an understanding. However, given that the United States is the premier external actor in the Gulf region and that Iran regards the United States as its primary antagonist, when we talk of Western perspectives, what is being discussed in significant measure is the ability of Iran and the United States to collaborate.

Solving the Middle East Problem: A Grand Bargain for the West?

Although conflict between Iran and the United States may interfere with the establishment of a regional security order, there are also other significant impediments. The West, on its own, has failed to reach consensus on a common policy towards Iran. As indicated at the outset, there are significant disagreements between the United States and its European partners on all major Middle East issues including the Arab-Israel dispute, the future of Iraq, and on an acceptable global and regional role for Iran. Thus, somewhat ironically, an understanding must be sought not only between Iran and the United States, but also between Washington and its Western allies on Iran as well as several other Middle Eastern issues. If this could be achieved and a genuine collective Western position arrived at, it might be less difficult for Washington and Tehran to arrive at some mutual understanding as well. The question is how to sequence this process of agreement and eventual reconciliation. To date, other than brief, functional collaborations, the United States and Iran have failed to identify a mutually acceptable and habitable common ground. Barring some sort of Libya-style Iranian submission to US terms, this common ground is unlikely to be identified. Or, put more directly, it is unlikely that the United States and Iran will actively seek or achieve rapprochement in the short run. Given that these deep-set suspicions and mutual recriminations will not soon be overcome, is there perhaps another way to secure a strategic and political outcome that is in the shared interests of Iran and the United States, but which neither country either independently or jointly seems able to achieve?

Although it is fashionable in some quarters to blame the West for all problems in the Middle East, it is quite clear that Middle Eastern governments themselves share much of the blame for their own plight. At the same time, the West, despite its major interests in the Middle East, has found it virtually impossible to formulate long-term and collaborative policies towards this uniquely challenging region. In some circumstances the West has imposed itself on this troubled area while at other times an array of Western powers have been invited by various Middle East interests to act as supporters, partners, or surrogates in a variety of failed political initiatives. Given that few Middle East problems exist in isolation from the broader array of issues challenging regional stability and progress as a whole, it might be helpful to Iran, as well as for the general stability of this tempestuous region, if the question of Iran's regional role was woven into the broader embroidery of Middle East issues. That is, much of the analysis in this chapter has emphasized Iranian concerns about issues on all of its borders and not simply those in the Gulf region. It is therefore reasonable to assume that in the eyes of Iranian policy-makers these issues are linked, thus creating a larger political universe within which Iran feels isolated, threatened and persecuted. Conversely, the West has interests throughout the entire Middle East region, many of which go far beyond Iran. The question is, are the broader areas of common interest so significant that they can surmount narrower areas of disagreement? In my view, the answer is yes, although this would necessitate a degree of political insight, imagination and commitment that to date have been largely absent in both the West and in the Middle East.

Areas where there is broad theoretical consensus in the West include first, the desirability of creating a democratic Palestinian state that will be designed so as to ensure Israeli national security; second, the completion of the nation-building task in Iraq so that a genuine indigenous Iraqi government will come to power and the role of external actors will be minimized; and third, the establishment of a new security order in the Gulf which will allow for free and open commerce and thus ensure stability. Although there are a variety of lesser although still important Middle East issues (e.g. Libya, Lebanon/Syria, Afghanistan, Sudan and Algeria) the three mentioned above are of paramount importance. Virtually all major regional and extra-regional actors share a common commitment to the achievement of these goals. Thus, consensus at first not between Westerners and Middle Easterners, but rather at the outset amongst Westerners themselves could in fact play a uniquely powerful role in ameliorating these conflicts. For this reason, an Atlantic alliance in the Middle East is advocated. Such an *entente* would permit common action amongst the Western powers seeking overall coordination and thus lead to the resolution of some, if not all, of the region's major political crises. Without delving into the precise linkages amongst all of these issues within this context, it is reasonable to assert that if both sides of the Atlantic could agree amongst themselves and speak with one voice to their partners and adversaries in the Middle East, then substantial subsequent progress could be achieved in dealing with the same major issues in direct partnership with Middle Easterners themselves.[15]

Let us make the case in an abbreviated fashion. In a general sense, while the United States has special sympathy for Israel, Europeans have a comparable special relationship with the Arab world. Yet all sides seem to agree that the creation of a democratic Palestinian state is a reasonable goal and if properly designed and executed, it could help to guarantee Israeli national security. Yet beyond the realm of rhetoric, there has been no meaningful Atlantic partnership devoted to resolution of the Palestine question. If Iraq were thrown into the mix, this would necessitate concessions by both the Americans and the Europeans. The United States would have to forgive and forget by allowing greater access to the Europeans not only in the area of Iraqi reconstruction, but also in helping to determine the future political design of Iraq. The primary benefit, if we accept the logic of the grand bargain, would not be isolated progress on any single Middle East issue but broad-gauged and powerful European and American collaborative action that would permit simultaneous resolution of more than one problem. It must be remembered that each of these issues is significant not only in their immediate environs, but also for the whole region. Thus, for example, resolution of the Palestine question would have a positive impact on Iran as it would remove a major stumbling block, albeit a symbolic and largely metaphoric one, that has prevented Tehran from dealing directly and openly with

Washington. Although Palestine is not an issue of major significance to the average Iranian, it is an important exemplar to the Iranian leadership of American attitudes towards and preferences in the region. At the same time, helping to bring Iran out of the "axis of evil" and back into the global political mainstream will have a positive effect on other regional issues including Arab-Israeli peace making. Each conflict needs to be seen as part of a whole, as part of a broader system of regional issues, and not merely as a smaller set of discrete political conflicts and challenges, which they are decidedly not.

The third and final piece of this grand bargain would involve Iran and its policies even more directly. Here there seems to be a general agreement amongst the Western powers about the desirability of Iran ceasing to develop its WMD and particularly its nuclear capabilities. Where the West differs profoundly is in its understanding of the intentions and aims of the Islamic Republic with the United States having quite negative expectations, in contrast to Europe which does not hold such a perspective. Here the United States would be expected to accept, partly at least, the European view that emphasizes the wisdom and desirability of engagement. At the same time, the Europeans would be expected to recognize and support American concerns vis-à-vis Iran's nuclear program as well as its support for terrorism and its handling of its Al Qaeda detainees. In short, the United States would have to be somewhat more flexible and tolerant in its views of Iran, while the Europeans would be expected to exercise some leverage over Iran so as to seek greater flexibility in Tehran's views of the United States. Again, what needs to be emphasized is that the role of Iran is not simply an issue affecting its immediate environs, but rather one affecting the broader region. A more constructive and positive relationship between Iran and the West in general and the United States in particular would contribute to greater Gulf security, to more effective nation building processes in both Iraq and Afghanistan, and to attenuation of a major source of instability that adversely affects the whole region. What is being argued here is that disagreements between Iran and the West can in part be ameliorated in the Western capitals themselves. If the West is able to speak with one voice on the major Middle East issues of the day, particularly as they relate to Iran, significant steps could be taken towards addressing some of the most debilitating problems of this tempestuous region. Whether the leaders in the many capitals that would have to be involved in this complex endeavor have the wisdom to go this route and to multilateralize long-standing bilateral conflicts that have defied resolution remains to be seen. Unfortunately, if history is a guide, then great optimism does not seem warranted.

12

The Impact of Iran's Nuclear Program on Gulf Security

Geoffrey Kemp

There is an irony concerning Iran's contemporary security dilemmas and the mixed signals it has given about its nuclear ambitions. During the 1980s, the most dangerous threat to Iran was Saddam Hussein's Iraq. Iraq had invaded Iran in 1980, and with considerable help from most of the Arab World, Russia, China, France and the United States, it was eventually able to inflict enough pain on Iran that Iranian Supreme Leader Ayatollah Khomeini had to end hostilities in 1988 on terms very favorable to Iraq.

Saddam Hussein's subsequent invasion of Kuwait in August 1990 resulted in the American-led First Gulf War, which saw Iraqi forces expelled from Kuwait. However, Saddam remained in power and, though greatly weakened, continued to be a source of anxiety for Iran, the neighborhood, as well as the United States and its allies. It was not until the War in Iraq in 2003, again led by the United States that Saddam Hussein's Iraq finally ceased to be a threat.

Iran's Changing Security Environment

During the 1990s, another neighborhood security threat loomed large in Iranian policy. The Taliban government of Afghanistan displayed increasingly aggressive tactics towards Iran, accusing the country of supporting groups hostile to the Taliban in the ongoing Afghan civil war. The Taliban used Iranian territory as a transit route for drug trafficking, its only real source of income. Relations reached a crisis point in the summer of 1998 when on August 8, eight Iranian diplomats were murdered by Taliban forces in a confrontation at Mazar-i-Sharif. The subsequent outcry led to a mobilization of Iranian forces. Upwards of 250,000 troops were sent to the border regions, and it looked like a war was imminent. It was prevented at the last moment, primarily because the Iranians realized that a protracted war in Afghanistan would be a very dangerous undertaking, and that they could find themselves bogged down in the same way as the Soviet Union had been during its futile confrontation in the 1980s.

In view of Pakistan's close ties to the Taliban, Iran formulated a very scary picture of the relationship between Islamabad and Kabul, and Iranians talked openly about their fear of the "Talibanization" of Pakistan. They argued that ever since Pakistan became a nuclear power in 1996, there was a possibility that at some point in the future, radical Sunni military officers in Pakistan would overthrow the government, and with access to nuclear weapons, would be in a position not only to assist the Taliban but also to dictate their will on their neighbors, including Iran. In sum, while the United States was always rhetorically regarded by Iran as the "Great Satan" and the most serious threat to Iranian interests, the reality was that it was Iraq and Afghanistan that posed immediate and direct threats.

The events of September 11, 2001 dramatically altered the strategic equation throughout the Greater Middle East. The first US military response to 9/11 was to wage war against the Taliban government in Afghanistan. In the weeks preceding the outbreak of hostilities, there were many dire voices worrying about the potential quagmire and the belief that the United States would not be able to do what Britain and the Soviet Union had failed to do in the past. However, as the war progressed and it was clear that the Taliban had lost, Afghanistan's neighbors including Iran calculated that it was better to cooperate with the Americans than to oppose them. As a result, Iran and the United States established an understanding about the nature and configuration of a future Afghan government. By all accounts, their interaction was cordial and business-like at the Bonn conference, which was convened in December 2001 to discuss Afghanistan's interim government. After all, the United States had removed a major thorn from Iran's side but from an Iranian perspective, the downside was that American troops were now stationed not only in Afghanistan but also in Pakistan and Uzbekistan. (The United States had demanded access to Pakistan after 9/11 and with Russian help had been granted permission to station military units in Uzbekistan).

A similar pattern followed the successful US campaign against Saddam Hussein in the spring of 2003. Saddam was gone, only to be replaced by US troops. Hence, Iran now finds itself surrounded on all sides by American military forces. The US currently has a military presence in Turkey, Iraq, Kuwait, Saudi Arabia, Bahrain, Qatar, the UAE, Oman, Afghanistan, Uzbekistan and Pakistan. Its ground force presence is backed up by sophisticated air and naval assets within easy reach of the Arabian Gulf from the Mediterranean, the Indian Ocean, as well as Europe and the continental United States. Furthermore, the last two Gulf Wars demonstrated vividly on television, in real-time, the effectiveness of American advanced conventional munitions, especially against fixed military targets. Consequently, Iranian military commanders and the conservative, hard-line *mullahs* know that, in terms of strict military power, they are in a very weak position vis-à-vis the United States.

However, there is another side to this equation. American forces in Afghanistan and Iraq are still fighting wars against well-armed and determined insurgents. The political stability of both countries is far from assured and Iran has great influence among certain groups, especially in western Afghanistan, and northeastern and southern Iraq. Iran could, if it wanted to, by direct and indirect intervention through its Revolutionary Guard Corps and its security services, cause the US forces a great deal of anguish. To this extent, there is something of a standoff between the two countries. Obviously, Iran's advantages in this balance of power will be lessened, if and when pro-US governments in Kabul and Baghdad achieve full control of their countries and have adequate security forces to assert control over insurgent groups, such as warlords or ethnic militias.

Iran's Nuclear Program

It is against this backdrop that the intense debate about Iran's nuclear weapons program, what it means for the region and what to do about it, is taking place. At about the same time that coalition forces were preparing for their assault on Saddam Hussein, in the winter months of early 2003, Iran was confessing to the world that its nuclear program was far more advanced than it had previously acknowledged. Iran was forced into this position because disclosures in the world press, based on information given to Western sources by the Mujahideen-e-Khalq opposition forces, showed that Iran had developed, among other things, a large uranium enrichment facility at Natanz, as well as a heavy water facility in Arak.

These disclosures must be examined against the backdrop of Iran's public posture on nuclear matters. The Iranian Republic has been a vocal and proactive advocate of international arms control agreements, especially the Nuclear Non-Proliferation Treaty and the Chemical Weapons Convention. As a state, it is party to both treaties. Its diplomats have been prominent in United Nations arms control negotiations at Geneva, Vienna and New York. Over the years they have become polished performers who understand the nuances of the subject and are well versed in the arguments on all matters relating to regional arms control.

For many years Iran has championed the notion that there are palpable "double standards" on the most basic elements of international security and the role that arms control is meant to play in enhancing the security of the less developed countries. Most egregious, in the Iranian view, is the fact that within the Middle East-South Asia region reside three nuclear weapons states – India, Pakistan and Israel – which have remained outside the nuclear Non-Proliferation Treaty (NPT) and are consequently not subject to any of the constraints that the treaty imposes on its members. Thus, from Iran's point of view, Israel is free to develop and deploy a large inventory of weapons of mass destruction without suffering any penalties, aside from endless carping from its neighbors. Yet Iran, despite its NPT membership, is subject to intense pressure from the United States and is denied access to nuclear technology by most friends or allies of the United States, including all G-7 members. Iran has many allies in the Arab world on this issue. Egypt, in particular, has been vehement in its insistence that the question of the Israeli bomb is central to any future understanding or more formal agreements on regional security and arms control.

Iranians are equally concerned about India and Pakistan, two countries now unofficially acknowledged "members" of the nuclear club thanks to a radical change in American policy towards both countries. In the case of India, the Bush administration from its early days made it clear that it would not chastise India on the nuclear issue despite the policies of the previous administration, which had put non-proliferation at the top of its agenda for the subcontinent.

Pakistan's case was different. Until September 11, US policy towards Pakistan was increasingly hostile but turned 180 degrees following 9/11 and President Musharraf's decision to side with the United States against the Taliban. Now the primary goal of American policy is to assure Pakistan's cooperation with the war on terrorism and the rounding up and containment of Al Qaeda and Taliban supporters who are hiding in both Afghanistan and Pakistan. On the nuclear issue, the primary US concern now is to ensure that the inventory of Pakistan's nuclear weapons remains securely under the control of Musharraf and his closest advisors.

Why does Iran Need to Build the Bomb?

Iran's desire for a nuclear weapons capability is based upon a number of established factors, including the perceived threat from the United States, the existence of other regional nuclear powers, the desire for status, and the bureaucratic momentum of a nuclear establishment within Iran's civilian and military leadership.

To determine how serious Iran is developing its nuclear program, it is useful to review the basic requirements that any country needs if it is to develop a nuclear weapon. Iran, like all aspiring nuclear powers, has had to come to grips with the basic chemistry and physics of nuclear weapons. Two elements can be used for the warhead in a nuclear weapon, uranium-235 or plutonium-239. Neither of these is found in a natural state. To produce uranium-235 requires a complicated process involving the enrichment of uranium fluoride, a gas derived from uranium oxide. Enrichment procedures are time-consuming and costly. However, once the U-235 has been produced, it is easy to handle and can be made into a workable weapon with a simple gun type device that shoots a sub-critical mass of U-235 at another sub-critical mass at the end of the barrel. They combine to form a critical mass and fission takes place.

The first atomic bomb dropped on Hiroshima in August 1945 was a uranium bomb. It had never been tested. The famous Trinity test a few weeks earlier on July 16, 1945 at Alamogordo in New Mexico was the first nuclear explosion in history. The test was made using a plutonium device similar to the bomb dropped on Nagasaki on August 9, 1945. While plutonium devices require less nuclear material than that of the uranium bomb, weapons grade plutonium is dangerous to work with—there is a high presence of the highly radioactive Pu-240 element and the plutonium isotopes are very unstable. Consequently, a plutonium device requires a much more complex design than that of uranium. Whereas uranium devices can use the gun type method, a plutonium bomb uses an implosion method, which is both highly complex and very difficult to engineer. In the implosion method, weapons grade plutonium is surrounded by high explosives, which, once detonated, send a shockwave of the fissile plutonium material into a super-critical mass thereby creating the nuclear explosion.

Where does Iran stand today with respect to its capabilities? It has acknowledged that it is developing a full fuel cycle. However, there are other ways it can obtain weapons grade material. It can purchase material from an illegal international source, or it can illegally divert spent fuel from the Bushehr power reactor, which the Russians are helping to build, and extract plutonium. It has decided to develop its own indigenous fuel cycle, which would provide it with the capability to produce plutonium or highly enriched uranium. The first route has the obvious advantage that the material would be fairly cheap to obtain in comparison to other options. However, the quantities would be limited, it would be a highly illegal transaction and it could trigger automatic sanctions under the NPT if Iran were caught. Furthermore, the quality of the material might be suspect (there have been many scams with front organizations from the former Soviet Union trying to make money pawning off weapons grade material to would be buyers.) However, if the purpose in Iran is to have one or two token bombs in the basement, this might be a preferred route.

The other two avenues have their own problems. The agreement Iran has with Russia to build the Bushehr reactor includes an understanding that Russia will provide the nuclear fuel and will, after it has been used, retrieve it for reprocessing or disposal in Russia. The Bushehr reactor would have to be inspected by the IAEA since Iran is a party to the nuclear Non-Proliferation Treaty, and therefore, diverting fuel would be a very risky option for Iran to consider since it would almost certainly be caught.

The third and the most secure and guaranteed way of obtaining a steady supply of nuclear weapons material is to "do it yourself." This means developing uranium mines with parallel conversion facilities, fuel fabrication plants, enrichment plants and all the other components that are necessary to produce nuclear weapons grade material. The process begins by mining uranium ore, chemically converting it to uranium oxide, converting the uranium oxide to gas and then separating the isotopes of uranium, which first provides low enriched uranium (LEU), which is relatively easy to produce or the more complicated and costly, high enriched uranium (HEU). With uranium dioxide, or low enriched uranium, fuel pellets can be fabricated for use in a nuclear reactor. Once in the nuclear reactor, some of the uranium converts to plutonium, which can then be extracted from the reactor by a chemical separation process. Thus, if a country chooses to go the plutonium route, it requires nuclear reactors to produce the plutonium byproducts and a separation plant to obtain the weapons grade material. It would not need the elaborate procedures necessary to produce high enriched uranium, but, as stated, plutonium is a very difficult material to handle and requires a more complicated engineering problem in designing the bomb.

To be truly independent, Iran will have to develop all the components that make up the fuel cycle, including the so called “front end” and “back end” activities. The front end includes mining, conversion, enrichment and fabrication of nuclear fuel. The back end refers to activities that occur after the fuel has been used—reprocessing of spent fuel to produce plutonium or disposing of spent fuel (itself a hazardous and costly operation). Spent fuel rods are extracted from the reactor and placed in cooling tanks for a number of months, depending on how they are to be disposed of. After cooling, the thick metal casing of rods is removed and the spent fuel is then placed in containers fitted with nitric acid and other chemicals. This chemical process separates out the plutonium and uranium from other waste materials. The plutonium and uranium can then either be used once more to become nuclear fuel or, if sufficiently pure, be used for weapons production.

Since the February 2003 disclosures, more and more information has come to light as a result of intense inspections by the International Atomic Energy Agency (IAEA) during the fall of 2003 and then the spring of 2004. Consecutive reports from these inspections, although couched in very bland, international, bureaucratic language – so as not to offend anybody – provide crystal clear evidence of Iran’s deception and the massive extent of its nuclear program.[1]

Iran has argued it is developing a nuclear fuel cycle for production of enriched material to be used in power plants because of its electricity needs. Certainly, Iran can argue that it has an electricity shortage and it would prefer to reserve its oil and gas treasure for sale on the external market. However, the evidence against Iran goes far deeper than the technology required for producing fissionable material. It is clear that Iran has been examining, if not fabricating, the components necessary for warhead design, and of course, the nuclear program must be examined in parallel to Iran's significant progress in developing a surface-to-surface missile capability with North Korean and Russian assistance.

Impact of an Iranian Nuclear Program on Gulf Security

To assess the impact of Iran's nuclear program on Gulf security, it is first necessary to distinguish between indirect and direct impacts. Indirect impacts refer to broader international and regional dynamics that would be influenced by an Iranian nuclear weapon and how these, in turn, will affect the Gulf. Direct impacts concern more specific links between Iran's nuclear ambitions and the possible responses of the Gulf countries and others, including the United States and Israel, in terms of multilateral, bilateral and unilateral national security strategies. The second requirement is to consider what differences will emerge, depending upon the scope and nature of an Iranian nuclear program. For instance, if Iran limits its program to a small, initially covert weapons program, one set of regional and international responses can be anticipated. However, if Iran embarks on a fully fledged overt program, similar to that developed by India and Pakistan, the reactions are likely to be different and have different long-term consequences for security arrangements.

Perhaps the most important variable is the nature of the Iranian regime, and the circumstances in which Iran crosses the nuclear threshold. A nuclear Iran – that is to say, an Iran with nuclear weapons – would, in all probability, be relatively isolated and most importantly, at odds, if not in a state of hostility, with the United States. In other words, the international circumstances that propelled Iran to cross the nuclear threshold would be reflective of a dangerous world and indicative of deteriorating relations between Iran and its neighbors.

There are dozens of scenarios that could be developed to provide a backdrop to the circumstances. The most obvious one is the continuation and entrenchment of the hard-line ideologues in Tehran, paralleled by worsening relations with the United States, perhaps over Iraq and Afghanistan, or confrontation between Iran and the international community at the IAEA or, in extremis, at the UN Security Council. It can be imagined that such isolation and pressure would reconfirm Iran's paranoid outlook and strengthen the hands of those in Tehran who believe that Iran must have the bomb as a deterrent to prevent a US-led attack to remove the regime. A more benign, but nevertheless worrying scenario would be one where an Iranian regime, perhaps even a reformer-dominated regime, calculates that the benefits of having the bomb outweigh the costs. In such a situation, although Iran would refrain from an aggressive foreign policy posture and might even consider a more moderate attitude on the Arab-Israeli issue, it will exercise its right to have the bomb, in part to ensure a level playing field and set the stage for genuine cooperative efforts to reach a regional security agreement on a nuclear (or WMD) free zone.

In between these two somewhat extreme scenarios there are other possibilities, some more frightening but all troubling in their likely impact on the neighborhood. The more benign the circumstances in which Iran becomes nuclear the more manageable the subsequent crisis is likely to be. It is therefore useful to assess the regional impact under a number of different assumptions. The following matrix outlines four different cases that are worthy of examination.

Table 12.1

Iran: Four Possible Scenarios

	Covert Bomb	*Overt Bomb*
Reformist Regime	Case A	Case B
Hardline Regime	Case C	Case D

Case A: In this case, a reformist regime is in power, but nevertheless decides to develop a covert nuclear weapon. In other words, the regime feels compelled to violate its NPT commitments for reasons of national security. This is perhaps the most unlikely scenario, because it would assume a reformist regime which had sufficiently poor relations with the United States, the European Union and its neighbors to be willing to violate its treaty agreements, thereby strengthening the likelihood of sanctions.

Nevertheless, there could be circumstances in which a reformist regime takes such drastic steps, especially if such extreme threats to Iranian security were apparent to the regime but were not considered relevant or legitimate by the international community.

Case B: This represents a more extreme version of Case A, but one that probably has more credibility. It assumes that Iran's security environment is sufficiently dangerous that the regime feels compelled to formally withdraw from its treaty commitments under the NPT and embark on a major, overt nuclear program. This could be triggered by a severe degradation in the regional balance of power, including, for instance, the reemergence of a strong, nationalist Iraq with renewed WMD ambitions. A Saudi or Turkish nuclear program or a radical regime assuming power in Pakistan, triggering Iran's nightmare of a Talibanized Pakistan could all dramatically change Iran's security outlook. Under these circumstances, the international community would have clearly failed to prevent proliferation and might regard Iran's decision as regrettable but understandable under the circumstances.

Case C: Two worst case scenarios are represented in Cases C and D. In Case C, it is assumed that Iran has a hard-line government which, by definition, is suspicious if not downright hostile towards the United States and its key neighbors such as Turkey, Iraq, Israel and Saudi Arabia. In case C, a hardline government develops a covert bomb. This would almost certainly trigger a major crisis with the United States and could lead to hostilities.

Case D: In this situation, it is assumed that the hard-line regime has formally withdrawn from the NPT, in effect announcing its decision to build the bomb. This, too, would lead to a crisis, including possible hostilities.

Impact on Saudi Arabia

How, then, do these different scenarios impact on Gulf security? Again, there are no simple answers, and it depends very much upon which Gulf state one is referring to. There is a major difference between Saudi Arabia and the smaller GCC countries, because of

Saudi Arabia's size, budget, infrastructure and regional aspirations. For instance, unilateral options open to the smaller Gulf states in the event of an Iranian bomb are very limited. Saudi Arabia, however, has the capacity and the wealth to consider some form of nuclear deterrent, most likely in cooperation with another country, such as Pakistan. Saudi Arabia already has Chinese surface-to-surface (CSS-2) medium range missiles in its current inventory. According to *The Military Balance, 2003-2004* published by the International Institute for Strategic Studies (IISS), these missiles remain in service, yet they are rarely discussed in the burgeoning literature on Middle East WMD proliferation.[2] It is not unreasonable to assume that Saudi Arabia could engage in nuclear purchases, either the basic fissile materials to make a bomb or a finished product. Furthermore, it is not only an Iranian bomb that could motivate Saudi Arabia to consider such an option. There is speculation that the propensity of Saudi Arabia to think about a nuclear option is very much related to the state of its relationship with the United States, which until recently, was always considered the protector of the Kingdom in the last resort. Also, Saudi Arabia's attitude would be strongly influenced by Egypt, a country with which it has had a historically tumultuous relationship.

In this context, the past must not be forgotten. When Egypt was waging war against Royalist forces in Yemen in the mid-1960s, its Russian-purchased bombers flew several combat missions over Saudi Arabia and hit Saudi outposts, because of Saudi support for the Royalists. Egypt also used chemical weapons during this brutal civil war. The traumas of the war forced the Saudi Kingdom to reconsider its defense needs and, as a result, a massive arms deal was struck with Britain and the United States to upgrade air defense capabilities. In the 1980s, Saudi Arabia started a new round of upgrades for its air defense forces and turned to the United States to be its primary supplier. The US agreed to provide very sophisticated aircraft, including the Boeing AWACs and the US Air Force's most advanced fighter, the F-15. However, major disputes with the Saudis erupted when the United States, under pressure from Israel, refused to provide long-range conformal fuel tanks that could extend the endurance and combat radius of the Saudi F-15s. The fuel tank issue was seen by the Saudis as an attempt by Israel to weaken its defense posture at the very moment when the Arabian Peninsula was threatened by Iran, which, at that time, was gaining the upper hand in its protracted war with Iraq.

Without informing the United States, Saudi Arabia negotiated with China for the sale of the CSS-2 missiles, which had a range to threaten all Middle East countries, including Iran. There was never any suggestion that the CSS-2 missiles would be armed with chemical, biological or even nuclear warheads, but the sale did highlight two very important lessons. First, that Saudi Arabia was prepared to "shop elsewhere" to get the defense technology and equipment it felt that it needed when the United States was an unreliable supplier. Second, there were willing sellers on the international arms market, who had fewer compunctions in making such advanced surface-to-surface missiles (SSMs) available. These SSMs clearly have no battlefield or tactical mission, and can only be regarded as deterrent forces, whether armed with conventional or unconventional warheads.

In the 1980s, Saudi Arabia and the other GCC countries were sufficiently worried by the dangers posed by Iran's revolutionary regime that they were prepared to take bold steps with their security. These steps included arms purchases as mentioned above, but also huge financial support to Saddam Hussein, without which Iraq could not have bought the modern military equipment which eventually gave its forces a major advantage in its war with Iran. The GCC also embraced defense cooperation with the United States, as the US intervened to protect Gulf shipping during Operation Earnest Will and become directly involved in combat with Iranian forces.

The next climax came in August 1990, when Saddam Hussein, two years after the end of its war with Iran, willfully and brutally invaded Kuwait, sending shock waves throughout the Arabian Peninsula, equivalent to those felt at the height of the Iran-Iraq War. The response of Saudi Arabia was decisive and resulted in the huge buildup of American and allied forces on Saudi territory in preparations for the liberation of Kuwait in the spring of 1991. Throughout the remainder of the 1990s, the US military presence in

Saudi Arabia was a key factor in determining the security of the Gulf. It was from bases in the Kingdom that no-fly zones over southern Iraq were monitored and Saudi bases were used on the several occasions that the Clinton administration launched major raids against Baghdad and military targets throughout Iraq. Yet, it was the American presence in Saudi Arabia itself that drew attention to and contributed to the rise of Osama bin Laden and Al Qaeda and the string of terrorist attacks against American interests, climaxing on September 11, 2001.

From that moment on, the relationship between Saudi Arabia and the United States has chilled, but both countries still depend on each other and, despite much public opposition, Saudi Arabia permitted its facilities to be used in the War against Iraq in 2003 and has recently begun to cooperate more openly with the United States on counter-terrorism activities. Therefore, how the Saudis will react during an Iranian nuclear breakout is difficult to predict. It can be suggested that any effort by Saudi Arabia to distance itself from the US umbrella would strengthen the bonds between the United States and the smaller GCC states, which in the absence of American protection, would be vulnerable to Iranian, Iraqi and Saudi intimidation. Alternatively, if the Saudis also decided to embrace closer ties with the United States, the smaller GCC states would be even less likely to object to an American presence for their own protection.

Gulf Concerns about Possible Israeli and US Actions

Aside from Saudi Arabia's reaction, the most likely initial response of the Gulf countries to the news of an Iranian nuclear weapons program in the works will be concern about possible US and Israeli preemptive military actions. The Bush administration and Israeli leaders have both made it clear that the Islamic Republic's possession of the bomb would be an intolerable threat. Consider the following statements by US President George W. Bush:

> The international community must come together to make it very clear to Iran that we will not tolerate the construction of a nuclear weapon. Iran would be dangerous if they have a nuclear weapon.[3]

> We…will not tolerate Iranian development of nuclear weaponry.[4]

> Iran…must abandon their nuclear weapons program.[5]

> The government of Iran is unwilling to abandon its uranium enrichment program capable of producing material for nuclear weapons. The United States is working with our allies and the International Atomic Energy Agency to ensure that Iran meets its commitments and does not develop nuclear weapons.[6]

As for Israel, Prime Minister Ariel Sharon has stated:

> Iran constitutes the main threat to Israel as it publicly calls for the destruction of…Israel. It is clear to everybody that Iran is trying to obtain weapons of mass destruction.[7]

Meir Dagan, the head of Mossad, has been even more explicit:

> Such [Iranian nuclear] weapons pose, for the first time, an existential threat to Israel.[8]

> Iran's nuclear program posed the biggest threat to Israel since its creation in 1948.[9]

However, since the Iraq war and the unreliability of western intelligence concerning Iraq's WMD programs, the case for preemptive war against supposedly proliferous states has become weak while the political costs of undertaking such action in future have become much higher. If the intelligence regarding an Iranian bomb is uncertain, the United States and Israel will have problems garnering support for military action. Even if the evidence is overwhelming and highly convincing (that is, Iran either tests a nuclear device or announces it is building the bomb), there will be reluctance to endorse US-Israeli military action for fear of the chaos this could bring to the Gulf and the region.

Despite their hard-line policies towards an Iranian bomb, how likely is it that either Israel or the United States could use military force against a nascent Iranian nuclear program? First, it is necessary to distinguish between the capabilities of the two countries. One, the world's only superpower, and the other, a relatively small, albeit technically advanced Middle Eastern country. Israel has a nuclear force built around a number of surface-to-surface missiles, including the Jericho I missile with a range of up to 500 kilometers, which is not sufficient to reach Iran, and the Jericho II missile with a range of up to 800 to 2,000 kilometers. It is estimated that Israel has around 200 nuclear warheads. The Jericho II could reach Iran but, unless it carried a nuclear payload, it would be little use against Iran's nuclear facilities, given the likely inaccuracies of the missiles over such long distances. The only conceivable circumstances under which Israel would attack Iran with nuclear missiles would be in retaliation against a nuclear or severe chemical or biological attack on its own soil by Iran. In other words, if Israel were to contemplate preemptive strikes against Iranian nuclear weapons facilities, it would have to rely on its conventional forces hoping for a repeat of the 1981 attack on the Iraqi reactor Osirak, which set back the Iraqi nuclear program by several years. The problem is that Iran is further away than Iraq and Iran's nuclear facilities are more dispersed and better protected than Iraq's facilities were in 1981.

The only instruments in the Israeli arsenal that could be used in a preemptive strike against Iran would be F-15I and F-16D fighters in the Israeli Air Force. These aircraft have the range to reach Iranian targets and return but it would be a very complicated and hazardous mission. Because of the long ranges involved, aerial refueling would be necessary. Israel has limited mid-air refueling capabilities and its tankers – the old Boeing 707 – are extremely vulnerable. Furthermore, Israel could hardly fly over Arab countries, including its ally Jordan, making the route even more difficult. In theory, Israel could fly to its targets from bases in Turkey or even from India, but it is extremely unlikely that it would be granted permission by either of these two countries which, although nervous about Iran's nuclear weapons program, are against preemptive action at this stage.

To successfully destroy the Iranian facilities, dozens of targets would have to be hit, probably repeatedly. Here, the real question is: how good is Israeli, or for that matter American intelligence on exactly where Iran's nuclear facilities are located? One knows from aerial photography where the big plants such as Bushehr and

Natanz are located, but what is important are other facilities, including the centrifuge production workshops and factories which may not be so easy to find, let alone destroy. One must also take into account that Iran has an air force and anti-air capability which, while no match for Israel one-on-one, could put up a formidable resistance for the few Israeli fighter-bombers that would make it over the long distance. Certainly, Israel would have to prepare for anti-air defenses, which should add to the burden and size of the mission. In view of the fallout from the Iraq war, Israelis themselves are re-examining their own intelligence capabilities, since the country, like the United States, Britain, Australia and others believed that Saddam had weapons of mass destruction (WMD) deployed and ready to use. Likewise, the extent of Israeli intelligence on the precise location of Iran's nuclear facilities may leave a lot to be desired, which is an additional reason why the operations might be too hazardous to contemplate.

What about the United States? Unlike Israel, it has an array of capabilities that could brought to bear very quickly against Iran, particularly in view of the huge presence in the Gulf, including Iraq, Afghanistan, Bahrain, Qatar, Saudi Arabia, Kuwait, Oman and the UAE. The United States has seaborne, airborne and land-based air assets, which could create havoc in Iran over a period of days and undoubtedly do much more effective damage than Israel could. However, this does not mean that even the United States could be successful in such an operation, given the complexities of the targets and the need to make absolutely certain that the program was stopped. What can be said with some confidence is that a sustained American air attack on Iranian capabilities would probably set the Iranian program back many years but would, at the same time, undoubtedly not end the program, since Iran would become determined to continue. Furthermore, Iran does have many means of retaliation against the United States, particularly in view of the exposed position of US forces in Iraq. Iran is believed to have infiltrated many of its Revolutionary Guard corps elements into Iraq, mingling with the Shiite community, and is quite capable of inflicting considerable damage on the United States and the nascent pro-Western democracy that the US hopes to nurture in Iraq.

Likewise, Iran has assets in Lebanon – notably, the Hizbollah – to use against Israel if it decided to conduct a preemptive strike.

Ironically, the one area where Israel, and certainly the United States, has a huge decisive advantage over Iran is the capacity to inflict extraordinary pain on the Iranian economy. Unlike its nuclear infrastructure, Iran's oil infrastructure is very vulnerable, easy to attack and its loss would cause immediate and long-lasting pain to the country's oil industry. Hence, there have been occasional warnings from Israeli officials that it is not so much Bushehr that Israel would attack, but the vulnerable loading facilities at Kharg Island and other offshore regions. While an attack on Iranian oil is a much easier military task, the international consequences would be considerable, including profound and disruptive implications for the oil market. It could even lead to Iranian retaliation against all oil shipments in the Gulf based on the principle that if Iran was going to suffer, then so should other countries. Iran certainly has the maritime capabilities to inflict great damage on Gulf oil facilities and could disrupt traffic through the Straits of Hormuz, at least for a period of time. The general economic consequences of such action would be global in dimension and so serious that it is difficult to imagine any likely scenario when this would happen, except as retaliation against Iranian actions considered so heinous by the United States and Israel that it deserves such a strong response. Obviously, the willingness to use attacks on Iranian oil would also be related to the overall status of the global oil market and the access to alternative supplies if Iran's production was rendered out of action for many months.

While these scenarios of Israeli and US attacks on Iran are somewhat far-fetched, they nevertheless highlight the reason why there could be great concern among the Gulf countries in the event of a serious crisis over Iran's nuclear program. Although, in the long-run, the Gulf countries have real reason to be concerned about an Iranian bomb, their short-term fears would surely focus initially on concern about what the United States would do. If the initial crisis was overcome and there was no military action and the United States relied more on international pressure on Iran through the UN

Security Council, the Gulf states would clearly be relieved and would almost certainly support any UN Security Council resolutions calling for economic measures to penalize Iran, though whether they would agree to an oil boycott is another matter. In fact, it is unlikely that an oil boycott would get full UN Security Council approval, given the close relationships between Russia and Iran, and China's growing dependence on Iran's oil for its own needs. However, lesser economic measures might well pass unanimously, in which case the Gulf states would concur with these.

The Impact of an Iranian Program on Iraq

One issue that the Gulf countries have to watch very carefully is the likely impact of an Iranian nuclear weapons program on the emerging regime in Iraq. Assuming that Iraqi sovereignty is re-established and that, over a period of time, elections and a binding constitution and a democratic government come into existence, how that government deals with its neighbors, particularly Iran, will be of critical importance. One issue that will clearly be uppermost in the minds of everyone in the neighborhood, including Iran, is the residual American military presence in Iraq after sovereignty has been regained. It is likely that the United States will have tens of thousands of troops there for the foreseeable future. These troops will be augmented by US air and naval assets based in Iraq and the Gulf. This will be necessary, first to provide an umbrella under which the new Iraqi forces can re-establish and re-equip themselves, and second, to provide protection against any belligerent neighbors.

In this regard, any move by Iran on the nuclear front would clearly strengthen the case for closer ties between the Iraqi government and the United States. Whether this would explicitly include security guarantees is as yet unknown. What is certain is that, without a US military presence, the new regime in Iraq will be very vulnerable to external manipulation and pressure. In the past, Turkey was considered the most likely country to intervene in Iraq's affairs. However, at least until the European Union (EU) has made a decision on whether Turkey should be allowed to apply for EU membership and is given a date to begin negotiations, the Turkish government is likely to be very careful about meddling in Iraqi affairs, since such action would poison the atmosphere in Brussels and a decision on Turkey would again be postponed.

Iran is another matter and it is already infiltrating Iraq with thousands of its people crossing the border, though there are debates about their ultimate purpose. What is clear is that no matter what regimes emerge in Baghdad and Tehran, both Iraq and Iran are intimately linked by reasons of geography, history and their respective Shiite populations. This need not be a threat to the new Iraqi government but it means that Iraq will need to have good relationships with its neighbors. In fact, talks are already underway between Iran and Iraq over building an oil pipeline from Iraq's southern oil fields to Abadan to facilitate the export of oil. The current capacity of Basra is limited, both because of infrastructure damage and the narrowness of the Shatt Al-Arab waterway. Thus, at one level, Iran and Iraq are likely to cooperate more than in past years, but at the strategic level, the Iraqis will clearly be wary of Iranian intentions and in this regard, the United States will remain their most important protector.

From an Iranian point of view, by far the most important utility of a nuclear force would be to protect its territory from outside aggression and deny the United States the option of an Iraq-type invasion. This would clearly be a significant deterrent to any American military adventures that directly threaten the survival of Tehran's regime. Since regime survival is of great importance to Iran's hard-line *mullahs*, this option clearly has its appeal, provided that the assumptions about American capacity for aggressive behavior are plausible. Are they? For the foreseeable future, the United States will continue to maintain a large military presence around Iran's borders, primarily to ensure the stability of Afghanistan and Iraq and to protect the GCC countries. Therefore, there will be many points of friction between the US and Iran that could lead to military encounters but much will depend upon Iran's proclivity for any intervention in the neighborhood that the US might find threatening. The test case of Iraq is still unfolding. Until now, Iran has shown considerable restraint, but stories keep emerging about embedded Revolutionary Guard soldiers in Iraq's Shiite community who, in extremis, could be used to wage guerrilla warfare against the United States, if the situation in the region deteriorates.

Gulf Security Ties with the United States

Such a situation would clearly impact on the Gulf, because it means the United States will have strong reasons to maintain its military presence in the Gulf states. This raises the question of what kind of relationship the smaller Gulf states will have with the United States in the event of an Iranian nuclear program. The nature and purpose of enhanced military cooperation between the US and the Arabian Peninsula could take many forms. The most important component would be a counter-deterrent to demonstrate to Iran – whatever else its clout with the Gulf countries might be – that any efforts to use nuclear weapons to intimidate or blackmail would be challenged by the United States. The credibility of this counter-deterrent would be linked to the vulnerability of US forces and US targets in the region to Iranian intimidation.

Iran is not expected to deploy an intercontinental ballistic missile capable of striking the continental United States for many, many years. By that time, the US and its allies plan to have deployed an anti-ballistic missile system as a component of a robust and multifaceted counter-proliferation strategy. For Iran to even consider an attack on continental United States (CONUS), it would have to be able to withstand a massive preemptive American strike using conventional warheads or, if Iran uses nuclear weapons first in the theater, a nuclear second strike by the United States. In other words, it is difficult to see, under what circumstances, Iran could use its nuclear weapons, except for some suicidal spasm—similar to the scenarios heard so frequently with respect to Saddam Hussein and his capacity for a glorious *Gotterdammerung* ending to his fiefdom.

In the absence of some catastrophic change in American policy – for instance a premature decision to draw down forces in Iraq and the Gulf – US security ties with the smaller Gulf states are likely to grow in the years ahead. The reasons go beyond Iran, though Iran must be regarded as an important factor.

The small Gulf states find themselves in a dilemma. By regional or even world standards they enjoy high per capita income, mostly energy and trade-related. However, they are surrounded by much larger countries whose social and political conditions are far from stable or reassuring. Iraq and Iran are not the only neighbors providing cause for anxiety. What happens to Saudi Arabia has immediate consequences for the smaller Gulf states, and as concern grows over the Saudi regime's ability to cope with its burgeoning problems, it is not surprisingly that the smaller neighbors become increasingly anxious. These circumstances provide all the more reason to work closely with the United States including defense cooperation, and to accept the inevitability that without some sort of US umbrella, their own security could be jeopardized. What happened to Kuwait in 1990 is an experience that none wish to see repeated.

The United States already has defense agreements with Kuwait, Bahrain, Qatar, the UAE and Oman. If it were decided that the threat from Iran had escalated to include nuclear weapons capability, what are the additional measures that might be sought? What would be the costs and benefits? If the Iranian nuclear threat remained in the realm of a nominal bomb or a nuclear breakout potential, probably no extra defensive measures would be needed except improved surveillance and intelligence capabilities. The United States has sufficient firepower in the region, or sufficient access to the region, to deter any conceivable non-nuclear posturing, or if one assumes that a nominal bomb would only be used in the event of an attack on Iran itself.

The exception, of course, would be the probability that a truly radical regime in Tehran might under some circumstances be prepared to transfer nuclear materials to terrorist groups, which could then use the material in an offensive mode against an array of targets in the Gulf and elsewhere. This possibility cannot be ruled out and would require intense vigilance if the Iranian regime appeared willing to engage in such dangerous behavior, but it would have to be considered an unlikely case. If the Iranian nuclear force is more robust, of the type described earlier, then a number of very expensive defense preparations would be in order, including the deployment of advanced theater missile defense systems, much greater funding for civil defense, and steps to protect civilian infrastructure by hardening redundancy and dispersal. Again, it is clear that the major deterrent to Iran from contemplating any use of nuclear weapons against the Gulf would have to be the overwhelming retaliatory capability of the United States. This deterrent is likely to be effective for many years to come.

To what extent will Iran's ability to influence the Gulf states be related to its nascent nuclear program? What difference will it make if the programs remain nascent or become fully fledged, including a nuclear force akin to that possessed by Pakistan and Israel? In some respects it does not matter, for as long as Iran has good relations with the Gulf states, the benefits of cooperation are likely to override fears about nuclear weapons. However, if relations sour over any number of potential issues, including territorial disputes, then Iran's temptation to use its nuclear potential as some sort of intimidation would be serious, especially since Iran has a significant conventional maritime military capability, which is where the US presence becomes vital. So long as the US sustains a physical presence in the Gulf, it will have the capacity to deter any Iranian military maneuvers that are designed to put pressure on the Gulf states. It would be foolish for Iran to threaten the use of force, including nuclear weapons, as long as the United States has the capacity to defeat any Iranian military adventures against its neighbors.

It is for this reason that the size and configuration of Iran's potential nuclear force would have little impact on the conventional balance of power in the Gulf, and the ability of the United States and the Gulf states to withstand any Iranian aggressive behavior. The size and configuration of an Iranian nuclear force would have impact if Iran itself is threatened or felt threatened by the United States. So it has to be assumed that the primary reason Iran would consider a nuclear option is for reasons of national survival, rather than as a tool for regional aggression. Under these circumstances, Iranian calculations as to the likely effectiveness of its deterrent would have to be influenced by the size and configuration of its nuclear force. Critics of an Iranian program say that nuclear confrontation with the United States would be madness even at the best of times, but if all that Iran possessed were one or two weapons, they might well be susceptible to preemptive strikes, malfunctions and active defenses by adversaries. Thus, the only sure way to survive a nuclear exchange with a superior foe would be to have a sufficiently large, hardened force that would be able to withstand a preemptive attack, and then be able to retaliate with high reliability against the adversary and be sure that a certain percentage of the retaliatory weapons overcame enemy defenses and reached their targets.

Conclusion

As the likelihood of an Iranian nuclear weapon grows, greater defense cooperation between the Gulf states and the United States will become inevitable. Even if Iran refrains from crossing the weapons threshold, its conventional force capability and its civilian nuclear infrastructure pose a challenge that no Gulf regime can ignore, nor counter without close cooperation with the United States. Whether this delicate relationship can be managed with both sensitivity and effectiveness will be the key test for Gulf-US relations in the years to come.

Section 5

IRAQ AND GULF SECURITY

13

The Impact of the New Iraq on the GCC States

H.R.H. Prince Turki Al-Faisal Bin Abdul Aziz

Currently, the Gulf Cooperation Council (GCC) states are facing challenging circumstances. At the same time, they have opportunities to surmount these challenges at both the internal and external levels.

Internally, there are a number of challenges:

First, there is an increase in the population of all the GCC states. This increase makes it imperative for the government to take measures and launch developmental projects that are consistent with and also anticipate this increase.

Second, there is the process of developing the system of government in such a way that it fulfills the aspirations of the citizens of these states by broadening the circle of political participation to include all citizens, both males and females.

Third, there is the process of developing defense arrangements so that these states acquire the capabilities to perform their duties with respect to defending their citizens and territories. This process involves developing the Jazeera Shield System so that it can carry out its functions with regard to warding off the dangers threatening the GCC states.

All these challenges are mentioned in the records documenting the proceedings of the GCC summit meetings, the press commentaries and expert views. However, it must be noted that with regard to meeting these challenges, we are racing against time.

Twenty years have elapsed since the foundation of the GCC. In comparing what the GCC has achieved with what Europe has achieved over the same period, the conclusion to be drawn is that there is a lot more to accomplish before the GCC can be considered as having achieved the same level of success as Europe.

I am not undervaluing the achievements of the GCC. The six states constituting the GCC have much in common. They believe in the same religion, speak one language, belong to one culture, pride themselves on one heritage, dwell in the same geographical milieu, and have the same ethnic origins. Yet they have been incapable of achieving a greater degree of conformity between themselves. This is a negative aspect that warrants further investigation and research.

Externally, there are a number of dangers confronting the GCC states.

First, the occupation of Iraq is the fountainhead of danger, because it has bred many possibilities. The most dangerous of these is the possible partition of Iraq, which will have ramifications for all. In this connection, we cannot forget what the Israeli Prime Minister David Ben-Gurion had written when he called upon Israel to seek the disintegration of Arab states into petty tribal and sectarian states, so that they would stop being a threat to Israel. He went on to say that the Arab states must become small, sectarian, ethnic states. We also read in the Israeli and American newspapers the writings of those who seek to justify and propagate the partition of Iraq.

The other possibility centers around Iraq turning into a focal point for terrorism—a magnet that attracts people from far and wide who find ample opportunity in Iraq to practice inhuman crimes that run counter to all divine laws.

There is a third possibility—namely, Iraq remaining an American colony, with a concentration of American forces. These forces would be ruled by a military policy the crux of which is a propensity towards unilateral decision-making dictated by American interests.

The second of the external dangers is the Palestinian Question. This represents a historical challenge—one whose ramifications extend not only to the GCC states, but to all other Arab and Islamic states. This is a challenge that has resisted the resolutions of the United Nations and made international hypocrisy a defining characteristic of our time.

The third danger relates to the general situation of the Arab world. This situation has resulted in successive disappointments, the first of which has led to the occupation of Palestine, whereas the most recent one has led to the occupation of Iraq.

In their totality, these challenges may provide the GCC states with an opportunity to conduct an in-depth examination of their problems, and enable them to find internal and external solutions to them.

Internally, they must avail of this opportunity to review their domestic conditions with transparency, without overlooking reality. Since His Royal Highness Crown Prince Abdullah Bin Abdul Aziz Al Saud uttered his famous words – in this country during the 1998 GCC summit – stating that the boom period has come to an end, a spirit of objectivity and impartiality has become the guiding light of the Rulers and people. Rapid development and reform have followed in each GCC state. The advisory council of the GCC summit has been broadened, and its duties have expanded. Hence we have to proceed in this direction, especially since our present position requires planning our economic activities to avoid duplication and squandering of money on similar projects. We have to concentrate on projects of an integrative nature. Economic activity alone can provide us with a labor market that will absorb the progressive increase in population.

We also have to take steps designed to develop the Al Jazeera Shield forces and unify arms as well as arms purchase contracts. This will considerably improve the GCC's current negotiating position.

The on-going dialogues, discussions and debates in the Gulf region attest to a sincere desire on the part of Gulf citizens to solve regional problems. These touch upon areas of discourse that, a few years ago, no one would have expected would be the focus of such intellectual activity. Since these dialogues focus on developing and improving education, they attest to the fact that the most important issue in our societies is being addressed. This is especially true because Gulf governments accept and encourage this trend. Needless to say, in aspiring to a prosperous future, education is the key issue.

Our interest in implementing internal reform will give us an opportunity to meet external challenges. Dealing with the Iraqi question gives us an opportunity to extend a helping hand to our Iraqi brethren, to support their aspirations to regain freedom and independence, to bless the quest of the Iraqi people to practice freedom of expression and to confront whoever tries to alienate or suppress this right. All this presupposes dealing with the Iraqi people as one nation, not as different sects, creeds and races. As soon as the situation calms down in Iraq, prosperity in the domains of economics, culture and development will envelop all. The GCC states can effectively contribute to wheeling the country towards greater prosperity.

The Palestinian Question remains a formidable challenge. We should not despair of achieving a just and comprehensive peace. However, this can only be achieved by communicating openly with our Palestinian brothers, providing comprehensive support to the Palestinian people, and seeking to halt actions that are devoid of any benefit and only attract condemnation from international circles opposed to suicide bombings. Such deeds victimize both the innocent and the guilty without differentiation. Both these challenges have a common denominator with which we must consider how to deal seriously and effectively—namely, the presence of the United States in Iraq as an occupier, and its involvement in Palestine as the patron of the peace process and the supporter of Israel.

The opportunity lies in supporting those American institutions and personalities that back our position. I am aware that there is a project, approved by many GCC summit meetings, which is aimed at supporting such American institutions and personalities. It has specified what each state is supposed to do. However, this project has not been implemented until today. How can we confront Zionist influence in the United States if we do not work there in the same way that Israel does?

As for the general conditions in the Arab World, the GCC states have a tremendous opportunity to meet the challenge represented by the deteriorating Arab situation. Despite expressing criticism of, and discontent with, the slow pace of development in the Arab World, the GCC states must realize that the Council constitutes a regional forum that still embodies hopes linked to the unity of people and ideas in the Arab World. The proposals of His Royal Highness Crown Prince Abdullah to reform the Arab League provide us with an opportunity to achieve our aspirations with respect to reform, development and Arab unity.

14

The Iraqi Economy: Aspirations and Challenges

H.E. Mr. Kamel M. Al-Kilani

Owing to the vast and diverse natural resources possessed by Iraq, especially its oil wealth, the Iraqi economy at the beginning of 1970s was viewed as a pioneering and leading economy in the Arab world and the region in general. Iraq has the world's second largest oil reserves (with proven reserves estimated at 112 billion barrels) after the Kingdom of Saudi Arabia. It also has an efficient labor force, highly qualified public administration departments, and advanced infrastructure. However, as a result of the mismanagement and misuse of public finance during the past three decades, per capita income from the gross national product (GNP) dropped sharply to its lowest level from US$ 3600 in the 1970s to US$ 770-1000 in 2001.

Economic activities also deteriorated sharply due to the hostilities of March 2003 and the subsequent looting and acts of sabotage, despite all the measures taken by coalition forces and Iraqi officials.

The Economic Legacy of the Former Regime

Policies Adopted by the Former Regime

The economic policies followed by the former regime have led to the systematic devastation of the Iraqi economy. A number of these policies are listed below:

- The centralized planning and management of the Iraqi economy leading to imprudent orientation and misallocation of financial resources.
- The unjust wars launched by the regime against neighboring countries, which claimed the lives of untold numbers of Iraqis, caused losses amounting to billions of dollars, and brought Iraq under the heavy burden of economic sanctions.
- The expansion of the public sector despite its weakness and total dependence on oil revenues to finance its activities, as well as the setting up of public companies for military production purposes.
- The diversion of Iraq's workforce to military activities, thereby limiting and weakening its productive capacities.

Problems and Consequences of the Former Regime's Policies

The policies and practices of the former regime have resulted in considerable negative economic and social problems and plights that have transformed Iraq from a rich nation to an impoverished country seeking assistance and donations from the international community and world finance institutions. Among these problems are:

- The overburdening of the country with such huge debts that the Iraqi economy is unable to repay, a situation that is bound to impede the revival and rehabilitation of the national economy.
- The deterioration of living standards, with most Iraqis now living below the poverty line, deprived of any benefits from the country's vast wealth.
- The deliberate, long-term neglect of vital infrastructure, severe shortage and degradation of public services.

- The misuse of economic resources by diverting them to non-productive fields such as military activities.
- The abuse of public funds, especially by the deposed regime's officials and followers.
- The reduction of the role and contribution of the private sector to the development of Iraqi economy.
- The collapse of the banking system, which has become incapable of performing its role.

Economic Policy Objectives of the New Iraq

Achieving a High and Sustained Rate of Economic Growth

The most effective strategy to boost the living standards of all Iraqis is to achieve the highest sustained rate of economic growth during the coming years. By attaining high economic growth rates, nations of the world will not only ensure economic welfare for their peoples and reduce poverty levels, but also establish social and political stability.

Iraq enjoys huge opportunities and capabilities enabling it to realize considerable economic growth in the coming years thanks to its enormous human and material resources. If these resources are rationally utilized and invested in pursuit of sound economic policies, they will certainly help to ensure continued growth and higher standards of living for all the Iraqi people. In the coming period, the estimated rate of economic growth that Iraq is expected to attain is 10.7 per cent.

The flow of external investment and assistance from donor states will provide great momentum to the Iraqi economy to achieve considerably higher growth rates than those previously reached in Iraq. This, in turn, could double the living standards of Iraqis in the near future.

Integration with the International Economic Community

Iraq seeks to restore its political, economic and financial position within the international community through creating constructive relations with all states based on mutual interests and non-interference in each other's internal affairs. It is also looking forward to establishing strong links with world financial institutions such as the World Bank and the International Monetary Fund, to obtain the support and assistance needed to guarantee equity and prosperity for the Iraqi people, and to rebuild its economy and political system based on democratic principles and respect for human rights.

To achieve such integration, Iraq will have to adopt clear, transparent economic policies, set up an active trading system founded on a legal customs regime and a unified tariff in the medium-term in order to build mutual confidence between Iraq and countries all over the world.

This process also involves the exchange of information and expertise with the above-mentioned financial institutions to win their trust and encourage them to extend financial and technical assistance to Iraq to help revive its own economic and financial institutions and achieve higher rates of economic growth.

Rapid and smooth integration with the international community will ultimately pave the way for Iraq to join the World Trade Organization (WTO) and conclude partnership agreements with other states in the region.

Economic Openness

Economic openness is a fundamental element of the market economy system as it allows markets to function properly and spontaneously away from interference by the central authorities. Thus, a favorable stage will be set for businesses, initiatives and enterprises to flourish and thrive, which in turn will create more job opportunities and reduce unemployment rates.

During the last decades of the past century, the Iraqi economy has greatly suffered from isolation, centralized planning and state interference, which had led to the obstruction of economic growth, the soaring of the unemployment rate, the expansion of an inefficient public sector dependent on subsidies from oil revenues, and the increase in the number of Iraqis living below the poverty level.

Hence, Iraq is eager to promote openness to other economies, regionally and internationally, and to eliminate all obstacles that would hinder this process in order to best serve the recovery and advancement of the Iraqi economy.

Building a Social Security Network

In view of the prospect of attaining high economic growth rates during the next few years, it has become imperative for Iraq to set up a broad and stable social security network to provide the necessary assistance to certain segments of Iraqi society that are highly vulnerable to the dangers of poverty, famine and illness—children, the elderly and those who have suffered greatly from the harsh effects of the transitional economic period.

This network constitutes one aspect of the moral responsibility that the state must shoulder in its endeavor to rebuild the new Iraq. It is also a step on the path to construct a solid social base and democratic system in Iraq to create equal opportunities for all Iraqis to benefit from, and share equitably in their country's wealth and resources without any discrimination.

Economic Policy Challenges

Rebuilding Vital Infrastructure

The wars waged by the former regime have resulted in the negligence of vital infrastructure and services and the postponement of maintenance needs, which have weakened operational capacities, leaving the majority of Iraqis with limited access to such services. Instead, they have no choice but to seek alternative services at extremely high cost, a situation that calls for radical improvements or the reconstruction of these infrastructures and services. This process will certainly play an important role in enhancing the quality of services, meeting the needs of both major and minor sectors, and strengthening security measures, all of which would attract more investments, develop the private sector and ensure the flow of strategic goods and services needed for the reconstruction efforts.

Iraq's Foreign Debts

In the 1970s, as a result of soaring oil prices and production capacities, Iraq was able to accumulate huge financial assets, both inside the country and in its accounts at international banks. However, the imprudent policies of the former regime, the diversion of most human and material resources to military activities, and the three devastating wars waged by the regime, have weighed down the country with such massive foreign debts that the Iraqi economy has been totally unable to endure and shoulder the burden. These debts remain an insurmountable obstacle that impedes the development of the national economy despite the financial aid extended by donor states and international institutions for reconstruction purposes. Indeed, Iraq's own resources are insufficient even to service these debts, let alone repaying them.

Setting up an Effective, Strict and Responsible Budgetary System

Rebuilding the new Iraq involves the adoption of a responsible financial strategy based on a strong foundation. Even a brief period without such a strategy would cause Iraq to suffer from such heavy financial pressures that the government might be compelled to resort to borrowings or issuing uncovered currency. These measures will certainly lead to a higher inflation rate, negatively affect national savings and investments, and directly harm the poorer segments of society.

Therefore, any financial strategy must abide by several fundamental principles, the foremost of which is the passing of a balanced budget, taking into account that any deficit thereof must be temporary and not permanent. Means of financing such deficits has to be determined in advance, and their negative effects on the economy have to be minimized. The ratio of consumption, investment and public debts to the gross national product (GNP) has to be balanced, taking into account the ability of the national economy to repay these debts.

Transparency is another important feature of financial policy, that is, the clarity of the assumptions upon which budget estimates (expenditures and revenues) are determined. Estimates of several previous years should be taken as the base for next year's expenditures and as a general framework for the coming years. Expenditures and revenues should be classified in a manner that facilitates analysis and follow-up, and a registration/reporting system has to be set up to ensure better implementation of the budget.

Creating an Attractive Investment Climate

The investment sector is one of the key sources for creating job opportunities that could absorb the unemployed in productive fields other than the oil sector, as well as a major means to stimulate and boost the national economy. However, the laws and regulations introduced during the last decades of the past century prohibited foreign investments, thus depriving Iraq of badly needed capital funds, projects and advanced technology. The unfavorable atmosphere prevailing at that time discouraged domestic investments due to many restrictions and constraints that hindered the development and expansion of the private sector.

Consequently, Iraq is making every effort to create an environment conducive for foreign and domestic investment by adopting macroeconomic policies, eliminating all obstacles preventing the flow of foreign capital to Iraq, and allowing Iraqi investors every opportunity to benefit from external capital markets and obtain credit facilities from international financial organizations.

Developing the Private Sector

The private sector constitutes an effective means to animate the national economy and create job opportunities. Hence, the creation of a favorable climate to bolster the private sector has become one of the challenges facing policy planners and decision makers.

During the last decades, the role and contribution of the Iraqi private sector to overall economic activities have been reduced and weakened as a result of the government policy of setting up state-run companies and substantially increasing their contribution to the economic activity despite their weak, inefficient administrative and financial performance that has made them dependent for their survival on the assistance they receive from oil revenues.

Thus, the authorities in the new Iraq are exerting strenuous efforts to provide the private sector with the necessary support and favorable environment to enable it to play a more meaningful role in advancing the Iraqi economy while the scope of the public sector is sought to be gradually reduced.

Promoting Public Sector Transparency

At the present time, the activities of state-owned enterprises constitute a significant percentage of Iraq's overall economic activity. Therefore, the work of these companies has to be based on clear and straightforward economic operating standards and responsible rigorous performance that are subject to checking and scrutiny by authorities concerned and the public in general in order to gain the confidence of both the investors and the citizens.

The acquisition by the government of public sector companies should not give it the right to interfere in the operations of these companies, as ownership and operational policies have to be separated from each other.

Generally speaking, in order to enable state-owned companies to function according to the principles of the free market system, radical reforms have to be carried out in these companies. These reforms include, among other things, conducting an individual assessment for each company in order to determine the type of support they need (taking into account the protection of the rights of workers and employees), and providing them with budgetary financial aids (in the form of operating capital or staff salaries) until they reach the stage of economical operations.

Expanding the role of the Iraqi private sector in economic activities will inevitably involve reduction of the role of state-owned companies. This can be done by encouraging the private sector to raise its share in these companies and lowering the share of the state.

The current ideas and proposals being debated for dealing with the public sector companies have not yet crystallized, except for the conviction regarding the need to provide qualified management and effective support to this sector to ensure fruitful and efficient performance and the retention of working staff.

Measures to Attain the Desired Objectives

Budgetary Allocations and Frameworks

- Officials of the Ministries of Finance and Planning prepared the 2004 budget based on specific assumptions to include the ministries' operational expenditures and those assigned for the reconstruction process, in addition to the 2004 general estimates of expenditures/revenues and thus provide a general framework for the years 2005 and 2006.
- The budget would be financed by available Iraqi revenues and not through foreign loans and assistance nor by issuing uncovered currency.
- For the first time ever, the budget included both current and investment expenditures, and covered all Iraqi governorates, including the northern ones.
- Budgetary revenue was estimated at Iraqi Dinar (ID) 19258.8 billion (b) (equivalent to US$ 12839.2m). Of this, ID 18000b (equivalent to US$ 12b) was the expected oil revenue generated from exporting 1.6 million barrels per day (mb/d) at US$ 21 per barrel, taking into account fluctuating oil prices. The remainder represented non-oil revenues generated from taxes, charges, services fees and earnings of state-owned companies (profits and returns of imported goods sold under the Oil-for-Food Program).
- Expenditures were put at ID 20145.1b (US$ 13426.5m), of which ID 19026.7b (US$ 12636.9m) are operational expenditures. Investment expenditures were estimated at ID 1118.4b (US$ 789.6m), with a deficit of ID 886.3b (US$ 587.3m) which was to be financed from unspent amounts of the Oil-for-Food Program).
- As a result of the anticipated increase in oil revenues, investment allocations are expected to double during the coming years.
- Budgetary spending included an amount of ID 5250b (US$ 3500m) allotted to finance ration system: ID 1117.5b (US$ 745m) to support local and regional administrations; ID 937.5

(US$ 625m) for construction projects; ID 750b (US$ 500m) to support public sector companies; and ID 300b (US$ 200m) to support the agriculture sector.

Restoring Vital Services

For the last two decades, due to the wars and sanctions imposed upon Iraq, the country's vital public services (water, sewage, communications, electricity and transportation) suffered from long-term negligence, rendering them technically inadequate to meet public needs. In addition, most of these basic services were further damaged, and some completely destroyed due to the war of March 2003 and the subsequent acts of looting and sabotage.

Therefore, efforts are being currently concentrated on restoring these services to their pre-war condition, and these were successful in restoring drinking water and sewage services facilities in Baghdad and many other regions within Iraq.

In the field of communications, most of the telephone exchanges are now functioning and efforts are underway to make the remaining exchanges operational. Contracts have been signed to set up three mobile telephone networks covering Iraq's southern, central and northern regions. Efforts also continue to rehabilitate power stations and distribution networks, and over 4000 MW of electricity are now available out of the total demand of 6500-7000 MW.

Massive funds need to be invested during the period 2004-2007 to finalize the process of rebuilding and to develop these services. According to estimates prepared by the International Monetary Fund and World Bank, some of these funds will be provided by donor states, US government grants and Iraq's budgetary allocations.

Creating a Favorable Investment Climate

On September 21, 2003 Iraq took a very important step to create a favorable investment climate—the enactment of a legislation regulating foreign investments. This law has put Iraq on the list of states most receptive to such investments in all types of projects in every economic field except those involving natural resources.

Under the law, foreign investors are allowed to set up investment projects, subsidiary companies (or branches thereof), and joint ventures together with local Iraqi partners. Foreign investors enjoy the same treatment and advantages offered by the law to national investors, and are allowed to transfer their profits immediately in any currency of their choice.

Also, the law gives owners of foreign projects the right to enter into long-term contracts to lease the lands on which projects are to be set up. The law, however, prohibits foreign investors from buying such lands. Besides, Iraq has taken several other steps towards improving the tax system, liberalizing trade and prices, modifying customs tariffs, and building up the banking system.

Building an Effective Banking System

Financial sector officials have exerted strenuous efforts to revive and operate payment systems and credit facilities. Most branches of the two state-owned trade banks have been rehabilitated and a communication system has been set up to facilitate payments and transfer transactions. These branches are providing loans and credit facilities to a large number of medium and small businesses.

One of the key challenges for these banks is to enhance their administrative potential and performance techniques in order to attain international standards. In this connection, Iraq is seeking technical assistance from international financial institutions, including the World Bank.

In November 2003, the Trade Bank of Iraq was established to finance export/import transactions through cooperation and coordination with world banks and financial organizations. Undoubtedly, such a step will expedite the integration of the Iraqi economy with the international financial markets.

The recently passed law on commercial banks will also create investment opportunities for foreign banks. Hence, these banks were approached to submit their offers under specific criteria laid down for the purpose of licensing six foreign banks. Local banks financed by the private sector were asked to increase their capital to match international standards and to expand their loan and credit potential.

The independent status which the Central Bank of Iraq currently enjoys, along with the amendment of its statute with regard to its supervision over banks operating in Iraq will have a positive effect in developing the functions of these banks.

Liberalizing Trade and Prices

By liberalizing trade and prices, Iraq is currently seeking to integrate its economy into the global trade and financial markets. Various practical steps have taken in this direction, including the abolition of import quotas, and the imposition of a unified customs duty (effective from January 1, 2004) of 5 per cent (the "reconstruction duty") on all imported commodities except humanitarian goods (foods, medicines, clothes, books etc).

The lack of flexibility in determining prices would cause diversion of consumption and investment trends and have negative effects on economic growth rates. Nevertheless, the process of price liberalization should be gradually implemented – especially those prices that considerably affect the limited-income class after taking into account its adverse social impact on poorer segments of society. Therefore, such a move has to be coupled with certain measures to mitigate this adverse impact.

Supporting the Private Sector's Medium and Small Units

Medium and small economic units set up by the private sector constitute an important strategic element to activate and boost the national economy in the coming years. The success and progress of these units will strengthen the efficiency of Iraq's labor force, increase the added value of the manufactured goods, encourage the trend to set up firms and factories involving intensive employment, advance economic growth and expedite the creation of more job opportunities.

Studies and surveys conducted by the World Bank have underscored the importance of advancing and supporting small and medium projects, given the strong relationship between economic growth and employment on one hand, and the development of these units on the other.

Iraq has taken a number of steps in this direction, including the streamlining of licensing and registration procedures for these projects; reforming the existing taxation system and enacting new laws reducing the maximum income tax rate to 15 per cent; and encouraging commercial banks, particularly state-owned ones, to expand their loan and credit facilities offered to such projects.

Moreover, Iraqi authorities concerned have approached the International Finance Corporation to provide these units with loans and financial aid. An amount of around US$ 18m is expected to be offered in support of Iraq's private sector.

Setting Up an Advanced Taxation System

Iraq has issued a new, transparent and streamlined taxation law that is expected to give strong, long-term momentum to the investment sector, thus creating additional job opportunities. From January 1, 2004, the income tax imposed on corporations and individuals has been fixed at a maximum of 15 per cent.

The new tax system aims at expanding the taxation base to include larger numbers of companies and individuals and limiting tax exemptions in order to ensure the continuity of reduced tax rates and thus maintain revenue stability.

The new law is expected to achieve several benefits including smoother implementation, lower collection costs and fewer cases of tax evasion, which would be much more if tax rates were higher. In 2004, substantial progress is expected to be accomplished in the fields of tax administration, estimation and collection techniques, and automation of taxation procedures. All these measures will reflect positively on the relationship between enforcement departments and tax payers, lending transparency to the work of the taxation authorities and curbing administrative corruption.

Issuing New Currency and Conducting Replacement Measures

On October 15, 2003, Iraq issued a new currency, which will be a cornerstone of the country's new financial/monetary system. It put an end to the existing dual-currency system—the use of the so-called

"Saddam dinar" in Baghdad and southern areas and the so-called "Swiss dinar" in the northern autonomous region.

The adoption of the new currency will certainly ensure the usage of a unified currency throughout the country; more protection against forgery; greater confidence by citizens; and higher consumption rates. The process of currency replacement was more successful than anticipated. A large number of bank branches were assigned solely for the currency replacement process and supplied with sufficient quantities of the new currency to ensure speedy transactions. Iraqi citizens were encouraged to open bank accounts in which old currency could be deposited both before and after replacement.

In order to increase the money supply in circulation, from October 15, 2004, public expenditures and salaries are being paid in Iraqi dinars instead of US dollars. The Central Bank of Iraq's auction sales of US dollars had a significant effect in improving the exchange rate of the Iraqi dinar against the US dollar, and in boosting the confidence of the people in their new currency.

International Assistance

Donor States Conference

In June 2003, UN and World Bank experts, in cooperation with their Iraqi counterparts, prepared a draft report on the estimated requirements of Iraq's reconstruction process. The report was discussed in the September 2003 joint meetings of the International Monetary Fund and World Bank held in Dubai and was later submitted to the Core Group meeting held in Madrid in October 2003 where it was agreed that a mechanism is to be examined and set up to deal with the UN/World Bank-run donor states fund(s).

The report was intended to advise donor states, during their October 2003 conference convened in the Spanish capital, of the projects and finance needed for reconstructing Iraq. Projects to be set up in 14 sectors at a cost of more or less US$ 35 billion were covered by the report. The Coalition Provisional Authority (CPA) submitted estimates for the requirements of 8 other sectors (including security and oil) at a total cost of around US$ 20 billion.

The Madrid Conference was attended by Kofi Annan the UN Secretary-General, Colin Powell the US Secretary of State and John Snow the US Treasury Secretary, along with a large number of donor-states' representatives. Iraq was represented by members of the Governing Council, ministers and experts, in addition to a number of advisors representing the CPA and the Coordination International Commission (CIC).

The Conference constituted a tremendous demonstration of moral and financial support for the new Iraq. It was the first time ever that such a large number of states had convened to express backing for a country that had been brutally oppressed by one of the most tyrannical dictatorships ever known to humanity. These states offered over US$ 23 billion in the form of financial support for the new Iraq.

Dealing with Foreign Debts

Under UN Security Council Resolution 1483, claims by Iraq's creditor states were postponed until late 2004. During this period, Iraq – in cooperation with many states and international institutions – has been exerting considerable effort to persuade its creditors to abolish or reduce their debts and to reschedule the remainder under more lenient conditions.

As a result of these efforts, coupled with frequent visits paid by the Chairman of Iraq's Governing Council and the former US Secretary of State James Baker to several capitals of creditor states, a number of these states have expressed their willingness to cooperate with Iraq by discussing the possibilities of abolishing or substantially reducing their debts.

Iraq is hoping that its foreign debts will be cut down to levels that the national economy can tolerate and service, thus enabling it to attain the high sustainable rates of economic growth desired.

At the same time, Iraq is gathering debt-related information and documents in order to verify the authenticity and the exact amount of its debts, to guard against any exaggerated debt claims by creditors.

Technical and Technological Aid

Due to isolation from the outside world during the last two decades of past century, and because of the three unjustified wars into which the former regime thrust the country, the majority of graduates from colleges and higher institutions of learning and other segments of the workforce were compelled to join the armed forces and fight in these wars. This has resulted in the gradual loss of their skills and qualifications, creating a big gap that has to be bridged through programs to rehabilitate and train civil servants and other workers with a view to improving their abilities and performance.

During the same period, training centers were incapable of fulfilling their functions due to the lack of training tools and equipment, the destruction suffered due to the March 2003 military operations and the rioting, looting and burning that followed. Therefore, Iraq has sought technical and technological aid from different countries, especially donor states. Several offers and promises have been made to organize training courses to rehabilitate Iraqi cadres and to provide some training tools and equipment, raising hopes that more training opportunities may be provided in the near future.

15

Post-Conflict Iraq and the Gulf: Ambivalent Effects of Change

Faleh A. Jabar

The US-UK led invasion of Iraq in 2003 will probably go down in history as a turning point in the destinies of the Middle East in general, and of Iraq and the Gulf in particular. The destruction of the old state power structures in Iraq has opened up the way for a radical restructuring of its political order, with far-reaching domestic consequences that will reformulate the state both as a system of governance and as a nation-state. This transformation will temporarily disturb the social and ethnic structures, and change the economic, political and cultural environment under which the social classes, ethnic and religious groups and forces have thus far acted and reacted. It will also change the geopolitical frameworks in the region.

The fallout of this domestic change will have deep consequences for Iraq's neighbors such as Iran, Turkey and Syria, but the strongest impact will be felt by the Gulf Cooperation Council (GCC) countries. The effects will have different outcomes for different players who are keeping a watchful eye on the current developments in Iraq. One of the greatest difficulties facing transition stems from the fact that the end of the Ba'ath patrimonial-totalitarian regime was not the outcome of home-grown, gradual change, but the product of an abrupt turn of US global strategy. Evolving as the "Bush Doctrine" in the aftermath of the tragic events of September 11, 2001 (9/11), this strategy is bent, among other things, on transforming the wider Middle East.

The basic hypothesis in this chapter is that the twin effects of the Bush doctrine, and the new reforms in Iraq that it envisages, will end up, by intent or otherwise, in the creation of a state weak in military and economic terms, but with strong political and cultural influence in the region. The challenge that Iraq will offer to the Gulf in general, and Saudi Arabia in particular, will shift from defense and security to the politics of change and regional alliances. It is also contended that while a reformed Iraq may offer neighboring Gulf states that are bent on reform a more friendly environment and wider opportunities for cross-border ties and economic cooperation, the same process will put additional pressure on hindered or slow reform in Saudi Arabia. However, as Iraq is not a closed box, its politics and society will be affected, in turn, by what regional players, the Gulf states in particular, would be willing to do or offer, given that Iraq is in a dire economic predicament, and that there is no shortage of Iraqi protégés looking for support and patronage across borders.

The focus of this chapter is, of course, to examine the multi-layered restructuring processes inside Iraq, and the uncertainties involved in these processes, in order to assess to what extent and in which ways the new Iraq may constitute a potential example (positive or negative) that can enhance or inhibit the drive for reform in the region.

Needless to say, reform is a frictional tool of change no matter how necessary it may be; hence policy makers hesitate long before embarking upon this course. Whatever the nature of change in Iraq may be, it will increase domestic, regional and global pressures on policy and decision makers in the Gulf countries to become part of a benign atmosphere in the region. The political, economic and cultural impact of Iraq's transformation will be tremendous and, to some extent, unpredictable.

The destinies and histories of Iraq and the other Gulf states, particularly Saudi Arabia, have been intertwined for millennia, and they will continue to be so in the future.

For the sake of clarity, the paper is divided into four major, interconnected sections:

- The US new global strategy and its consequences for Iraq, Saudi Arabia and the Gulf in general.
- The ongoing restructuring processes that the Coalition Provisional Authority has undertaken to reshape Iraq.
- An outline of Iraq-Gulf legacy of animosity and mistrust throughout the twentieth century.
- The regional fallout of the three above-mentioned factors on the Gulf region.

New Problems, New Strategy: Bush Doctrine and Removal Campaign

Launching Operation Iraqi Freedom signaled a new phase in US strategy vis-à-vis the world, the Gulf region and Iraq. Globally, this was a shift from mutilateralism to unilateralism in global politics, embedded in the concept of preemption, a new and aggressive means of projecting US military might and readiness, in the post-Cold War era, to defend its interests, allies and friends. Under the administration of President George W. Bush, unilateralist preemption is specifically geared to combat global terrorism, pariah (or "rogue") states known as members of the "axis of evil," and prevent the proliferation of weapons of mass destruction (WMD).[1] These components have been almost unanimously recognized as constituting the core of the new strategy. Other elements in this global strategy have been the subject of controversial assessment, and need to be discussed.

The 22-page national security document of the new US administration[2] tempted a purely military interpretation. A more nuanced reading,[3] however, "suggests a more complex and even conflicted agenda" that goes beyond emphases on military power, military superiority, unilateralism or preemption.[4]

Preemptive, unilateral war on global terrorism is incorporated into a set of value-system objectives, such as promotion of global economic growth, enhancing market economies and building the infrastructure for democracy. Preemptive, unilateralist power is thus wedded with "democratic idealism."[5] The shift away from the previous standard of national interests-based strategy to one deeply delved into social, political and economic engineering was encouraged by the end of the Cold War and the emergence of the US as the mega-superpower, the advance of globalization, and the precedence of the humanitarian interventions in the 1990s in countries like Haiti (1994) Bosnia (1995) and Kosovo (1999). In these instances, limits were set on state sovereignty; states could no longer have the exclusive rights to sovereignty over their land and people, if they were found in breach of certain universal legal norms and standards.[6]

In the aftermath of the September 11 attacks, the US inclination to subsume national sovereignty under additional scrutiny was finalized. In other words, the evolving "Bush Doctrine" and the war on terror imply far-reaching consequences as far as the sovereignty of nation-states in the Middle East or the developing world is concerned. If need be, this sovereignty would be curtailed or even neutralized. In the opinion of Toby Dodge, economic globalization has already reduced the autonomy of governments, now the Bush Doctrine is targeting political autonomy. The logical outcome is that pressures would increase on targeted governments to comply with certain rules such as the suppression of all terrorist activity on their territories, the transparency of banking and trade arrangements, and the disavowal of WMDs.[7] Certain political reforms, it may be added, would also be imposed on the agenda of authoritarian or primordial, autocratic governments, such as the establishment of the rule of law, expanding or creating participatory mechanisms, enfranchisement of women and the like. The US strategic framework had overt implications for Iraq, but covert implications for the Gulf and beyond.

Viewed from US strategic perspectives, Iraq symbolized the marriage of authoritarian rule, WMDs and possible links (right or wrong) with Islamist violent groups. The Bush administration saw America's former containment strategy of Iraq as a total failure. Under Clinton's administration, this policy functioned on three main levels: first, sanctions, which impacted enormously on the civilian population, a burden only somewhat alleviated when the Oil-for-Food program was agreed upon in 1996; second, a humiliating and enforced disarmament, supervised by the UN special committee – the United Nations Special Commission for Iraq (UNSCOM); and third, reduced sovereignty (with the imposition of 'no-fly' zones in northern and southern Iraq). In retrospect, Iraq fell victim to its own successful survival through the loopholes it exploited in the sanctions regime. The Ba'ath regime's survivability in the 1990s prompted neo-liberal, hawkish voices to demand regime change long before the 9/11 attacks.[8] Nevertheless, it was only after 9/11 that such calls were to become part of the new strategic planning.

Shortly after the collapse of the World Trade Center in New York, the push for regime change in Afghanistan and Iraq took precedence. The swift military success scored by the US in Afghanistan provided the first proven test case in removing the fundamentalist menace, and boosted the propensity to further experiments in regime change.[9] How then did Iraq and the Gulf fit in this new strategy? The answer is to be found in the US interpretation of the 9/11 attacks. These attacks have been the subject of several interpretations in the West and elsewhere, and will continue to be debated and analyzed for a long time to come.

The rise in religion-embedded violence has been phenomenal since the 1980s. Clandestinely organized political violence targeting civilians – otherwise known as "terrorism" in the political jargon – has been justified equally by secular and religious ideologies. Almost all religions contributed, one way or another, to this steep increase in ideologically motivated non-institutional violence, but the notoriety fell on one small group, Al Qaeda, and one man, Osama bin Laden. This was a paranoid overstatement of reality. In 1980, the US scarcely registered one single religious organization involved in terrorism; in 1990, 30 terrorist groups were registered, half of which were religious. The proportion of religious terrorist groups increased from 16 out of 49 in 1994, to 26 out of 56 in 1995.

In that sense Osama bin Laden and his Al Qaeda, or other Islamic terrorist groups, were not loners.[10]

The fact, however, that Al Qaeda directed its strikes at the heart of the US economic life – the twin towers – and symbols of power – the actual attack on the Pentagon, and the aborted attempt on the White House – shook the US tremendously, public and administration alike. Most interpretations of 9/11 relate to the Middle East, Islam, radical splinter-Wahhabism and Saudi Arabia.[11] The fact that a majority of the perpetrators, 15 out of 19, were Saudi nationals and were ideologically motivated by militant strands of Islamism of the same Salafi hue, Wahhabism included, was deciphered as an accusation of the country of origin and its creed.[12] Wahhabism and other conservative strands of Islam had been ideological allies of the US during the Cold War. They were deployed against the pan-Arabist, radical Nasser of Egypt in the 1950s and 1960s; against the Soviet Union in the 1970s; and against communist-led Afghanistan in the 1980s. Following the 1991 Gulf War, the 'ally' sustained schisms, and splinter fragments turned against its originator, Saudi Arabia, and against an old patron, the US. As a result, US-Saudi relations were 'poisoned' by mistrust and Saudi Arabia was viewed as a haven of violence, needing to be reformed and/or replaced.[13]

In this strategic and regional context, Iraq fits in a dual manner. First, its destruction as a hostile "pariah" government and a threat, in the sense defined by the Bush Doctrine, would be a showcase for other pariah states. Second, if reconstructed along liberal-secular lines, Iraq might serve as a beacon of reform in the region. From the American standpoint, then, the new Iraq would constitute an antithesis to a conservative Saudi Arabia affected by fundamentalism:

> "The Saudis…were increasingly perceived by some Republicans as unreliable allies, whose style of governance and subordination to conservative clerics was bound to bring them into conflict with America's regional agenda. A liberated Iraq seemed the ideal substitute for the sour partnership with Riyadh."[14]

For Iraq to serve such grand designs, three preconditions must be achieved in the restructuring of this country: viable democracy, secularism and market economy.

New Iraq: Uncertainties of Transition

The speed and ease with which the seemingly monolithic totalitarian Ba'ath system collapsed one month into the invasion nurtured great expectations in Washington that nation rebuilding in Iraq would be as smooth and swift as the military conquest. The neo-conservatives in the Pentagon, who embrace a maximalist approach of full-fledged liberalization of Iraq under US tutelage,[15] succeeded in taking full charge of Iraq's transformation. The less radical and more cautious State Department, who advocated moderate reforms in Iraq, was sidelined. Institutional tribalism in the US administration is notorious. Rivalries between the Pentagon and the State Department have and will continue to affect the future of Iraq. If the Pentagon maximalist designs of full-democratization fail, the State Department might step in with its minimalist, strong-authoritarian government. Indeed these are the two main conflicting scenarios available—the maximalist option is now in operation, the minimalist is dormant. We will, of course, focus on the maximalist and how it is changing Iraq.

When the Pentagon was authorized by National Security Presidential Directive No. 24 (January 20, 2002) to take control of post-war Iraq, it set off on a new trajectory with vague conceptions of how Iraq should and could be transformed. One crucial outcome of this victory scored by the Defense Department is that the studies prepared by the eighteen workshops of the "Future of Iraq Project"[16] under the auspices of the State Department, were cast aside. The State Department staff have substantial knowledge of Iraq and the Middle East, but little enthusiasm or courage to embark on radical reform; the Pentagon staff, by contrast, lack genuine knowledge of the region but have the will to initiate radical change. The paradox was apparent as soon as General Jay Garner, and later, Administrator Paul Bremer would set foot on Iraqi soil.

Shortly before the invasion, the Office of Reconstruction and Humanitarian Assistance (ORHA) was formed under General Jay Garner to provide humanitarian assistance and civil administration. According to Garner's team, the political options for civil administration were not clearly defined.[17] The US had several options in Iraq: direct military rule, coalition civil administration,

Iraqi civil administration, or Iraqi interim government. Garner seemed to negotiate a transition from option one to option four.[18] Following UNSC Resolution 1483 of May 22, 2003, the Coalition Provisional Authority (CPA) replaced ORHA on June 1, 2003. In June 2003, Administrator Paul Bremer reversed the political process from option four to options one and two to take full control of the situation. Barely six months later, Administrator Bremer had to change course again and opt for transfer of sovereignty to be effective on July 1, 2004. The reason behind these backward-forward moves was a matter of failure. The rise of Shi'a militancy was aggravated by anti-coalition violence and the collapse of essential services—the situation seemed to go out of control.

What went wrong was a multitude of factors, including the lack of proper planning, blunders and misconceptions. A major miscalculation in this regard is an underestimation of Iraqi nationalism, and of security or the lack of it. No proper planning had been made for the deployment of international constabulary forces, or to quickly install an Iraqi interim government to control the situation. Lawlessness, criminal violence and political violence shook the daily lives of Iraqis and overshadowed the transitional phase from day one and the situation deteriorated even further during the summer and autumn months. These developments made uneasy reading in the White House in a sensitive and delicate election year.

Thus, the processes of total liberalization initiated by the CPA proceeded under unfavorable conditions to dismantle and replace the old power structures, liberalize the economy and manage a plethora of social and political actors.

Old State Agencies "Melting Down"

Citing the US experience in post-war Germany and Japan, drastic measures were taken by the CPA to dissolve five crucial state agencies: the ministry of defense, the ministry of the interior, the presidential guard and special security formations, the ministry of information, and the Ba'ath party organs, with no proper arrangements to vet the staff of these organizations, prevent the proliferation of weapons, and provide for the livelihood of that section of the professional staff that had no criminal record.[19] The irony is that party and state functionaries, including the agencies of state-violence, had already vanished. Part of the staff was ready to cooperate; another part was planning to wage clandestine, guerrilla warfare.[20]

Disempowering the Ba'ath party is legitimate and necessary for any meaningful democratization. However, the methods used to achieve this were ill-conceived, alienating much of the talented and rich technocratic class who numbered around 250,000.[21] Dismantling the army and security formations was problematic—it mobilized ex-servicemen,[22] and worsened the security situation.[23]

Almost the entire arsenal of weaponry and ammunition fell into the hands of defeated regime loyalists. Iraqi dailies criticized the coalition authorities on this score, saying that "the livelihood of these people has been cut off and they have no other means of making a living."[24] The CPA did some self-correction. A total of 200,000 Iraqi military personnel, up to the rank of colonel, have been put on the CPA payroll.[25]

The CPA controls security and defense tasks, including selection, training and financing of the new Iraqi army. Plans have been laid to train and equip 40,000 soldiers by the end of this year, and 120,000 by the end of 2005. The Bush administration earmarked $2.1 billion for the New Iraqi Army (NIA), and a similar amount for other security agencies.[26] In addition, 40,000 policemen of the old police force have been re-commissioned, as well as 5,000 border police, 6,600 Iraqi Civil Defense Corps and 20,000 officers in the Facilities Protection Service. Security improved, but political violence increased [27]

The bottom line is that the central government will not be strong enough to impose its will and will have to accept compromises, power-sharing and a certain degree of chaos. On its own, Iraq will remain militarily fragile, compared to its strong neighbors, Iran or Turkey, and will be unable to pose any threat to its neighbors in the foreseeable future. This situation may encourage regional security cooperation, for example, between Iraq and the GCC. It may also encourage strong military cooperation with the US to enhance national defense. The final status of the coalition forces is under negotiation and the outcome will determine the future of the US military presence in Iraq.

Native Governance

Facing power vacuum and bitterness among the public at large, the CPA proceeded to create domestic governance structures, such as municipal councils in the provinces and major cities, reform and run governmental departments, and commence negotiation to form an Iraqi interim governing council. Creating municipal councils was not an easy task, nor was the selection of provinces' governors. In many cases, the public rejected ex-Ba'ath officials being re-instated, and wished to install domestic figures. Indeed, several provinces (*mahafazat*) changed hands more than once. Municipal councils have long been part of Iraqi political tradition, but they were previously tightly controlled by the central authority. Now, they have been set free from central fetters but lacking resources and manpower, they have little means to govern. Nevertheless, they may contribute to spreading democracy from below once nationwide normalcy is reached.

On the other hand, the CPA took direct control of the departments of government. Success was scored in rehabilitating the educational system and reforming the justice department; but the CPA failed to quickly rehabilitate essential services or maintain security, much to the resentment of the public. The CPA was also now willing to share power with Iraqis in a controlled and calculated manner, in the face of resurging Iraqi nationalism and the failure of the CPA to provide security and essential services.

Viewing Iraqi society as neatly divided into Shi'as, Sunnis, Kurds, Turkomen and Assyrian, the CPA determined the selection of the Governing Council (GC) along these ethnic-sectarian and religious lines, with some balance sought between the exile and native elements. The GC is composed as follows: 13 members (52%) are Shi'a, 11 members (44%) are Sunnis, and one is Christian (4%). Along ethnic lines, Kurds, Turkmen and Assyrians got respectively: five seats (20%), one seat (4%), and one seat (4%).

Establishing inclusive and participatory mechanisms on the basis of these criteria will end for ever the monopoly of power by thin and narrow elites, mainly drawn from provincial localities and their tribal networks. While participation has now been expanded, the method of quota-distribution of seats implies the danger of entrenching such communal-ethno-religious norms in a Lebanese type of political-institutional communalism. Concern over such an eventuality has been voiced by the burgeoning political class in Iraq, as well as by regional and external observers and political elites.[28]

Another alarming feature in the GC is a precarious balance between the Liberal-leftist bloc and the Islamic camp. The Liberal-leftist has seven members (3 Sunnis, 3 Shi'a, 1 Assyrian). The Islamic bloc is large, comprising eight members who are segmented in ethnic and sectarian terms. The Shi'a Islamic bloc, the most radical by ideological standards, is predominant but will be more or less isolated from other groups. Secularism and non-sectarian politics do not seem to have promising perspectives. This ideological-political dividing line was clearly seen during the debate on the constitution, and the question of early elections.[29]

Like Sunnis and Kurds, Shi'as do not constitute a homogenous social and political space. A multitude of social, political and clerical forces compete in this space, which involves radical and moderate currents.

Market Economy in Question

Prospects of democratization in Iraq will be equally determined by sound mechanisms of nation-building, enhancement of the rule of law and the creation of a market economy. In modern polities, no democracy has ever developed without a market economy. Liberalization of polity, as envisaged by the US-led coalition, requires the creation of a market, or freeing it from the hegemony of the state. The Ba'ath regime's political economy rested on oil rent and a command economy. Private capital was patronized and controlled by the state. Such weak crony capitalism is hardly a starter in any democratic project. The CPA and its junior partner, the GC, embarked on deregulation and economic liberalization reform, that fell short of radical privatization. Oil rentierism, on the other hand, has been left intact.[30]

Generally speaking, economic liberalization involves privatization, deregulation of the financial sector and of prices, trade liberalization, incentives for foreign investments, labor market reform and social security, among other things. Given the political difficulties facing the constitutional and transitional processes, the CPA steered clear of privatization and confined its plans to institutional reform. Several decrees were issued to open up the market to foreign investment, allowing foreign capital 100 percent ownership in every sector other than natural resources. The private banking system also received a boost, while government-owned industrial complexes, already obsolete and outdated, have yet to be privatized.

The future of the oil industry is not yet known. National ownership may continue. Privatization seems out of question. Other solutions may be considered. Following the handover of sovereignty on July 1, 2004, these issues may be raised again. Much of the future of the democratic development of the country will depend on how the problem of oil rentierism is addressed. Although oil may be an instrument of authoritarian rule, of state patronage and "clientele capitalism," it is also an instrument of development. Even the liberal CPA has to rely on it for running the state machinery and has to expand production, which is still very low and barely meets Iraq's financial requirements.[31]

Economic liberalization might also be hindered by the weakness of the Iraqi entrepreneurial classes, including some 24,000 businessmen,[32] who rely heavily on government contracts which they used to acquire through kinship and party networks. Rehabilitating the upper and middle business classes seems to be imperative to the successful operation of any market economy.[33] Iraq's notorious indebtedness and the staggering cost of reconstruction are another impediment to prosperity, stability and, by extension, democratic perspectives. Debts stand at a staggering $170 billion, compared to Iraq's projected annual oil revenue of a meager $6 billion.[34] Furthermore, the cost of reconstruction is estimated at $150–$170 billion.

Democracy requires a market economy, and a market economy is required to yield prosperity. Prosperity is the only leverage that generates consent simply because democracy rests on consent but does not generate it. It is likely that privatization will continue under the forthcoming Iraqi government scheduled to be formed on July 1, 2004. No apparent political force envisages economic hegemony by the state, and the market economy may receive a radical boost.

New Forces, New Polarization

The self-deceiving myth that the Ba'ath regime had built a monolithic, homogenized Iraqi society ended abruptly in April 2003. A plethora of social, institutional, economic and cultural forces emerged in full force to compete with already active political and social forces in the diaspora or at home. More than 140 organizations and associations based on culture, ethnicity, gender, education, human rights, and countless issues of national or local significance, have emerged. Ending state monopoly of the public space, at least 170 daily, weekly, bi-weekly and bi-monthly periodicals have come into existence.

The most striking feature is the recurrence of all politico-ideological trends that existed under successive regimes from the inception of the monarchy in 1921 to the present, such as liberal monarchists, old liberal outfits, old parties embedded in the politics of notable Baghdadi houses, the Qassimites, or the Nassirites, expanding the already existing ideological spectrum of Pan-Arabist, Kurdish Nationalists, Marxists and Islamists. Another feature is the resurgence of ethnicity-based politics beyond the Kurdish movements (the Turkomen and the Assyrians). The rise of new, modern, issue-oriented civil society groups and associations is another characteristic. All-inclusive pan-Iraq tribal associations and leagues also came into being, reflecting the strength of the tribal factor that staged a comeback in the 1990s.

However, most important of all are faith-based movements and institutions, such as the informal institution of Shi'ite marja'ism (a Shi'a notion to name the supreme religious authority), centered around Grand Ayatollah Ali Al-Sistani, or the *Hawza* movement (an idiom denoting the supreme religious authority, including the institutions of religious learning) led by the young cleric, Muqtada Al-Sadr, on the Shi'a side, and various *Salafi* societies and associations of *ulama* (a Sunni term for doctors of religion), on the

Sunni side. Unlike the previous forces, these faith-based forces are segmentary and particularistic. One way or another, all these forces are engaged in peaceful negotiations with the CPA to retrieve national sovereignty and prevent the return of the old Ba'ath regime.

In the opposition are clandestine armed groups emerging from several sources, such as the remnants of the defeated regime's party and institutions of violence, from loyal clans, or from some native or imported *Salafi* organizations. The polarization that divides these forces has shifted from supporting or opposing the Ba'ath regime, to supporting, engaging, opposing, or fighting the Coalition. The majority is bent on peaceful, institutional or extra-institutional politics. Fringe currents are waging armed resistance. The mainstream, peaceful trend is a conflictual space in which secularism competes with Islamism, centralism with federalism, liberalism with statism.

As has been observed earlier, thus far, Islamists on both sides of the sectarian divide are the strongest. Most Shi'a Islamists steer clear of an Iranian-model of "guardianship of the jurist," but envisage a strong role for the clergy and some form of Islamization of polity and society. Islamists seem to have vast resources, well-disciplined militant manpower and religious-communal symbolic capital. Secular liberalism is not wanting, but is fragmented. Most probably, Iraq will emerge as a moderate liberal-Islamic polity. The contours of such a polity will take shape gradually. The constitutional process is unfolding slowly and has yielded a 100-member Iraqi National Council, set up in August 2004. The socio-political composition of the forthcoming National Assembly due to be elected by January 2005 is still an open question. The direction of this change will be crucial in shaping Iraq's political system and the future of US–Iraq relations. These outcomes, in turn, will impact on the legacy of Iraq–Gulf relations. To assess these effects, we shall now turn to the Iraq–Gulf legacy.

Iraq-Gulf Relations: The Legacy of the 20th Century

For the most part of the twentieth century, Iraq's relations with the Gulf (inclusive of Saudi Arabia) have been characterized by mistrust, animosity and antagonism save a short-lived phase of rapprochement during the eight years of the Iraq-Iran war. Much of this legacy stemmed from the transition of the Ottoman Empire to the new reality of modern nation-states. Iraq emerged under British mandate as a sovereign state in 1921. Following decades of successful military campaigns, Prince Abdul Aziz Bin Saud, founded the newly unified Kingdom of Saudi Arabia in 1933.[35] The Gulf emirates, then under British control, joined the community of nation-states in the second half of the twentieth century. Prior to the modern era, the whole region, more or less, shared a cultural space united by common religion, language, tradition, tribal organization and trade ties. Their political development as nation-states, however, drove them worlds apart. Political systems diverged, political cultures clashed, territorial limits were disputed and old animosities revived.[36]

Under British mandate, for example, a liberal, constitutional monarchy developed in Iraq in sharp contrast to the patrimonial political system of the Saudi dynasty, whose lands had never come under any colonial control. A sharp contrast also built up relative to the traditional Gulf emirates, where primordial, traditional polities were not disturbed by a century or so of British military control. Accordingly, the political institutions and discourses in these three parts (Iraq, Saudi Arabia, and Gulf emirates) set off on different trajectories. Like the other Arab countries in the Levant, Iraq would embrace modern ideologies and modern forms of mass mobilization, in the form of Arab nationalism, socialism and Marxism. Moreover, these ideologies would make inroads into Saudi Arabia and certain parts of the Gulf, causing a considerable degree of instability.[37]

Primordial polities usually invoke tradition and rely on kinship networks. Modern polities, by contrast, are embedded in the concept of nationalism and citizenship. The novelty of the system of nation-states into which Iraq, Saudi Arabia, and later, the rest of the Gulf polities, were integrated, was that it required, for its functioning, the full recognition of the state as the exclusive owner of the national space, or what is known as the Westphalia Peace principle. Much of the strain and antagonism between the political entities of the region stemmed from the difficulty encountered in bringing this Westphalian principle into effect. In other words, the thorniest legacy of the transition to the post-Ottoman era was the recognition of the borders or even the very existence of local "nation-state."

Questioning the legitimacy of the political entities in the region triggered border disputes, and border disputes questioned the legitimacy of the new entities. These clashes linger today. Iraq's borders were disputed by most of its neighbors; Iraq in turn challenged the borders of some of its small neighbors. The Kuwait-Iraq border dispute, and the Iraq-Saudi neutral zone that existed, for more than six decades on the map, are cases in point. Latent irredentism, fueled by nationalism (pan-Arabist or, in this specific case, Iraqi nationalism) surfaced on several occasions, under the monarchy (1921–1958), during the republican era (General Abdul-Karim Qassim 1958–1963), and under the Ba'ath regime (1968–2003) leading to several regional crises in the 1930s, 1960s and a war in 1991 with destructive results.[38]

Another legacy of this transition was the cultural-religious rift between Iraq and the rest. The Hanafi-Shafi'i Hashimite kingdom of Iraq with its secular political system contrasted sharply with the Salafi interpretation of Islam in Saudi Arabia, known among its adherents as "*muwahidoon*," that is, Unitarians, and among its rivals as Wahhabism. Iraq's rapidly urbanizing society also differed greatly from the traditional tribal predominance to the south. Even under the Ottomans, Iraq was a cultural buffer zone vis-à-vis the ambitious Saudi plans to expand northward. The Ottomans were keen on compromising and discrediting the Saudi school of puritanical Islam.[39]

Alongside the Hanafi Ottomans, the Shi'a of Ottoman Iraq had an added interest in opposing Wahhabism. The shrine cities of Iraq, vulnerable to Saudi raids (Karbala was attacked and plundered in 1802), converted the tribes of the south to Shi'ism, throughout the 19th century, creating thereby another cultural barrier.[40] Both Hashimite and Shi'a inheritance overshadowed the thinking of political elites, Shi'a and Sunni, old and new. In part, this legacy, *mutatis mutandis*, lingers on to the present day. The resurgence of forceful Shi'a groups in post-conflict Iraq has evoked this history in symbolic ways that may taint politics in the years to come.

Throughout the first half of the twentieth century, the competition between the Shafi'i Hijazi Hashimite monarchy and the Najdi, Hanbalite Saudi dynasty, was not only a case of two houses in pursuit of power in a zero-sum-game, but also a case of two nations in pursuit of regional influence, a competition that has become part of the geopolitical contention between the two countries. As republican Iraq tilted in favor of the Soviet camp during the last decades of the Cold War, the Iraq–Saudi competition intensified.[41]

Contrasting ideologies were another factor. Arab nationalism, socialism and Marxism in the wider Arab world, or indeed in monarchic, republican or Ba'ath Iraq, which professed "progress" and modernization, have been viewed as perilous and blasphemous in the non-ideological Saudi and Gulf environment. The limited proliferation of such ideologies in the fifties and sixties was a source of security concern in Arabia. During this period, the Gulf region was almost a net importer of modern ideologies from the rest of the Arab world. Tensions reached the level of war fought directly or through proxies in Yemen in the 1960s.[42] In the eyes of Arab radicals, Saudi Arabia was the bulwark of pro-Western conservatism in the region; and the activities of the Muslim Brotherhoods against radical Arab regimes, which were largely native, were conceived as part of the Saudi foreign policy onslaught. With the rise of militant, fundamentalist Islam in the 1970s to the foreground, reciprocal ideological influences became the rule.

Iraq–Gulf rivalries were also accentuated by oil competition. As from the 1960s, Iraq was much inclined to assert some form of radical oil nationalism, which conflicted with the moderate and accommodating oil production (including embargos) and pricing policies pursued by its neighbors to the south, who acted as swing producers to moderate the oil markets. The nationalization of oil companies in Iraq in 1972 heightened "oil nationalism," much to the embarrassment of its moderate neighbors.[43]

In foreign policy, republican and Ba'ath Iraq steered a course hostile to the US, the major ally of the Saudi Kingdom. Iraq also pursued radical policies towards Israel and the Palestinian question, an issue that added to the differences between the two parts of the region in terms of means rather than ends. In political, social and cultural terms, Iraq was the anti-thesis of the rest of the Gulf. The cooperation between the adversaries was confined to the eight years of the Iraq-Iran war. This short-lived marriage of convenience was an effort to stem the tide of radical Shi'ism in Iran, which posed a national security threat to Iraq and its neighbors. This cooperation was the exception not the rule. Needless to say, without covert and overt support from the West and the Gulf, Iraq would not have been capable of sustaining its war effort as long as it did, nor impose a ceasefire on Iran in July 1988.

Iraq emerged from the first Gulf War a military giant, but an economic dwarf, and thus the protector of the Gulf turned into its persecutor. The invasion of Kuwait was, in retrospect, the beginning of the end for Ba'ath Iraq.[44] Two devastating wars and thirteen years of debilitating sanctions have reduced Iraq to an impotent nation for some time to come. Perhaps this explains the genuine concern in the Gulf over the fallout from the US campaign to remove the Ba'ath regime. The Saudi Foreign Minister, H.R.H. Prince Saud Al Faisal expressed the Gulf's concerns over the possible fragmentation of Iraq, a clear reference to the rise of communal and ethnic politics, but above all, to the unprecedented resurgence of the Shi'a factor in Iraqi politics.[45]

The menaces of the old regime are gone, but potential new hazards are looming, while latent benefits that may arise from the new regime are not yet in place. The question now is how the unfolding realities in Iraq would affect the Gulf, given the long legacy of mistrust.

The Regional Fallout

In view of the processes set off by the US-led coalition, the contours of the present and near-future Iraq are beginning to take shape. If the US policies for reshaping Iraq go ahead uninterrupted, Iraq may well emerge as a moderate Islamic state – a liberal, pluralistic polity, with a vibrant civil society – with a transformed market-based economy and possibly strong links to its patron, the US.

However, Iraq will be consumed by the tensions of change and transition. It will emerge as a very weak nation in military terms. Thus far, Iraq has only managed to train 700 soldiers, and plans to create an army of 40,000 by the end of 2004, an unrealistic target, given the slow pace of the process. Even if Iraq managed to reach the ultimate target of 120,000-men, this army will be weak relative to its neighbors: Iran (420,000), Syria (310,000) and Saudi Arabia (190,000). (See Table 15.1 on military factors).

Table 15.1
Military Factors (1999)

Country	*Size of Armed Forces (AF) (in thousands)*	*AF Per 1000 People*	*Military Expenditure (ME) (in million $)*	*ME/GNP (%)*	*ME Per Capita ($)*
Bahrain	9	14.5	415	8.1	666
Iran	460	7.1**	6,880	2.9	106
Iraq*	420	19.1	1,250	5.5	57
Jordan	102	21.1	725	9.2	150
Kuwait	21	11.0	2,690	7.7	1410
Lebanon	58	16.4	653	4.0	185
Oman	38	15.5	1,780	15.3	726
Qatar	12	16.7	1,060	10.0	1470
Saudi Arabia	190	8.9	21,200	14.9	996
Syria	310	19.5	4,4507	7.0	280
UAE	65	27.9	2,180**	4.1	935**
Yemen	69	4.1	374	6.1	22
USA	1,490	5.4	281,000	3.0	1030

*Pre-war figures. Now the Iraqi army is in the process of reorganization, and is planned to reach 40,000 men by the end of 2004.
** Estimates based on partial or uncertain data.

Source: *World Military Expenditure and Arms Trade,* 1999, released on February 6, 2003. (http://www.state.gov/t/vc/rls/rpt/wmeat/1999_2000/) accessed on January 7, 2004.

The only compensation Iraq will have for its weakness is the reduction in per capita military expenditure, which, by pre-war levels, was among the highest in the region. Freed from this burden, it may follow the Japanese or German model—more focus on civilian development.

Iraq will also remain weak in economic terms. With its per capita income sinking to $2,400, the lowest in the region, except for Yemen with $800 (See Table 15.2 on economic factors). Add the staggering indebtedness, which again is the highest in the region at $120 billion (this does not include Iraq's debts to the GCC countries), and the nation's ability to rehabilitate its economy and provide essential services will be severely limited. If oil *rentierism* is curtailed, and federalism established, the central government that will emerge will be less powerful than its predecessor. In other words, this central government will be more controlled by the new checks and balances that the liberal political system will offer. Hence, the central state that will emerge will have less freedom of action. Paradoxically, Iraq's main asset will be its liberalization experiment. This experiment will be conditioned by US pressures for full-fledged liberalization, domestic counter-pressures to create a strong state and regional pressures that seek to minimize the fallout from Iraq's transition.

In this new polity, both Sunni and Shi'a political Islam will be as prominent as ethnic and local political forces. Traditional and modern elements will interact, peacefully or otherwise, to define the post-occupation course of the country. Nation re-building will be slow, at times painful, to redefine the mechanism of power-sharing. As in every journey over uncharted land, uncertainties exist. Unpredictable and uncontrollable forces may disturb the steady flow of the transitional changes to a post-conflict, independent Iraq. Such changes and the uncertainties involved in them, have already given rise to great expectations and edgy apprehensions in the region.

Table 15.2
Economic Factors

Country	*GDP per capita (purchasing power parity 2002 est.)*	*External debt in billions of $ (2002 est.)*	*Oil production in barrels/day (2001 est.)*
Bahrain	$15,100	$3.7	43,000
Iran	$6,800	$8.7	3.804 million
Iraq	$2,400	$120.0*	2.452 million**
Jordan	$4,300	$8.2	40
Kuwait	$17,500	$10.4	2.117 million
Lebanon	$4,800	$9.3	0
Oman	$8,300	$5.7	963,800
Qatar	$20,100	$15.4	864,200
Saudi Arabia	$11,400	$25.9	8.711 million
Syria	$3,700	$22.0	522,700
UAE	$22,100	$18.5	2.566 million
Yemen	$800	$6.2	438,500
USA	$36,300	$862.0	8.054 million

* Current figures uncertain due to 2003 war and reconstruction.
** Production was disrupted as a result of the March-April 2003 war. Estimate of production capacity in August 2003 was just under 1 million barrels/day (http:// www.eia.doe.gov /emeu/cabs/iraq.html).

Democracy and its Aftermath

Although still in its infancy, political liberalization in Iraq has thus far set free civil society forces and institutions that will, if balanced and stabilized, have strong demonstration effects on societies in the GCC countries. Broad political participation, a universal ballot, parliamentary democracy, free press, transparency, women's rights, human rights and open, though contentious, debate would hardly pass by without a regional trace. Such a radical change would galvanize civil society in the region and heighten pressures and demands for more reforms, notably where reform has long been overdue. Reform-minded elites would be encouraged to step up their efforts in pressing for change, inasmuch as conservative elites and groups would intensify resistance against change, and the decisive moment to make crucial choices would arrive sooner rather than later. The GCC countries, then, have to take this change of political culture and reality into account with well considered reforms.

The Shi'a Factor

Introduction of all-inclusive participatory mechanisms in Iraq have created a politically and ideologically pluralistic Governing Council, cabinet and other new state agencies, reflecting the pluralistic character of Iraq's society. The deposed regime was monistic—one leader, a single party, one ideology. With the end of totalitarian monism, the Shi'a demographic majority translated into a majority in the governing bodies and will, of course, enjoy parliamentary majority when elections are held (See Table 15.3). Shi'as may command some 50-54 percent of the vote.[46] All other ethnicities and sects, taken individually, will be *de facto* minorities. Rule by consent and majority rule are the basics of democracy. Nevertheless, in multi-ethnic, multi-cultural polities, additional constitutional constraints are required to protect the rights of minorities and put checks and balances in place to curb the power of sheer superior numbers.

However, what is meant by "Shi'a majority?" In this paper as in previous works, we do not conceive of ethnic, religious or any other Iraqi communities as monolithic entities, imbued with a mono-dimensional sense of political direction and unity of purpose. Such monoliths exist only in oversimplified representations of Iraq's society. Among the Shi'a community, Islamist, liberal, leftist, nationalist and sectarian trends compete for power. Thus far, Islamist groups seem in better shape and may dominate the political scene in the Shi'a provinces for a period of time. Nevertheless, they are divided into several political parties (five thus far), and schisms split the clerical class in Najaf and several other centers of religious authority. Furthermore, the Sadr camp fiercely opposes the Al-Sistani and Hakim camps, and other local religious dignitaries in Kadhimiya and Karbala hold independent positions.

Another social cleavage separates Shi'a tribal chieftains, to some extent, from clerical and ideological co-religionists, placing the tribal factor, at least in certain districts, on its own and in contest against other divides.[47] That said, the Shi'a factor, in the form of Islamic parties and the informal religious authority, will have an essential role to play now and in the foreseeable future. Informal institutions of the marja'ism, the networks of *Husaynias* (places of worship intended for gatherings to commemorate the martyrdom of the third Imam, Hussein), religious visits to shrine cities (religious tourism), networks of religious *madrasa*, and other institutions, will constitute a regional and supra-regional civil society that will systematically spread these effects to the region. Not only will the GCC societies be on the receiving end, but Iran will also not be able to avoid the effects of such democratization in neighboring Iraq.

As the "Shi'a factor" is itself made up of diverse and conflicting groups and trends, the forces it represents will exert diverse influences. Islamist Shi'ism is only one among many. Ideologically, the rise of moderate Najaf under Al-Sistani will pose a counter weight to radical Shi'ism in Iran. Moderate Shi'a thinking as represented by Grand Ayatollah Ali Al-Sistani, stands in contradistinction to the radical doctrine of *"velayat e-faqih"* (guardianship of the jurist), applied in Iran. Najaf's moderating influence is already felt in Iran's Qum.[48] A similar result could come about in the Gulf countries, leading local Gulf Shi'as to shift from radical Khomeinism to moderate Sistanism, or encouraging the rise of influential moderate trends. Iraqi political Shi'ism would also enthuse and back up the inclination among Shi'a communities in the region to organize and mobilize.

In business, Iraqi Shi'a entrepreneurs could also expand their activities through communal associations, facilitating cross-border entrepreneurial rapprochement. The political consequences of such ideological, institutional and commercial links would increase local Shi'a motivation in the GCC countries but could moderate their methods. With their large Shi'a constituencies, Bahrain, Kuwait and Saudi Arabia will be the most affected by this increased activism. Demands for better representation, if not for wider claims or greater reforms, will be enlivened.[49] Such demands may take peaceful forms.

The Sunni Factor

The Sunni factor is being strongly agitated in Iraq and, perhaps, in the region as a whole. After half a century of uncontested supremacy, Arab Sunni elites find themselves in the minority. Theological and juristic differences between law schools (singular *math-hab*, plural *mathahib*) that mattered little before, have now been politicized by extreme Salafi groups. While "sectarian" discourses and motifs were instruments of Shi'a grievances in the past, at present it is Sunni Salafi and Islamic groups that are inclined to use such motifs. Fragments from the deposed regime are trying hard to ignite or heighten sectarian feelings. Some attacks were deliberately directed against Shi'a or Sunni mosques in Baghdad, Najaf, Ba'quba and beyond. Yet, no sectarian clashes materialized.

Foreign observers, who overestimate the communal (*ta'ifi*) factor, have poor understanding of the nature of Iraqi nationalism. In the past, sectarianism was a matter of government policies, but inter-communal relations were never strained. Now, they face such a potential threat. Much will depend on the ability of the political class, above all the liberal and moderate groups, to strengthen the sense of Iraqi nationalism and have it embedded in equality before the law—wholesale participation of all Iraqis. In ideological, political and social terms, Salafi agitation and organization in Iraq has been on the increase since 1991. The deposed regime had already encouraged communal polarization to sustain its principle of "divide and rule."

The GCC countries have expressed deep concern over the marginalization of the Sunnis, and some governments may well be tempted to support their co-religionists in Iraq, irrespective of the latter's discourse. This course could worsen rather than ease embryonic communal tensions. The best guarantee against communalization, or "Lebanonization" of Iraqi politics, is the liberal-moderate alternative. Indeed, the voting patterns in the Iraqi Governing Council show how far Sunni-Shi'a liberals, moderates and leftists are united against communal themes. The GCC countries can contribute positively to make (or negatively to break) a pan-Iraqi, moderate outcome.

Regrettably, the overwhelming representation of Shi'a political forces in the Iraqi governing council has been misread as being a communal rather than democratic feature, and triggered fierce reactions among militant Salafi activists in Saudi Arabia, who verbally threatened Saudi Shi'a with physical elimination. Similar sentiments can be detected in the Gulf. A sound consideration of the communal factor should be made.[50] All in all, both the Shi'a and Sunni factors will have both positive and negative cross-border influences.

Table 15.3
Demographic Factors

Country	*Population (July 2003 est.)*	*Muslim (Sunni/Shi'a) %*
Bahrain	667,238	30/70
Iran	68,278,826	10/89
Iraq	24,683,313	32-37/60-65
Jordan	5,460,265	92/<2
Kuwait	2,183,161	70/30
Lebanon	3,727,703	70%*
Oman	2,807,125	Ibahdi 75%
Qatar	817,052	95%*
Saudi Arabia	24,293,844	100%*
Syria	17,585,540	74/16
UAE	2,484,818	80/16
Yemen	19,349,881	*
USA	290,342,554	–

*Total Muslim percentage. No breakdown among sects given.
Source: *CIA World Factbook*, 2003 (http://www.cia.gov/cia/publications/factbook/index.html).

A New Political Elite

The old ruling political elite was drawn mostly from the provincial and lower middle strata of society and confined to a narrow regional-clan group, with little or no academic education, little knowledge of the world, strong anti-urban culture, cruel values and primordial cohesion. The members of the emerging political class, by contrast, hail from middle and upper urban classes, possess good knowledge of the world and high academic credentials. This class is also more diversified in ethnic and religious terms, possess civilized norms and culture, and a mixture of modern and traditional organization. The best way to assess the new political class is to engage it in a fruitful dialogue.

Military Factors and the US Link

As we have noted Iraq will emerge as a state weak in economic and military terms. Its planned armed forces will not exceed the 40,000 threshold by 2005, and 120,000 by the end of this decade. Its projected economic growth, its high indebtedness, the cost of its reconstruction and the low per capita income would cripple any attempt to remilitarize the country beyond modest margins. In other words, Iraq poses hardly any direct military threat to its neighbors, whether individual or collective, in the foreseeable future. A weakened Iraq, however, may disturb power relations in the region in favor of Iran, possibly stimulating an Iraqi-GCC alliance, if governments sympathetic to such rapprochement were formed in several capitals. Otherwise, Iran might well attempt to take advantage of this situation, and the US would respond as it has done in the past by enhancing direct military presence and cooperation with friendly governments.

Whether or not the future Iraqi state will have strong ties with the US remains to be seen. However, US military presence on Iraqi soil, or strong Iraq–US cooperation, would augment Iraq's strategic importance and diminish that of major players in the GCC.

Oil Swing and Oil Nationalism

With the second largest proven oil reserves in the world, Iraq's significance as an oil-swing producer would increase at the expense of Saudi Arabia. Oil rivalry would prove unaffordable as both the Iraq and Saudi economies are running into difficulties, Iraq out of its dire need to reconstruct, Saudi Arabia out of the need to alleviate the pressures of demographic growth and expand employment opportunities, among other things.

Patrimonial Versus Share-Holding Oil

As has been indicated, oil wealth is not only an economic factor of stability and development, but is also the political economy of governance. There are four forms of oil ownership: patrimonial (as in most GCC countries), national (as in Iraq and the rest of oil producing countries), private (as in Western Europe and the US), and societal share-holding (as in Alaska). Iraq's oil wealth is still national, but it may shift to private or share-holding society forms. This shift is considered by the US administration and is recommended for Iraq. The reason for tabling this recommendation is the fact that oil rent is the basis of authoritarian forms of governance. Any viable democratization requires a synchronized market-based economy rather than an *étatist* form. If Iraq privatizes the oil sector, or, better still, if Iraqi citizens in their entirety become shareholders of their national oil wealth (the Alaska option), the government's political economy will be simply dependent on taxation. Separation of the economy from politics, the *conditio sine qua non* of democracy, will become a novel reality. A reform in this direction will put unprecedented pressure on the governments of the GCC countries where oil wealth is still patrimonial.

Market Economy

Iraq's open-market economy, if not restricted or partly reversed by future elected governments, would have tremendous impact on the economic environment in the Gulf. It would, first of all, offer the GCC countries ample opportunity for investment. The countries that would most benefit are those with a developed banking system, deregulated economy, and financial surpluses to invest. The United Arab Emirates, Kuwait, Saudi Arabia and Qatar, are best situated in this respect. In addition to their burgeoning entrepreneurial classes, cross-border cultural ties among common tribes and houses constitute a secured channel to encourage business. Intensified private economic cooperation would improve the political climate and encourage practical and diplomatic compromises.

However, reformed market economics in Iraq could also have other influences. In particular, greater radical deregulation would become necessary in the GCC countries to synchronize them with the economic realities upon their doorsteps, and with those effects already produced by globalization.

16

Defining Success in Iraq and its Implications for the Gulf

Frederick D. Barton

Lost in the midst of the diplomatic exchanges, war planning and combat of 2003 was any strategic effort by the United States and its coalition allies to visualize a post-conflict success in Iraq. The result has been a stumbling and humbling start to Iraq's reconstruction with unreliable public safety, insufficient Iraqi ownership, unclear political choices and muddled communications.

Now, with thousands of lives lost and nearly US$ 200 billion later in the enterprise, the likelihood of a felicitous outcome has diminished and the enormity of the task threatens to overwhelm the coalition. A calamitous result is not likely, but a no win–no chaos scenario is probable with considerable ramifications for the region.

How did the US Reach this Point?

There has been a lack of direction from the beginning. Within the United States administration there was an ongoing debate between a group of Utopians, with a triumphant view of the war and a euphoric view of regional transformation, and the pragmatists, who felt that a short war should be followed by a prompt exit. The latter group had long opposed nation-building, feeling that the United States had no proven skill in this area.

Those baseline prejudices were bolstered by a series of existing industries – defense, diplomacy, development, intelligence and humanitarian – all of which eyed the post-war period with wariness. In each case they hesitated to think about the challenges to be confronted because these were of secondary importance to their traditional work. The military was reluctant to prepare for post-combat public safety; diplomats were not ready with the cadres of civilians needed to lead the political work on the ground; development specialists disliked the inherent instability and the requirement for speed within the setting; intelligence did not have the local knowledge to complement their high technology approaches; and the humanitarians felt overwhelmed by the politics which threatened their neutrality. Structural flaws would compound other weaknesses.

Foremost among these was a bureaucratic fight to determine who would be in charge. The US State Department had convened working groups on "The Future of Iraq," and prepared a series of papers, yet was in no position to make those ideas operational. The Department of Defense (DoD), which had not thoroughly studied many of the prospective challenges, was prepared with tens of thousands of soldiers on Iraq's borders. When the moment of truth arrived, the President chose to go with the default route, DoD, and a "headless horseman" effect resulted. There was an active fight for leadership roles, while ducking critical components.

Another distressing dimension was the international and multilateral argument vs. the US-led and unilateral view. Both spoke as if the contest was between the hare and the tortoise, with real capacity at the ready on both sides, when the available tools suggested something more akin to a race between two tortoises, neither too swift.

Into this mess rode former General Jay Garner, a leader who arrived late—just two months before the war. Given a murky mandate, little time, and no team, he faced a very tough mission.

While others had been thinking about the war for months and gathering the talent they would need, Garner spent much of the little time he had having to improvise as he went along. A scapegoat was found where a leader was needed.

General Garner's work was made more difficult by being subordinated to the military's needs in the field, while being dependent on them for all logistical and security support. The awkward arrival in Baghdad of the civilian team confirmed the difficulty of that arrangement.

Finally, despite the importance of the post-conflict reconstruction issue, the critical public, both in Iraq and in the coalition countries, were ill-prepared for the huge undertaking at hand. In many ways, post-conflict reconstruction was being treated as a nuisance or sideshow.

If the elements of success are a clear sense of direction, a solid team and constant communication, the US-led coalition lacked all three elements from the beginning. The ongoing internal argument between the euphoric Utopians and the keep-it-simple anti-nation builders, with all others being sidelined, has produced a confused and confusing sense of direction. The team was not only late in selecting the civilian political leadership but also replaced it on the run, leaving it still staffed at the 50 per cent level. Communications from the Coalition Provisional Authority (CPA) have not reached out to the Iraqis, the region, the world or the American public in a self-confident manner.

We can only hope that the New England phrase, "bad start, good finish," holds true in the case of post-conflict Iraq.

What Should the Vision Be?

If the Utopian and the minimalist views are inappropriate, what approach makes sense? The idea of nation-building is both colonial and paternalistic and a poor fit in most places. This is especially true in Iraq with its strong local traditions and proud history. Given the huge coalition footprint and the statements of President Bush and others, making a smooth exit with a public relations-driven declaration of success is an equally unattractive option.[1]

Post-conflict challenges require a catalytic model that is centered on core, universal values, whether based on the Universal Declaration of Human Rights or on major religious traditions. This idea was described by Immanuel Kant: "Act so that the principle of your act can be a universal law." To be an effective catalyst for a fresh, national start, there must be a clear sense of direction, full engagement of the Iraqi people, and clear, consistent communication. These have not been strengths of the coalition forces in their early days in Iraq.

Multiple definitions of success have emerged from all quarters, further confusing an Iraqi public that is already spinning from the many changes. In an October 2003 planning document titled, "Achieving the Vision to Restore Full Sovereignty to the Iraqi People" the Coalition Provisional Authority spoke of a "free Iraq governed by a representative government chosen through democratic elections." That vision was supported by a mission statement and an "End State" that was described this way:

> The ultimate goal is a unified and stable, democratic Iraq that: provides effective and representative government for the Iraqi people; is underpinned by new and protected freedoms for all Iraqis and a growing market economy; is able to defend itself but no longer poses a threat to its neighbors or international security.

L. Paul Bremer III, the Director of the Authority, addressed the issue at the start of Ramadan on October 24, 2003. He spoke of preventing "the rise of a new tyrant," ending advantages "arising from religion or ethnicity," and achieving "equality of opportunity for all men and women." He concluded with these words:

> Your future is full of hope. In this blessed time, we share a common vision that: You will live in dignity; You will live in peace; You will live in prosperity; You will live in the quiet enjoyment of family, of friends, and of a decent income honestly earned; You will live in an Iraq governed by and for Iraqis.[2]

Both of these visions are considerably broader than the one put forward by the Undersecretary for Policy at the US Defense Department in February: "it will be necessary to provide humanitarian relief, organize basic services, and work to establish security for liberated Iraqis."[3] The British Ambassador to the UN, Jeremy Greenstock emphasized governance: "An internationally recognized government, representative of all Iraqis, working for the benefit of all Iraqis."[4] Anthony Cordesman, a seasoned regional observer explained it thus:

> The most the US can hope for is to leave Iraq having created conditions that give Iraq real hope *if* new leaders emerge, *if* they can work together and towards the national interest, and *if* the Iraqi people are willing to follow. The minimum that the US and its allies should strive for is to create conditions where they gave Iraq these opportunities, and it is clear that the resulting failure is Iraq's and not that of the US and its allies.[5]

The formulation that we proposed last January was for "a promising future for a prosperous Iraq at peace with itself and its neighbors." The focus was on a non-intimidating environment, open participation, rule of law, and relief from the financial obligations of Saddam Hussein's regime.[6]

Today's vision should borrow from many of these statements but ring with clarity for the Iraqi people: *An open, safe, and non-intimidating country that respects the basic freedoms of its people and invests in their well-being.* This is both ambitious and achievable, and has the advantage of being driven by three basic freedoms: speech, assembly and movement.

It is also possible to imagine what it would look like in practical terms: safety on the streets; a political process that is not captive to any armed group or system of threats; free flow of reliable information; and a clearly articulated political transition.

The second critical element is the delivery of tangible progress that reinforces the avowed direction. While much has been done by way of governing councils, repaired schools, restoration of utilities and numerous other initiatives, the Iraqis have not been shown how these good works are part of a cohesive whole. Jessica Mathews expressed it in this way:

> CPA is also letting the best be the enemy of the better-than Saddam, employing US. contractors in needlessly expensive projects that strive for US-level technology…Iraqis, with a fraction

> of the money and sometimes with help from their original suppliers, could make it go…and a priceless sense of ownership for them.[7]

It may be a heavily laden airplane on a long runway that is off to a bumpy start, but the passengers must know where it is going in order to accept the slow takeoff and the nervous early moments. So far, that sense of direction is missing.

It seems that the fear of prospective failures is better understood by most Iraqis: chaos with intercommunal violence; a failed state with jihadist control; a playground for regional mischief makers; a weak US puppet; or three separate states. With negative views more dominant, public support trend lines are likely to be pessimistic—the case in Iraq right now.

Establishing a broadly understood, unifying and positive definition of success and convincing Iraqis that progress is being made on a daily basis is the central challenge of the post-conflict reconstruction. The coalition is rediscovering that war should not be fought if the peace cannot be won.

Where Are We Today?

Iraq is in the midst of multiple, simultaneous transitions, with only a few being digestible at a time. While progress is being made, there are four vital areas where improvement must be felt to reach success: public safety, political direction, Iraqi ownership and communications.

The greatest problem remains that many parts of Iraq are not yet safe. The 25 or so daily attacks on coalition forces, assassinations of Iraq's developing leadership and the larger bombings of mosques and police stations have been well documented These have been carried out in a way that suggests organization with the intent of spreading doubt and dissension. A story filed by Edward Wong from Karbala at the end of December 2003 described the effect:

> But a day after four suicide car bombs killed seven soldiers of the occupation forces and seven Iraqi policemen and injured more than 100 others, Karbala is a city transformed, joining the ranks of other wounded places in Iraq where security can no longer be taken for

> granted. Here, in what were once relatively peaceful streets, the rumble of Humvees and the loading of Kalashnikovs are sounds that are becoming as familiar as the calls to prayer.[8]

It is not a huge surprise that a well-orchestrated effort is underway where once there were thousands of "elite Republican guards"– few of whom surrendered or were captured in the war – and where arms are stored all over the country. Coalition forces were unprepared for and slow to recognize the likely developments, from systematic looting to organized attacks. In return, there has been a general skittishness among, and loss of confidence in the capabilities of the occupying forces in parts of the country. The Oxford Research International poll released on December 1, 2003 showed that of 11 institutions tested, public confidence was lowest in the US and UK occupation forces, with just 21 per cent expressing confidence. On the other hand, 54 per cent indicated confidence in local community leaders and 70 per cent expressed confidence in Iraq's religious leaders.[9]

Less attention has been paid to the daily lawlessness facing Iraqis. In a telling *New York Times* article in September 2003, Neil MacFarquhar reported on a visit to the Baghdad Central Morgue:

> Dr. Faiq Amin Bakr, Director of the Baghdad Central Morgue for the past 13 years, reels off the grim statistics that confirm to Iraqis that they have entered what they see as a terrifyingly lawless twilight zone: 462 people dead under suspicious circumstances or in automobile accidents in May, some 70 percent from gunshot wounds; 626 in June; 751 in July; 872 in August. By comparison, last year there were 237 deaths in July, one of the highest months, with just 21 from gunfire.[10]

A simple extrapolation of gunshot wounds, assuming that the fatalities were only a percentage of those being shot at, suggests that thousands of Iraqis have been victimized or been witnesses to gun violence in the past few months—a state of disorder that will not encourage the citizenry to step forward and assume leadership. "Regaining public security in the country" is far and away the top priority for Iraqis, cited by 67 per cent in the Oxford poll.[11]

These inherent dangers are further complicated by a political transition that has been unclear. While progress is most visible in various localities, the national order is confused by regular changes and the absence of skilled Iraqi practitioners; not surprising in a country where the best were often killed because of their talent. Now, with the handover from the CPA to the Iraqi interim government already completed, the election of a National Assembly scheduled to take place by January 31, 2005 and with ratification of the constitution and permanent government elections due by the end of 2005, official guidance is nearly in place.

What the rules of engagement will look like, what should be done with political groups that have their own militias or protection forces, and how to implement the many operational steps are enormous tasks that will require full public support.

Lagging Iraqi ownership in the process of change is hampering progress. This can be ascribed partly to the physical dangers that Iraqis encounter when they embrace change publicly. Yet, there has been a spectator quality from the beginning that may be due to other factors: years of passive engagement in their communities, the complete lack of public trust, fears of a return to the practices of the past, and the foreignness of the mostly military presence. However, none of these factors prevent the Iraqis from complaining about the performance of the "occupiers" or even of speaking of the attackers as the "resistance." Both terms do not bode well for the partnership that is needed.

A nervous and controlling style of leadership by the coalition exacerbates the reluctance of Iraqis to engage. Bunkered down in Saddam's palaces, "a mostly American administration…remains largely inaccessible."[12] At a focus group in July 2003, Iraqi participants complained of the shutting down of entire sections of Baghdad, with bridges being closed, roads barricaded, and traffic disrupted. One local resident told us, "We could get closer to Saddam than we can to the coalition. He shut down a quarter of the city. The Americans have doubled that." The image of the coalition taking over and basing their operations in the despot's palaces also rankles, suggesting that these national assets will not reach the people.

Communications continue to be blurred. Without a central message, the coalition has been worried about *Al Jazeera* and

Iranian TV. The new national TV network, *Al-Iraqiya*, is Pentagon-sponsored and seeks "to cast the US occupation in the most favorable light…(but) may actually be losing the war for viewer's hearts and minds." Seen as an official mouthpiece or source of propaganda, the station captures a modest audience with a style that is not thought to be competitive.[13]

In our 2003 conversations with Iraqis, there was only one idea that seemed to be understood by all we spoke with: the US$ 25 million bounty for the capture of Saddam. Where local TV stations were operating, there was a greater sense of connection between the coalition forces and the Iraqi people, but there were few stations in most cities.

Many of these problems are the result of a war where Iraq's people recognize that Saddam and his loyalists were losers, but their own win is still unclear. The Oxford poll asked the open-ended question, what was "the best thing which happened to you" in the last 12 months. "Far and away the top choice" was a 42 per cent response that the demise of the Saddam regime was the most positive news. When asked the "worst thing," 35 per cent cited, "war, bombings and defeat," again the top choice by far.[14]

What Must Be Done to Succeed?

A radical departure from business as usual will be necessary to achieve a greater success in Iraq. This will require a non-institutional, non-bureaucratic and entrepreneurial approach, something that is not easy for a US government-led effort. Just doing things on a big scale will not be enough, rather, new ways of doing things must be featured, because post-conflict situations require management by chaos vs a strong, controlling hand. In addition, the effort must maximize the role of Iraqis, strike a good balance between the military and civilians, and reach out beyond the coalition countries to recruit the talent for the job.

The four areas of greatest challenge are: public safety, Iraqi ownership, communication and funding.

1-Creating a Sense of Public Safety

This remains the primary responsibility of the coalition forces. While there has been an evolving concept of operations, there is still a real need to fit the force to the task of making the people of Iraq feel safer.

In order to do that, the military will need to move out of the palaces and bases, leave the garrisons and cantonments, curtail the convoys and heavy patrols, and curb the use of retaliatory strikes. Now is the time to move into the community, reassure the neighborhoods, and establish a higher level of local confidence. There are significant parts of the coalition that have tried a lighter, faster footprint, but the dramatic suicide bombs and daily attacks, have caused a reversion to heavier responses and to increased force protection measures.

A January 7, 2004 article in the *Washington Post* described the thoughts of US Marines who are preparing for a second tour of Iraq. "I'm appalled at the current heavy-handed use of air (strikes) and artillery in Iraq. Success in a counterinsurgency environment is based on winning popular support, not blowing up people's houses." Citing experiences in Vietnam, the Marines compared their "clear and hold" operations, which were seen as "one of the few success stories in Vietnam," with the Army's "search and destroy" approach. The Marines "are also aiming for more restraint in the use of force and intend to limit the use of heavy weapons, using bombs and weapons as a last resort."[15]

As we traveled around Iraq in the summer of 2003, we did not see a reassuring presence of authority in most places. On the streets of Baghdad there were entire areas that were secured because of coalition bases, but there were only occasional Humvees, often driving quickly through the streets or parked and isolated in busy places, and few visible street patrols. On the one hour drive from Baghdad to Hillah, we saw one Humvee patrol, leaving wide areas of a busy part of the country exposed. In Basra, the British were based out of town at the airport, and they too could not cover most of the city or area.

The result of this profile is that the coalition has little ability to prevent attacks or crime, no real rapid response when something goes wrong, and soldiers doing work which they are not prepared for. Much of the problem is caused by the size of the country, but the military's aversion to work that is more akin to policing has a compounding effect.

The new look will have to include some of the following elements:

- *SWAT teams*: Special Weapons and Tactics (SWAT) teams from police departments are needed. These should be active duty squads, working in US cities and not retirees. Such teams will be required to root out organized crime in Baghdad and elsewhere.
- *Improved intelligence*: The local councils could be engaged in recruiting thousands of young people, shopkeepers, and others who are active in the community and outfitting them with cell phones and two-way radios. If massive switchboard and computer profiling operations are set up to back such an operation, to evaluate individual efforts and the quality of the information received, the interdiction of threatening events is more likely. The often repeated premise is that Iraqis want to see an end to the violence. Unless the public is engaged in preventing and identifying crime and attacks, there is little likelihood of US law enforcement alone succeeding. Only if it is done as part of a community watch program can the system move from a history of intimidation and fear to one of open communication in support of a peaceful environment.
- *Constabulary forces*: Several European countries have this capacity to deal with organized crime, anti and counter terror, crowd and riot control, and the protection of key officials. There are many places in Iraq where such forces could be immediately and usefully deployed.
- *Specially trained small street units*: The Marines plan to expand this approach in the coming months. As the US Army reviews its approach, a special effort should be made to expand their capabilities in this area.
- *Special force interdictions*: There are hot spots in the country that will need the unique skill sets of independent, quick strikes. This is a critical way to reverse negative trend lines in difficult places.

- *Securing arms and armories*: Potential supply depots exist throughout the country. Many have been identified, but they need to be guarded and the weaponry destroyed.
- *Mobile forces*: These forces can move from one hot spot to another. Greater mobility and surprise are needed to keep ahead of the guerrilla tactics of Saddam's loyalists.
- *Realistic expectations*: There must be more realistic expectations of the 160,000 new Iraqi police, guards and military. While the handover of responsibility is desirable, there is considerable evidence that patience in this area is required. The initial reports of high desertion rates among the new Iraq Army recruits should be expected, as should the on-the-job learning of police who have not had experience with street patrols, human rights, or even arrests.
- *A review of force levels*: More forces are likely to be required in order to increase their visible presence in the community and to deal with the insurgents in difficult areas. The next few months are critical to establishing the law and order that makes everything else possible. There is a feeling that the reluctance of the US Administration's civilian leadership to encourage this idea has made it difficult for the military leaders on the ground to request an increase of troops.
- *A status of forces agreement*: There is general Iraqi acceptance of the need for coalition forces to remain in Iraq for some time. A clear statement of rights and responsibilities, much as existed in Japan or Germany, is necessary to prevent the continuing development of private militias.

On the issue of unrealistic expectations regarding the Iraqi security forces, Senator John McCain expressed the following opinion:

> When the United States announces a schedule for training and deploying Iraqi security officers, then announces the acceleration of that schedule, then accelerates it again, it sends a signal of desperation, not certitude…Prematurely placing the burden of security on Iraqis is not the answer. It is irresponsible to suggest that it is up to Iraqis to win this war. In doing so, we shirk the responsibility that we willingly incurred when we assumed the burden of liberating and transforming their country, for their sake

> and our own. If the US military, the world's best fighting force, cannot defeat the Iraqi insurgents, how do we expect Iraqi militiamen with only weeks of training to do any better?[16]

Success in Iraq will be driven by the establishment of public safety. The number of attacks on coalition forces needs to drop to a few per month rather than the current 25 or so per day. Iraqi civilians must feel free to move around their cities and the countryside. Women and children will have to feel safe on the streets. Citizens should not feel the need to keep a personal armory to protect their families. When a crime is committed there should be some confidence that there will be a positive response. This kind of progress will depend on a series of unconventional choices—most of which will have to be made by the dominant force, the US military. "The moment has arrived for United States forces to reconsider the tactical ways and means they are using to remake Iraq."[17]

2-Accelerating Iraqi Ownership of the Reconstruction Enterprise

The US operating mentality in Iraq has been that of a wholly owned subsidiary, creating landlord vs. tenant, headquarters vs. field, and occupier vs. resistance relationships. None of these relationships benefit either the coalition or the Iraqis. As Senator John McCain explains it, "The United States is treated as an occupying force in Iraq partly because we are not treating Iraqis as a liberated people."[18]

Under Saddam, individual responsibility was replaced by survival instincts. People felt that they made it because they were lucky, clever or through divine intervention or protection. Trust did not exist and relationships were not the building blocks of society. This situation creates a difficult starting point, since it requires reaching out to individual Iraqis instead of traditional groupings. The temptation to look for ethnic, religious and tribal relationships should be resisted, not only because of their complicating effect on the future governance of Iraq, but more importantly because they are not strong enough to carry much weight.

Another guiding principle is not to embrace any promising parties too closely. In a process that is forced to mature more rapidly than it is capable of, there is a natural rejection of anyone that is seen as a puppet. Those whom the coalition wants to encourage will need to be given space to make their own mistakes and to express their independence. This is a doubly wise choice, because the coalition is not so popular as to be able to win support for its chosen friends. A Baghdad resident asked us: "If Saddam's idiots could get the electric power and water going weeks after the Gulf War, why is it taking the US months?" Such attitudes are evident everywhere.

The adoption of responsibility by the Iraqis will require several initiatives:

- *A clear handover of governance from the CPA with a practical working calendar*: The present agreement on the political process, which leads to a fundamental law, which leads to the selection of an interim government, which determines how those who will draft the constitution will be elected, which produces a constitution by the end of 2005 is just too complicated and fluid for the Iraqi people to rally around. "After eight months of debate and delay, the United States this week will formally launch the handover of power to Iraq with the final game plan still not fully in place," said the *Washington Post*.[19] A starting point would be to leave the complex questions to the constitutional debates: what of type of government; the role of religion; and the nature of local rule. That would relieve the weak central Governing Council of that responsibility, simplify the calendar, and allow for more time to expand the political class. Real authority and less process should be the rule.
- *Be clear with all parties that there will be no favoritism or pre-selection of winners*: Each ethnic and religious group seems to be behaving as if it deserves or will soon receive special treatment. Confirming that oil will be a national resource, with a wealth sharing formula; that minority rights will be the norm; that new geographic boundaries will not be drawn for the governorates; and that decentralized rule will encouraged, could reduce private demands or rash behavior.
- *Make the constitutional selection a truly bottom-up, grassroots process*: Overcomplicating or denying popular participation at

this stage will guarantee the long term failure of the country. Finding ways to broadcast the proceedings, showing the sharp differences of opinion and the resolution of disagreements will do more to mature the body politic than dozens of other initiatives. This must be credible and the skepticism about the present course has the appropriate ring of caution.

- *Emphasize the local councils and enhance their ability to deliver services*: All spending decisions should move to the local Iraqi level, giving nascent political figures a chance to face choices, hear voices, make decisions, and deliver benefits to their people. The standards of coalition oversight should be maximum community participation in the decision-making, a local contribution (of labor or materials), transparent handling of the funds, and a local group that will take ownership of the project. This is the next stage of what started out as a hearts and minds campaign by the coalition forces; now is the time to make the shift to Iraqis clear.
- *Encourage Iraqis to take the lead on the trial of Saddam*: International standards can be met in Iraq. Furthermore, it is the domestic market that most needs the lessons of a fair trial, the full story, and the experience of putting the complex puzzle together in front of the people who lived with the atrocities. Where expert assistance is needed, it should be available—something that could come in many forms, from hybrid courts with external members to technical advisors for the prosecution and the defense. The UN has shown it is capable of delivering valuable assistance in this area, from helping to shape the trial structure to the daily elements of a fair procedure. Up to now "there has been too little effort to engage Iraqi society in a debate about the best way to confront its past."[20]
- *Engage the youth market*: Start off by hiring thousands of university students and recent graduates to serve as community organizers, promoters, and representatives of the coalition. This is the critical demographic group that must be involved for the transition to take flight.
- *Avoid ideological overreach*: While privatization, a pure market system, foreign ownership and flat tax rates all have

appeal, there is danger in overloading the near-term agenda. These are issues that a new Iraqi government will be capable of addressing, in good time. Under the present circumstances they are not likely to be implemented, which is another reason to hold back.

There is considerable evidence that Iraq is a land of capable and resourceful people. Nesreen M. Siddeek Berwari, the Iraqi minister of municipalities and public works expressed confidence: "Every time I look around I see Iraqis doing things. We think that the same expertise that exists will be the backbone of any success on the reconstruction of this country in the future."[21] Only when Iraqis are fully engaged in every element of their country's transition will success be within reach.

3-Flood the Market with Reliable Information

The national appetite for visible evidence of change, progress, threats, and complications is not being met. Opportunities abound, but they will require fresh approaches.

Rumors, word-of-mouth, and official messages from the center continue to dominate Iraq's communications channels. While there has been a blossoming of newspapers, satellite dishes, and official organs, there is still an unmet, pent-up demand from the public for more news. The years of fabrications have left people wanting to see what is happening on a minute to minute basis—this is why television is so important.

There are new directions that must be encouraged in Iraq to meet this challenge:

- *Expand the number of television outlets*: The coalition's concern that the media it has helped to establish might be too independent should be replaced by an effort to create new stations and networks throughout the country. Multiple voices need to be heard during this time of accelerated development. The strength of their signals and their reach are not as important as giving voice to the range of feelings and thoughts that need expression.

- *Create a wide range of new programming*: More of everything is needed, from crime watch shows that involve home viewers in the solving of crimes, to call in shows, to Cable-Satellite Public Affairs Network (C-SPAN) type coverage of trials, public meetings and civic events. Engaging the public is the first step to ownership.
- *Provide film of daily events throughout the country*: It would be possible to hire 500 youthful camera crews, equip them with mini camcorders, and ask them to film events that are happening in their cities and towns; Editing shops set up throughout the country could help to develop and circulate these daily insights throughout the land.
- *Feature the ultimate reality TV*: Show the state of Iraqi decay that was brought on by Saddam's regime: from tours of the palaces, to the Basra water system. Develop a schedule of national hearings for people to speak about their personal suffering. So much happened in the past 30 years and the public record of these events should be one that people can see, tape and discuss.

Television is the medium of choice in Iraq. If this is approached in a bold enough way, Iraq's experience will not only be positive, but it will set a standard for the many transitions which face difficult pasts.

4-Make Sure there are Enough Funds for the Reconstruction

Getting Iraq to the point of becoming self-supporting is critical to its future. The early days of reconstruction were aided by almost US$ 2 billion of seized and found Iraqi money. That allowed for quick dispersal of funds without the usual US government constraints. Those original funds have run out and the next stages will continue to be expensive.

At this stage the available funds are mostly coming from the US Treasury, with its Congressional and Office of Management and Budget (OMB) restraints. Reports from the field are that there is less available for "hearts and minds" expenditures (quick repairs, potholes, local events) and that the bureaucratic constraints are great.

A liquidity crunch was always over the horizon because of the encumbrances of Saddam's prior financial obligations and due to the inevitable delays that would be faced in oil production. Those who thought that oil would allow Iraq to support itself right away were wrong, though its promise continues to give Iraq reason to be more optimistic than most post-conflict nations.

There are several positive choices that could be made in order to give Iraq a better chance of success:

- *Disencumber Iraq of Saddam's debt, reparations and contracts*: It is important to think of these together, because each offers an element of flexibility. Some countries will forgive some debt and all of their claims from the Gulf War; others will forget the debt but want a preferred position on the contracts in exchange. What is clear is that Iraq will have no financial future if it tries to honor more than a token amount of the US$ 200 billion that still hangs over its treasury. It could probably deal with US$ 30-35 billion, especially if payments on any refinancing were delayed for a few years. If Iraq is going to have a future it will need to clear the past. Former Secretary Baker should set the ceiling and conclude the discussions as quickly as possible. No debt was going to be paid as long as Saddam was in charge—so why should those expectations change now?
- *Make the oil business in Iraq transparent*: While production is now nearing 2.5 million barrels per day, the information flows, the expectations, and the international oversight boards are still not reassuring Iraqis that their wealth will benefit them.
- *Prepare a new oil wealth sharing model that will reach individual Iraqis*: If only in a small way starting in a few years, some oil payments must be enjoyed by Iraq's citizens. That is the best chance to encourage public interest, circumvent official waste and an immature leadership, and create a unifying force that binds every citizen to a single nation. Most will still go to fund the central and local governments, but the feeling of having a piece of the national patrimony will matter.
- *Accept some of the immediate realities of a broken down, command economy*: Yes, there are too many non-competitive

state enterprises, agricultural price supports, and padded payrolls. Iraq is not Poland, so the adjustments should be calibrated so as not to overwhelm the political events.

The standard practices of international financial institutions and the oil industry will need to be stretched for Iraq to approach self-sufficiency. External assistance will be needed for some time, but making it possible for Iraq to fund most of its recovery will go a long way towards reaffirming a positive national identity.

Since it is unlikely that a large and often ponderous institution like the US government could make so many significant changes, an ambiguous outcome is most likely in Iraq over the coming few years. Neither chaos nor triumph should be expected, rather, more of a muddle.

A note of caution seems appropriate at this juncture. As Milt Bearden, a 30 year veteran of the CIA warned:

> There were two stark lessons in the history of the 20th century: no nation that launched a war against another sovereign nation ever won. And every nationalist-based insurgency against a foreign occupation ultimately succeeded. This is not to say anything about whether or not the United States should have gone into Iraq or whether the insurgency there is a lasting one. But it indicates how difficult the situation may become.[22]

Iraq is likely to be safer than now, but still intimidating; more pluralistic, but still overly influenced by bullies; much more Iraqi-run, but fractious; with ongoing US and coalition support, but much less money.

If this situation is correct, what does it mean for the Gulf?

What Might All This Mean for the Gulf?

Underlying any projections of regional opportunities is the assumption that post-conflict Iraq will be in an ambiguous state for some time. Iraq will go through a gradual, long term evolution, with many missteps, neither winning nor descending into chaos. Today's dangerous trend lines may be mitigated by committing massive resources, but that will not be enough to fully check various non-pluralist voices and a confused mass of youth.

Nevertheless, significant changes will be felt in the Gulf region. Iraq will, in one form or another, challenge the status quo. Whatever happens inside Iraq, the Gulf will be changed. It will be important to think of Iraq as an opportunity, not as an irritant.

Of particular interest are those changes which test conventional wisdom, show new ways of addressing challenges, and reach the people in progressive ways. There are eight changes that bear discussion.

1- The security situation is rapidly moving from an unofficial to a de facto reality under NATO's protective umbrella

Starting with the response to the attack of Kuwait and now broadened by the war in Iraq, the coalition has made clear that neighboring countries cannot meddle militarily in this part of the world. Combined with the overwhelming technical superiority of the United States and NATO forces, there is little likelihood of local military incursions. Three traditional beliefs in relation to the Gulf region deserve to be tested:

- *A large force is needed to protect against possible invasion:* Iraq's plan for a 70,000 strong mix of armed forces should be sufficient to guard its borders from illegal traffic, while at the same time Iran should be able to reduce its 520,000, Turkey its 514,000, and Syria its 319,000 strong military. Much of these forces were premised on a belligerent Iraq. Without Saddam and with a NATO policy of zero tolerance of invasions, all these countries should be able to redirect vital resources.

- *Many of the region's militaries are intended for internal control as much as for external threats*: Until those dual functions are separated, regional arrangements will be difficult to advance, and the dependence on foreign forces will continue.

- *Large and permanent foreign bases are necessary:* Recent war experiences have shown that having large and numerous foreign military bases may be a dated concept. Not only are they expensive and distracting to the countries where they are

> located, but they are a less effective deterrent than a long range missile that has a sniper's accuracy. The willingness to use force and its devastating effects has now been seen, from Serbia to Afghanistan to Iraq. The growth in the number of US military bases in the Gulf region is neither wise nor necessary and its visibility produces an unwelcome, meddlesome and threatening profile. Agreements that cover mutual protection through landing rights, staging, navigational enforcement and joint procedures, would be better than more permanent bases. Such a shift would benefit an overextended US military and also the Gulf states that have provided direct financial support for certain operations.

Because a large force will need to stay in Iraq for at least two years, the bases will not move out soon, but that planning should begin right away.

2- Markets will be more open

The exaggerated acceleration of the moment will not last forever, but Iraq is now more a part of the world than it has been for decades. While it is a poorer country, due in part to the rapid growth of its population, it remains a force because of size, promise and location. With a modicum of stability, it should attract outside investment—most especially from the Gulf with its proximity, familiarity and external view.

The weakness of Iraq's former command economy will continue to mean high unemployment and low productivity, but a real opportunity for the region to introduce more modern methods and equipment. Some of those will be in more open systems, such as loans that will be based on collateral as opposed to friendship. In our summer travels through Iraq, we heard of local manufacturers who were eager to update their technology, while in Hillah and Karbala regional real estate speculators were reported. Movements of people and goods were increasing, information was flowing more freely, important regional hubs within Iraq were attracting business, and 25 million Iraqis were savoring the chance to be part of the intellectual, business and political bazaar of the Middle East.

3- The aftermath of the war will produce dramatic globalization effects

While it comes as a shock wave, the war's introduction of tens of thousands of foreigners with the many demands that they make on the market and the informal exchanges that are generated, will produce dramatic changes in the region.

The countries around Iraq are already seeing the surge, as Taiwan, Thailand and South Korea did during the Vietnam War. A great economic benefit will be felt by the neighboring countries, but so will the growing presence of outsiders. The return of sophisticated Iraqis from around the globe will add to these phenomena.

At the same time, it will require governments in the region to increase investments in building up the skills, infrastructure and global outlook of its people. One advantage that the Gulf states could gain is to pursue a strategy that makes its citizens the most international in the world. There are multiple reasons for this and there are few national competitors, apart from the Nordic states. Among the primary reasons for a government to exist is to help its public to become more competitive—and the war will heighten such pressures.

4- Iraq will be more pluralistic and that will radiate in the region

This will be felt most directly with women, youth, and within ethnic, tribal, and religious groups that are seen as monolithic.

The new leadership in Iraq, will by definition be younger and more diverse than most of its neighbors. Saddam's elimination of a generation of leaders means that the next group is more likely to be in its 30s, 40s and 50s. It will also be more diverse. In the past, Iraq has involved women more than other countries and because of the many needs of the reconstruction period, combined with the destruction of the talent base by the past regime, will mean that women will have more opportunities than tradition would provide.

Youth will be more influential because of their demographic dominance—with 40 per cent of Iraq's population under 18 years of age. This group will require special remedial attention because the last two decades saw the lowering of educational levels in Iraq. Further complicating the political roles of youth will be the higher expectations that come with external influences and expanded information flows. A reasonable expectation is that they will assume an anti-establishment role, due to frustrations during the transition. Women and youth are among the most active communicators, so the changes that take shape in Iraq will spread like a contagion.

The growing diversity of opinion within traditional Iraqi groups, the creation of coalitions, and the way that these more complex forces begin to influence Iraq's political future, will have a powerful demonstration effect in the region.

5- Governance lessons will be generated by Iraq

Chaotic as they might seem, the constitutional process and the setting up of strong, local governments could serve as democratic models for the region. If Iraq's constitutional assemblies bring together disparate parties in an open, inclusive and ultimately productive exchange, it could serve as a powerful example and help to comfort regional insecurities about change.

Since it will be awhile before Iraq's central government is functioning, the most progressive environment is likely to be the governorates and municipalities. Moving beyond the strong, central system of maximum control to one of local initiatives will benefit the region.

6- With Saddam no longer a distraction, a global unity might coalesce on the Palestinian-Israeli issue

The Palestinian and the Israeli people deserve a solution and the region would benefit from the resolution of this issue. A two-state model (which only Iran opposes) is likely, but there will be a cost. First, there will be a steep price to pay for the transferring of the right of return and the relocation of settlers. These are practical matters which will require all in the region and globally to contribute. Second, when this issue is decided, it will no longer be an excuse for foot-dragging on internal reforms.

Much of this may have to be supplemented by non-official channels. The recent Geneva accords have shown a clever way to circumvent obdurate parties, though the next step will be to design a practical way to achieve the political goals that many recognize but that regional leaders have been unable to embrace.

7- The region's own vision of an affirmative global future will be far more important than that of the United States or any other external party

The inability to create a regional policy to deal with Iraq's last regime was evidence of a core weakness. Now is the time for the region to express a positive vision of its own future, not premised on being opposed to others but on how best to prepare its people and its infrastructure to compete globally.

Opportunities abound for collective structures that will help the nations of the region weather changes, challenges and disagreements. General Anthony Zinni's call for a common statement of goals and the open formalization of agreements that go beyond the bilateral would elevate the diplomatic and security standards of the Gulf to those enjoyed by most areas of the world. Whether those should include non-regional allies would be for the region to determine, but it would give the military cooperation more of a feeling of local direction and ownership, and could reduce the volatile quality that is attached to a foreign presence.

The wise and public use of the oil resource, a work ethic that emphasizes rigor over inheritance, and an open, transparent and meritocratic set of rules is another area where local efforts could lead to significant transformation. Where the region has tried a forward-looking initiative, it has been rewarded. Such moves must be expanded. Development depends on three elements: size, wealth and stability. By creating a unifying vision, the region could enhance its natural advantages in the first two elements, while projecting a forward-looking and stable system.

The Iraq war is likely to be the highest level of US involvement in the region until there is a Palestinian-Israeli agreement. Even then, the role of the US military, the largest element in Iraq, will not be substantial. At its core, American foreign policy is based on enlightened self-interest, and the US public is divided on the merits of the kind of preemptive strike policy that was put forward in the Iraq war. The significant loss of lives and the US$200 billion expense make the future American role more diplomatic and economic than military. Another war would be deeply unpopular and Congressional opposition would be greater. If the US is moving back to a position of peaceful persuasion, then the region has a chance to avoid the perceived dominance of the superpower.

8- Multilateral institutions may be revived as a result of the war

Smaller, less powerful states depend on an international order for their well-being. While the UN's stature was apparently diminished by the Iraq debate, war and reconstruction, it could be strengthened in important ways.

The enormity of the Iraq undertaking has overwhelmed the appetite of the United States for such ventures and shown that partnerships are vital. It has also confirmed that alliances and multilateral approaches are necessary to achieve credibility and to complete the task. In the short-term the UN appeared to be sidelined and less relevant. As time has passed, its indispensable nature has been reinforced by the struggles of the coalition.

Furthermore, the two most valid reasons for going to war seem to have been the enforcement of multiple UN Security Council resolutions and the removal of a human rights-abusing despot. If the US means to stand behind the votes of the Security Council, this will boost the credibility of the UN, which has been undermined for some time. If the UN chooses to take on the issue of abusive sovereignty, progressive forces around the world will have gained a substantial edge. Helping Iraq's people to deal with the trial of Saddam Hussein could be a valuable contribution. All of these changes could work to the long term benefit of the region.

Conclusion

Iraq is fraught with challenges but also ripe with opportunities. For the huge commitment of human lives and material resources made so far, a greater likelihood of success in Iraq should have been possible. Armed nation-building combined with little sustaining capacity makes the challenge ever greater.

Most commentators believe that success will depend on American resolve. Would the US be willing to stay the course when its soldiers and civilians were being killed? Would the US taxpayers support such a distant cause? Despite indications that growing numbers of Americans are skeptical of the official story in Iraq, the political support is sufficient.

Could the Iraq reconstruction succeed without the United Nations and a broader alliance? Opportunities remain, but the US obduracy has made the potential UN political leadership in Iraq unlikely. There is plenty of work for all in Iraq, but perhaps the allies who were not part of the coalition should be focusing more on Afghanistan, where progress is also desperately needed. The next stage of rebuilding in Iraq is going to be Iraqi-led. While this may be premature in some respects, it has simplified the process of the US handover.

The total lack of preparation for post-conflict reconstruction and the slow start in Iraq have produced a greater obstacle to success. If the US, the coalition, the Iraqis and to a lesser extent the international community, are unable to change their normal, predictable patterns, Iraq is going to flounder–perhaps not sink, but merely stay afloat.

Unconventional approaches and an affirmative vision are now needed both within Iraq and in the region to produce the progress that the sacrifices of so many deserve.

Section 6

PROSPECTS FOR THE GULF

17

Political Reform in the Gulf: Society and State in a Changing World

Bassam Tibi

In identifying the challenges of the future from an Arabian Gulf perspective, it must be noted that these challenges emerge as much from "external events" as they are related to regional developments and internal trends. Whilst seeking proper "strategies and policies that will guide the Gulf countries in successfully managing future developments"[1] the present study suggests a framework for reforms that can help to meet these challenges. At the outset, one needs to put the pending challenges and issues into the overall global context. This is a basic requirement for grasping the real challenges, enabling appropriate responses and adapting Gulf societies and states to their international environment.

In the present age, a global network links all parts of the world. This is an age characterized by the intensifying processes of globalization[2] which are reducing the world to a global village. To cope with the great challenges confronting the Arabian Gulf it is necessary to create special conditions. With this insight as the starting point, this chapter examines the Arabian Gulf from the perspective of three pertinent and intertwined aspects:

- First, the position of the Gulf area as a sub-systemic region in the international system[3] of the present global age
- Second, the meaning and implications of reform, which requires a cultural underpinning
- Third, the cultural commonalities that need to be established for the whole of humanity from the perspective of a success culture shared by all civilizations.

The problem is double-edged and was earlier addressed by the Oxford International Relations scholar Hedley Bull who stated that the process of shrinking of the world to a global village labelled as "globalization" has contributed to a shrinking "of the globe while it has brought societies to a degree of mutual awareness and interaction that they have not had before." However, he adds, this "does not in itself create a unity of outlook and has not in fact done so."[4] It follows that this lack of "a unity of outlook" either leads to a cultural diversity accepted by pluralism, or it results in a conflict-laden cultural fragmentation.

Given this potential for cultural fragmentation accompanying the ongoing globalization process, there is a need to search for the commonalities that underpin cultural pluralism. This fragmentation is among the side effects of globalization. In my work I have coined the formula of "simultaneity of structural globalization and of cultural fragmentation."[5] I maintain that globalization is placing humanity in one structure, but does not and cannot create a universal world culture. In fact, there is no such thing. However, the challenge for humanity is to participate in a globalizing world whilst finding ways to link its cultural fragments without abandoning or forsaking cultural diversity. I believe that this inter-cultural bridge could be built up by a culture of success.

Reforms and the Significance of the Arabian Gulf in the 21st Century

Economically, the Arabian Gulf is an area of particular significance among the world's regions.[6] It is therefore pivotal not only to world economy but also a region of the Middle Eastern subsystem[7] that is crucial to world politics and to the future of the international order. Thus, the political reforms required and recommended for meeting pending challenges also matter to the entire international community. This insight is based on the assumption that the promotion of a healthy world economy requires political reforms in all parts of the globe.

Certainly, this line of argument touches on political questions. An older approach recommends separating economy from politics to avoid dispute and controversy. This approach has been adopted earlier by some sections of the Western business community. In this regard, I recall a German-Kuwaiti meeting in the mid-1980s which I attended as an expert consultant. When Kuwaiti participants addressed some pending political issues, the response of German businessmen was that business, not politics was at issue. Two decades later, poised at the beginning of the 21st century, in the aftermath of three wars in the Gulf region, it is impossible to address Gulf challenges properly while separating economy from politics. Today, we live in a changing world in which economy and politics are clearly interlocked. Nevertheless, a balance needs to be established between both aspects to avoid falling into the trap of politicization, because such politicization creates more severe problems that are best avoided. Nevertheless, the challenges of the Gulf must be confronted without overlooking the linkages between economy and politics. At this point it is necessary to highlight the particular significance of the Arabian Gulf and identifying regional challenges in order to find appropriate ways to deal with them.

Political and opinion leaders in the Gulf need to be aware of this political-economic interlinking in the emerging challenges to recognize the high risks looming over the Gulf region. The challenge for these leaders is to make the Gulf an exemplary place in order to play a pivotal role in the Arab world in particular and the Muslim world in general. However, there is a need to know the limits and constraints, while meeting the challenges of the future.

The basic challenge for Arabian Gulf states is to develop a vision for the future beyond the oil economy. This involves enriching the oil economy by adding other dimensions like free trade in a modern market economy. In the UAE, these efforts are ongoing and they belie the prejudiced views conveyed by some Western media that Arab and Islamic cultural patterns are at odds with the essentials of a modern economy. Appropriate political reforms could contribute to a smoother development beyond the *rentier* economy based primarily on oil resources. It is advisable for the West, as represented both by international companies and governments, to support the UAE model of diversification and the related liberalization of the economy. Similar accomplishments should be promoted throughout the Gulf region as a positive vision for the future. How can this be done? There are some avenues for pursuing reforms. However, the inherent feasibilities for this development potential are subject to regional and geopolitical uncertainties and are linked to security policies. As these matters are dealt with elsewhere in this book, the particular focus of this chapter will be on the political culture and its scope for evolving a reform approach that takes into account the significance of the Gulf in the present world, and how to sustain this in the future. In the global age, these envisaged reforms should also be dictated by the changed environment requiring both "new security" and cultural geopolitics, as both these needs apply to the Arabian Gulf.

Political and Institutional Reforms: The Case of Democracy in the Gulf

In our age, democratization and liberalization – in state, economy and society – are considered as a means to promote world peace and economic prosperity as well as political stability. In a globalizing world – however diverse or fragmented it may be in cultural and civilizational terms[8] – the call for a "global democracy"[9] cannot be based on a monolithic pattern. Of course, all democratic systems have some basic common characteristics, but they can and do need to be based on different foundations, particularly when it comes to establishing a cultural underpinning for democratization. Democracy can only be successful and be considered legitimate if it has such a cultural underpinning in the society in which it exists.

Therefore, an imposed democracy could never thrive and gain legitimacy. Taking this into consideration, opinion leaders in the Arabian Gulf could base any future democratization effort on the Islamic concept of "*shura*" (participatory deliberation) that could be further developed to become the Arab-Islamic name and face of democracy. In this case we may speak of a "*shura* democracy." A modern interpretation of this Qur'anic concept can provide a smooth path to modernity for Gulf societies. The Holy Qur'an teaches us that the political culture and the public life of the *ummah*, being the expression of Islamic society, have to be in line with *shura* (the participatory deliberation model of Islam). It may be noted here that there is no mention of the state or "*dawla*" in the Holy Qur'an. While the Holy Qur'an provides only the instruction, the progress of reforms depends on the scope and opportunities provided by the relevant political culture.

In this regard, democracy is the political culture for promoting reform. In keeping with reform needs, it is possible to promote a political culture open to democracy and to democratization, in which maintaining one's own cultural heritage is combined with learning from other cultures and their reform experiences.[10] There is a record for this kind of cultural adoption in Islamic history as indicated in the saying attributed to the Prophet (Peace be upon him): "*wa utlibu al-ilm wa law fi al-Sin*," which means "search for knowledge even in China." In line with this instruction, Muslims during medieval times engaged in learning from other civilizations, above all from Hellenism. For this reason, historians have coined the term "Hellenization of Islam" to describe a process from which the Islamic tradition of philosophical rationalism evolved. That tradition promoted science and philosophy in Islamic civilization. In fact, medieval Muslims accorded high respect to Aristotle, even ranking the Islamic philosopher Abu Nasr Al-Farabi, also known as *Al-Mu'allim Al-Thani* as second in greatness to Aristotle.[11]

The reference to the above-mentioned historical record is also relevant to come to terms with present and future challenges. In this connection, a two-track strategy is needed:

- keeping faith with one's own cultural heritage[12] while reinterpreting it to legitimize the building up of an open society
- to develop democratization efforts

I believe that democracy is a global issue that also matters to the Arabian Gulf. Certainly, in the 21st century democracy has moved center stage and has become a pivotal issue for the whole of humanity. Since the end of the Cold War the idea of a "democratic peace"[13] among nations and civilizations has moved to the forefront of current debates. The substance of the idea is that democracies do not wage war against one another, because they resolve their conflicts peacefully. From my experience as a Fellow at the World Economic Forum in the years 1999-2001, this observation also applies to the business community, which associates the success of a market economy with democratization. The market economy issue is also a recent challenge for Arabian Gulf societies. Reiterating my view that no order can be imposed, whether it is market economy or democracy, I do not propose any backdoor imposition of "democratic peace," a concept which, in its essence, goes back to the German philosopher Immanuel Kant.[14] In drawing on the idea of "democratic peace" I refer equally to the democratic order that Kant recommends and to the Islamic heritage—Al-Farabi's vision[15] of "*Al-Madina Al-Fadila*" in which the *ummah* is led by a reason-based philosopher. Seen from this perspective, no *Al-Madina Al-Fadila* could wage war against a similar order, because a perfect state ruled by a leader who complies with reason in his political decision making is both peaceful and democratic.

In addition to its contribution in meeting security concerns through war prevention, democracy also seems to be more favorable for the development of a market economy. From the security perspective, looking back to the contemporary history of the Arabian Gulf, it is likely that this region would have been saved from all three recent wars[16] had Iraq and Iran been democratic states that would have resolved their conflicts peacefully through negotiations and thus refrained from imposing their models on others, whether it be the principles of the Islamic Revolution of Iran[17] or the Baathist idea of Pan-Arabism.[18] The Gulf states do not need any of these models, because they aim to develop their own version. To pursue this, they need suitable reforms rather than the emulation of unsuccessful models. Among these reforms are the proposed efforts at democratization as well as education policies for promoting a success culture. This observation leads us to the third major part of this chapter in which I would like to introduce the concept of success culture. It is based on three pillars: politics, culture and economics, all intricately connected to one another.

Success Culture is the Key to Modernity

What are Cultural Reforms?

In acknowledging that culture matters[19] to the economy and politics, the pertinent idea is that a success culture should be introduced for the purpose of underpinning and promoting the reforms necessary for meeting the challenges of the future. People from Islamic and Western civilizations are different and have traditions and worldviews that vary from one another. Nevertheless, it is possible for both to share a universal, or cross-cultural set of values related to what may be termed a success culture. Two issues are fundamental—first, democratization and second, education in success culture. These form the core concepts for promoting political reforms. I have already addressed the first issue and would like to deal at greater length with the second, while relating the idea of a success culture to the Gulf area.

At the outset, let me express the view that the existing economic embedding of the Gulf in the international economy would receive a positive boost through reforms related to the two issues of democratization and education in success culture. Efforts in this direction can be pursued both in state and society. The ongoing modernization process in the Arabian Gulf can be upgraded into a model worthy of emulation by other countries. The preliminary requirement is to implement institutional reforms in the political system that smooth the way for educational reforms that promote the transmission of values, worldviews and conduct, all of which are collectively integrated in the formula of a success culture. This section will focus on two steps—first, outlining what is meant by cultural reforms and second, introducing the concept of success culture as a reform in itself. It is assumed that there is an interplay between social and cultural change.[20]

Reforms are not an end in itself, but rather a means to guarantee and promote the well-being and prosperity of a society, helping it to meet looming challenges while maintaining its social fabric. A success culture is introduced as a means to serve this end. To reiterate a basic insight: The first requirement of a success culture, which may be termed as the "culture of modernity"[21] is that it should never be imposed. Without establishing a local cultural underpinning to support such reform, it cannot thrive successfully. At this juncture, it is pertinent to underscore the basic premise—namely, how much the spirit of Islam, if understood well, is in line with political and cultural reforms that could serve as a bridge between diverse civilizations in embracing modernity.[22] In the globally changing world, democracy and success culture could provide the necessary elements needed to bridge the gap between the Islamic world and the West and thus strengthen commitment to world peace, as earlier outlined. The Arabian Gulf could enhance its existing economic significance by demonstrating its ability to play a new, pivotal role as a cross-cultural bridge in the present age—one that is marked by the return of civilizations as a defining force in the arena of world politics. This revival of the forces of civilization implies the risk of a clash, but there are ways to avert such an occurrence.[23]

Those civilizations that promote the success culture on the grounds of a common understanding of cultural and political reforms would not engage in a clash, but would rather contribute to a democratic peace. The global significance of the Arabian Gulf in this regard is rapidly growing. If the UAE model – aimed at adding further political, cultural and economic aspects to the importance of the oil economy – manages to succeed with spill-over effects, there would be promising future prospects for the entire region. In a world where everything is subject to change, the oil economy is no exception. Prudent Gulf leaders are aware of the fact that the region cannot maintain its significance forever merely by restricting itself to the oil industry and oil-related revenues.[24] A reform strategy for the Gulf based on the triple task presented at the outset of this chapter would promise a continuing significance for the region beyond the age of oil. To elaborate on this vision it is necessary to explain the implications of political and cultural reforms.

In the ensuing discussions I shall focus on a new aspect of the idea of reforms. During the Cold War the significance of the Gulf region was related politically to Western security concerns and economically to the dependence of Western economies on Gulf oil supplies.[25] One of the remaining obstacles to the development of the Gulf was Saddam's Baath regime—a legacy of the Cold War, which was persistently trying to impose itself on regional states, even through war.[26] This regime has now been toppled. What are the far-reaching consequences and implications of this political development?

To answer this question, the issue needs to be placed in a broader context. At the Annual Meeting of the International Studies Association in Chicago, 2001 the term "Cultural Turn" was coined for the new post-bipolar age that followed the end of the Cold War. In overcoming the burdens of that Cold War as well as its remnants, the world is facing new challenges, both current and future, which will impact broadly on state and society, particularly on the economy and politics of the Arabian Gulf. In this new age of the Cultural Turn, it must be noted that culture matters, given that some of these challenges are politico-cultural in nature. The introduction of reforms to adapt the Arabian Gulf culturally to the global changes taking place in the post-bipolar age seems to be the proper response to these challenges. If this path is given priority then there would be promising prospects for this region, which, as outlined earlier, is one that is pivotal to the international system. Within this general context, the following section proposes that success culture is part of the reforms that can be introduced through education and training. The shift from the political burdens of the Cold War – to the Age of the Cultural Turn, the removal of Saddam Hussein's regime, and the weakening of the Iranian quest to export the "Islamic Revolution" – are altogether positive elements that facilitate a greater focus on reforms in the light of receding obstacles and obstructions.

Success Culture as a Reform in the Arabian Gulf

What is a Success Culture?

Education is the most important sub-system in society, because it determines the worldview, values and the related behavior of elites.[27] If the educational system of the Arabian Gulf states accepts the introduction of success culture as being a framework committed to raising elites educated in the related worldview, values and behavior, then this would lay the groundwork for momentous reform in state and society. Political elites, decision makers and the business community would then be in a position to adjust according to the conditions of a globalizing and rapidly changing world. The seeming contradiction between the universal claim of this success culture and the global realities of cultural and religious diversity could be overcome in the Arabian Gulf, if a cultural underpinning for the worldview, values, and the related behavior of a success culture could be established as a common foundation. A preliminary effort towards this end will be made in this section. A comparison with the universality of individual human rights and their application within Islam[28] provides a clearer view.

In explaining the concept of a success culture as a reform, I would like to refer to a project conducted in the United States jointly by the Harvard Academy for International and Area Studies and The Fletcher School of Law and Diplomacy. The current project entitled "Culture Matters," is chaired by Lawrence Harrison and builds upon the work completed earlier and published as a book under this title.[29] As a project member, my assigned task is to study Islam and the Middle East[30] and I am therefore keen to introduce the concept to the Arabian Gulf. All project members recognize the following principles:

- First, cultural reforms in state and society are subject to political and economic constraints, but still require a vision of change.

- Second, the scrutiny of any culture must take into account its cultural and civilizational specificities, such as the existing worldview and value-related differences between Western and Islamic cultures. However, it should neither overlook the potential for commonalities nor over-emphasize the differences as some authors do.

In acknowledging these two principles and the related requirements, the project "Culture Matters," argues that, next to economy and politics, culture also matters. In this context, there is a success culture in business and economy that can be shared by elites belonging to the whole of humanity despite the existing, normal cultural differences. The results of this ongoing project are quite pertinent, both in general and in particular terms, in that they touch on the present deliberations regarding the kind of reforms necessary in the Gulf. The major assumptions will be presented as the basis for a model that is suitable for emulation. To avoid any alien imposition, efforts have been made to find and uncover a cultural underpinning suitable for the introduction of success culture into the Arab world. In exercising caution about any cultural imposition, this cultural underpinning deserves serious consideration. In our project we employ the following 23 variables developed by the two professors, Roland Inglehart and Larry Harrison. I shall present these variables in the following section, while adding to my own deliberations in order to relate them to the subject matter under discussion—political reforms in the Arabian Gulf. At the outset, the members of the project engaged themselves in a major distinction between factors of a progress-prone culture and those of a progress-resistant culture. We relate this distinction to three issue areas: worldview, values and economic behavior. Does this distinction also apply equally to the Arabian Gulf? The ensuing discussion aims to answer this question.

Variables Relating to World View

In the first area, worldview, there are five variables, which may be identified as follows:

1-Religion: Religion is an essential part of any culture and thus it is pertinent to shaping reforms. It could nurture rationality and achievement but if it is based on scripturalism, it could nurture irrationality and inhibit promotion of material pursuits. In the first case, religion promotes a progress-prone culture while in the second case, it does not. In fact, the understanding of religion leads to different ways of dealing – a focus on this world (pragmatism) or on the other world (utopianism) – creating tensions between the two approaches. In reading the Holy Qur'an and studying its precepts in the light of existing realities, I observed two things. First, I found in Islam a deep commitment to rationalism and achievement as well as to the pursuit of worldly affairs,[31] and second, I missed this spirit among contemporary Muslims who pay lip-service to the Holy Qur'an but are lagging behind in the practice of what it calls for. The second variable makes this contention clear. John Waterbury, the well-known expert on the Middle East tells us what he learned after his year-long research in Morocco:

> I would reiterate that I am particularly concerned with that aspect of political culture which relates to patterns of behavior rather than manifest belief systems...In this sense it is more important to understand what Moroccans really do and why they do it than to understand what they think they are doing. It should also be noted that certain patterns of political behavior persist well after their structural underpinnings have begun to decompose. There is thus a behavioral lag that develops (like cultural lag in value systems) and continues until the social context for politics has so changed that further adherence to the old patterns becomes clearly ill-advised and politically non-productive.[32]

2-Destiny: This variable is related to worldview. A success culture requires a worldview suggesting that a man can influence his own destiny in contrast to an attitude of fatalism and complete resignation. There exists a false, but established perception that Islam promotes fatalism. In contrast, I read in the Holy Qur'an: "Whatever good befalls you, man, it is from Allah: and whatever ill from yourself"[33] This phrase makes clear that humans are responsible for their deeds, and for the negative consequences that might befall them if they fail in their actions. The Holy Qur'an is against fatalism, but the fatalist worldview can be observed working in reality, even though belied by the Islamic revelation.

3-Time: Cultural attitudes with a future focus promote planning. In contrast, present or past focus discourages an orientation committed to punctuality, defers fulfilment and planning. I would like to quote an example for such a worldview. In Egypt, people joke about themselves and local attitudes, ironically referred to as the Middle Eastern "IBM." The letter 'I' stands for "*Inshallah*" (God willing) the letter 'B' for "*Bukra*" (tomorrow) and the letter 'M' for "*Ma'lish*" (It doesn't matter). In contrast to this attitude, a success-prone culture requires a behavior that would be in line with the Islamic value of "*hayya ala al-falah*" (stand up to achievement today, not tomorrow).

4-Wealth: This is a product of human creativity and the existing zero-sum is expandable to a positive sum. The oil-derived wealth of the Gulf states is a reality, but this wealth will not last forever. The needs of the future are pressing and therefore, there is a need for reforms in order to discover and secure new sources of wealth. As part of these reforms, a new worldview on wealth is also required.

5-Knowledge: Knowledge may be viewed in two ways—either it is practical and verifiable, which means that facts do matter. Or it is viewed as cosmological and hence not verifiable. This distinction has been made already in the *Muqaddima* of Ibn Khaldun in which he distinguished between religious and historical knowledge. The same distinction may be found in Ibn Rushd's concept of *al-haqiqa al-muzdawaja* (double truth). I recommend the revival of the Ibn Rushd legacy for a better future, based on rationalism.[34]

Variables Relating to Values

In moving to the second issue area of values and virtues, three variables may be presented as follows:

1-An ethical code: Such a code could be rigorous within realistic norms and thus somewhat elastic, providing possibilities to adjust one's behavior to a changing world. In contrast, the ethical code could be dogmatic and thus inflexible. The attitudes of success can be learnt in a reform-oriented education related to a more elastic ethical code that is close to reality.

2-*The virtues of responsibility*: A well done job carried out in a proper manner demonstrates that punctuality matters. So too, courage in taking risks (no risk, no gain) should be recommended in contrast to the mentality of seeking hundred percent security and safety.

3-*Education in the values of success culture*: Such an education is indispensable for promoting independent thinking, dissent and creativity, in contrast to the scriptural orthodoxy of mind that considers change as a source of deviation.

Variables Relating to Economic Behavior

In the third area of economic behavior, there are fifteen variables that can be either based on an orientation of success (progress-prone culture) or conversely, based on a progress-resistant culture:

1-Work matters: This variable suggests that work is important and behavior is led by the orientation that "we live to work, because work leads to wealth." The opposite mindset suggests that work does not lead to achievement and wealth—work is for the poor and wealth is provided by destiny, which we cannot determine. The *rentier* economy promotes this kind of behavior and there is a need for a new mindset in the Gulf in preparation for the post-oil age.

2-Frugality: In most Western economies, frugality is considered to be the mother of investment. However, some consider this orientation not only as a threat to equality, but also think it could hurt the prosperity of people. As a Muslim who is open-minded about cultural reform, I believe that the Holy Qur'an also teaches frugality.

3-Entrepreneurship, investment and creativity: The foregoing observation applies also to the variable concerning entrepreneurship, investment and creativity, as well as to the related behavior.

4-Moderation: This variable concerns the propensity to be moderate, to abide by law and to avoid even occasional excesses or adventures. The Prophet (peace be upon him) called upon the *ummah* to be moderate (*ummat al-wasat*).

5-Competition: In a modern society, competition leads to excellence, as opposed to envy and aggression, which threaten equality. Privilege must be based on competitive behavior. In Islam, envy and aggression (*udwan*) are among the sentiments viewed as despised (*makruh*).

6-Rapid adaptation: For a mindset promoting reform, one needs to be open to rapid adaptation, and not to be too suspicious of change, because such an orientation results in slow adaptation. The acceptance of this variable by Muslims requires a new connotation of reform, which should not be viewed as *bid'a* (change detrimental to religion), since this is not the intended impact of cultural reforms.

7-Merit-based achievement: In a modern society, advancement can be based on merit and achievement, not on family relations, patronage, connections or what is termed *wasita* in Arab culture. Here, major reforms are needed to change the patterns of advancement in Arab societies. In this regard, economic behavior is also linked to socio-cultural behavior.

8-Rule of law: Willingness to abide by reasonable laws is necessary to avoid corruption, which is based on the sentiment that money and connections do matter. Corruption should be prosecuted, rather than tolerated.

9-Identification with society: Social behavior is also related to a stronger identification with the broader society rather than the narrow community (family or clan). This is very important for establishing a modern polity in the Gulf.

10-Association mindset: An attitude related to a broader society, promotes the mindset of association, trust, identification, affiliation, participation and inspires higher achievement.

11-Balancing individual and collective identities: In the Arab world, there is a conflict between notions that underline and emphasize individual identity and those that emphasize collective identity and the group. Western individualism is alien to Arab-Islamic culture. A middle way between both patterns of behavior could be possible through reforms—finding a balance between the identities related to the individual and those that are collective or group-oriented.

12-Nature of authority: A sensitive issue is the one relating to authority: Is it centralized, unfettered, often arbitrary and inherited? Or is it based on consensus with the result that authority should be accountable and subject to checks and balances? This is a basic area of political-cultural reform.

13-Role of elites: The role of elites and their responsibility to society also matters to success culture. How do elites conduct themselves with regard to the acquisition of power and revenue? It is a fact in the Arab world that the present crisis among the ruling elites is leading to the emergence of destabilizing counter-elites.[35] A balance is urgently needed to rectify the situation.

14-Relation between state and religion: A very basic issue is the nature of the relation between religion and state (*din/dawla*) addressed in the West as church-state relations. This factor plays a major role also in the contemporary Gulf. Representatives of political Islam claim the unity of "*din wa dawla*" while other sections reject this perception and prescribe the promotion of a civic sphere, the latter standpoint being the more promising alternative.[36]

15-Gender: This final variable is perhaps the most sensitive of all. In reality, can there be equality between man and woman? Are women being subordinated to men? What does Islam[37] say on this issue? It may be noted that gender inequality would be inconsistent with the value system of a success culture. A modern society needs the skills and services of all of its members, men and women equally, for its prosperity. This need was stressed by Muslim women speaking at the ECSSR Ninth Annual Conference, during the panel discussion on the changing role of Gulf women in the context of globalization.

In short, these twenty three variables incorporated into the three issue areas – worldview, values and behavior – reflect an understanding of a success culture that could be supported by a religious-cultural underpinning that is rooted in a positive interpretation of Islam. Therefore, a success culture can be embraced by Gulf societies in their bid for political reforms.

Conclusions

The Gulf region is considered central both in terms of international affairs and in terms of the world economy. At the same time, this region has a special position as the core area of Islamic civilization. In this capacity, the Gulf can relate itself to the vision of becoming a cross-cultural bridge between Islam and the West. This cross-cultural bridging can be promoted through a series of reforms in political institutions and in society, based on a new educational approach that is oriented towards a success culture. To reiterate an earlier point, Islam prescribes *shura* and is thus compatible with the spirit of democracy, which forms the crux of reforms. Islam asks its believers to be successful in their life and holds them responsible for their own shortcomings. Thus, the Holy Qur'an prescribes also a mindset of achievement-prone attitudes and related behavior. It follows that a dynamic understanding of Islam would promote both the participatory culture of democracy and the achievement-oriented culture of success.

A properly understood and appropriately interpreted civil Islam[38] is fundamentally compatible with democracy, civil society and success culture. Yet, why is this basic compatibility misunderstood? I believe that the political, cultural and structural constraints underlying this unfortunate situation in the Arabian Gulf can be altered through reforms in state and society. The promotion of a success culture that is prone to progress will also help to prevent the

"Clash of Civilizations."[39] An international morality based on cross-cultural bridging would also promote reforms. The Arabian Gulf could become a progressive place, committed to reform in a changing world, if decision makers would embrace and implement the political and cultural reforms needed to promote the region by equipping it with the means necessary to successfully meet the challenges of the future.

18

The Political and Economic Dimensions of GCC Foreign Relations: Prospects and Challenges

Saleh Abdulrehman Al-Mani

For the past one hundred years, the Gulf states have been dependent on outside powers for security, production and trade. In terms of security, Britain, since the eighteenth century, provided the required manpower and firepower to keep outside contenders away from the Gulf region. However, from time immemorial, the Gulf enjoyed a remarkable geographic location linking the Indian Ocean with the Arabian Peninsula and East Africa. The Gulf was the linchpin of a regional trilateral trade route that benefited from the seasonal trade winds that pushed cargo *dhows* from East Africa to the southern tips of the Arabian Peninsula and onward to the western shores of the Indian subcontinent. Such trade has flourished from the dawn of history till the fifteenth century in one form or another.

An important aspect of that trade specialization was not limited to profit generation, although this was an important aspect, but it also allowed the smooth flow of people and goods through these ports. Arabs of the Gulf, the Persians, East Africans and Indians all profited from these exchanges. The historical legacy of such trade left an important mark on the culture, language and religion in each of these societies. The Arabic language left an important imprint on Swahili and other East African languages, just as Arabic itself borrowed many words and idioms from Farsi, Hindi and other languages of the subcontinent. Later on, Islam in its many variations and *Madahibs* (sects) became the predominant religion of all the countries of the Arabian Sea, and parts of the Indian subcontinent as well. Later on, Arab merchants of Hadhramaut and other Gulf countries carried the message of God with them as far as Indonesia, Southern China, the Philippines and other South East Asian countries. An Australian historian has suggested that if the continent had been flourishing at that time, it would not be too far-fetched to say that Arab traders would have visited its shores as well and established Islam as the official religion of the new continent.

What is important to note is that regional and sometimes long distance trade created harmonious relations in which, not only goods, but people, language and culture could truly become transnational. Historically, during the Middle Ages, the city-states of Southern Europe and Italy such as Venice and Florence, could forsake the adventurism and pillaging of the Crusader Wars in the Near Orient because of their cultural and trade ties with Arab and Muslim societies. Those trade links forbade them from wielding their swords against their trading partners. Unlike today's oil trade, the old regional trade was more pacific. The trade of the past century was externally based. Greed necessitated the suppression of local culture, values and people to the point that regional wars were often waged in order to protect and maintain the trade for oil and other natural resources.

The change from what I call regional trade to hegemonic long distance trade gave rise not only to regional wars and conflict, but also to internal discord with the governing elites of the Gulf. Thus, it was not too uncommon here in the Gulf for the British High Commissioner to replace an Emir with his son or his nephew. Such manipulation of local power holders was a clear negation of the general notion that Britain was a power oriented towards upholding the status quo. Such manipulation was accepted at the time as a tool to achieve grand hegemony.

Historically, the Gulf did not have a huge commercial output but local commodities and products were of high value in adjacent countries and/or segments of the population living in those countries. Noteworthy is the pearl trade of the Gulf, the small but valuable production of high quality tobacco in the valleys of Hadhramaut, and the dates and gum incense of southern and eastern Arabia. Although the supply of these commodities was in little quantities, it generated enough money to finance sea travel and purchasing of goods of commercial value that were then brought back from China, India and East Africa. Even the fish catch of the Gulf and Southern Arabia could create good demand in the interior of the Arabian Peninsula and in other parts of the Indian Ocean basin.

The early era of British domination was focused on attaining imperial glory and keeping other potential rivals away from the region or away from India—the Jewel in the Crown of the British Empire. It was after this period that the region witnessed the rise of oil as a basis of long distance trade. Many writers have written extensively in the seventies and the eighties on the economic nature of hydrocarbon production, as contributing to the rise of the so-called *rentier* state. Looking back at about half a century of oil production in Gulf region, another important aspect of the oil trade may be noted—it was not harmonious, but contributed to a rise of conflict, competition and outright regional wars. Thus, unlike the old harmonious regional trade, the new long distance trade in oil, a single commodity, contributed to the growth of competition among regional and external powers and inflicted tremendous suffering and misfortune upon the area.

Propensity of Oil Trade to Create Instabilities and Wars

The phenomenon linking long distance trade with war and instability seems something of a contradiction. The massive amount of funds generated through such trade may offer one explanation for this phenomenon. The strategic value of the commodity itself may be another important factor, particularly since this commodity is the energy driving modern industry and transport.

Perhaps, it might also be linked to other factors, including the military strength of the purchasers, vis-à-vis the relative weakness of the producers, the high value of the product itself, and the ease of its acquisition.

Another important aspect of this phenomenon is total dependence of the Gulf states on a single external power. The "mother country" hypothesis may explain the longing of many Gulf elites to return to their maternal womb, even if they are not always welcome there. Gulf reliance on "the mother country" stems from psychological as well as mundane reasons. For a long time, the lack of adequate local capital and manpower necessitated the existence of international companies to manage, produce and deliver economic goods and services in those societies. This whets the appetite of defense planners in major countries to deploy their power and force in these lands in support of their mercantile interests. The classical three G's of imperialism are very much at play in the region. "Gold, God and Glory" manifest itself not only as motivators to the grand designers, but as real outcome of an external power maintaining a reasonably low price for energy, seeking business contracts for its corporations, suppressing local culture and religion, and building a grand empire commanding allegiance from all.

Imbalance between Long Distance and Regional Trade

Many political economists[1] have shown that trading states tend to be more pacific. Richard N. Rosecrance has indicated that during the American Civil War, New York merchants and financiers were close to the trading cities of the confederacy. He adds, "Hamburg traders were closer to their English trading partners, in the late 19th century, than they were to the imperial class of Berlin."[2]

My earlier research has shown that the city-states of Northern Italy were in close partnership with the Islamic cities of the Eastern Mediterranean. Such a network of trade relations whether local or long distance, tends to create more harmonious relations among trading partners. However, the main issue here is the imbalance between the trading partners themselves and also between the volume of regional trade and the volume of long distance trade conducted by these states.

An examination of current trade relations within the six GCC countries or between them and neighboring regions, indicates that intra-GCC trade amounts to only ten percent of their total trade with the outside world. The volume of GCC trade with other regional states is even lower and does not total more than three percent. Given such discrepancy between the size of those states and their long distance partners, as well as the opaque relation between regional trade partners versus international trade partners, the end result is a situation of inordinate size and trade terms, as well as an increase of dependence and sensitivity towards international trade.

The mix becomes more contentious when we add perceptions, cultural and religious issues. In my view, the major difference between today's international system and the old British system is manifested in the field of culture or cultural politics. Although the British faced tensions with some religious movements within their domains, whether in India or in Sudan, they remained largely sensitive to local cultural values and tried to maintain and perhaps even strengthen the ideological status quo.

New international powers of the day are really different. They resemble the ideologues of the French Revolution who preached a new ideology pertaining to reinventing the state, re-engineering society, and re-framing local culture and religion. This new anathema to local traditions, culture and religion is truly an invention for recreating a regional system. The ultra-right or what journalists have called the empire builders are attempting to put in place their vision in Iraq, but their manipulation and experimentation may not be limited to that country whether they succeed or not and regardless of the human costs. Given their zeal, those builders are apt to repeat their political and cultural experimentation in other parts of the region—Syria, Iran, or even the GCC states. I think the challenge that faces the region, as a whole, including the elites of the GCC, is to make sure that the Iraq adventure in cultural manipulation does not succeed easily in order to prevent similar regional catastrophes.

The Gulf states, Europe and the old liberal America can in one way or other find a new alliance that champions the trade systems versus the new war system propagated by the New Right. There are encouraging signs that the so-called "empire-builders" are recognizing that their aversion to harmonious relations among states and cultures does not fit well with the established major trading concerns of the United States itself.

What is needed here is not merely a return to the old Wilsonian paradigm of respecting the universal right of self-determination, but also the re-emergence of the old liberal cultural paradigm in which the people of the world vie with one another by building up cultural self-respect, not by cultural and religious aggrandizement more characteristic of the pre-enlightenment era.

Perceptions of Competition Within and Competition Without

When we think of the disharmonious by-products of long distance trade, we also tend to think of external competitors vying to lay their hands on a natural product. However, the emerging perception is of a competition to own the output of the land, between local expanding populations and externally hegemonic powers.

During the Cold War, it was fashionable to speak of a race between the Soviet Union and the United States to control the warm waters of the Gulf. However, successful geo-strategic positioning in itself did not necessarily ensure acquisition of the commodity called "black gold."

The old conflicts of the past were largely external in their dimensions, causing little internal impact. Today the situation is completely different. Although couched in religious-cultural terms, the current perception of conflict is between external powers attempting to use all their military might to acquire as much as possible of the available finite hydrocarbon, oil and gas reserves and a rising number among the local population attempting to regain those same high value goods.

Local elites are, therefore, caught in the middle between aggressive international powers that are not only interested in usurping the region's natural resources but also imposing their cultural and religious vision on those territories. Unfortunately, conflict and unrest would be the resulting outcome.

The greater the pressure from the external power, the greater is the internal backlash. The positive visions and hopefulness of the nineties, or what I call the liberal decade, have unfortunately been replaced completely by prophets of conflict and unrest on all sides. Regional peace is not preached by the powers-to-be, rather, it is chaos and managed instability that is the daily norm. Within this new political and strategic milieu, it is no coincidence that regional Gulf elites may find themselves caught between two shifting pressures—one external and the other internal. When the demise of local culture and religion is advocated by theorists of the new international war system, the local reaction by extremist groups is also likely to take a bloody turn.

The Gulf: Between the Trading System and War-like System

There is a general agreement between many scholars that the Arabian Gulf has always been a trading post where the goods of many countries find their destination and provides a meeting place for entrepreneurs in the region who can pick and choose those goods with little tariff and redirect its flow to their respective ports. Trade has always been a hallmark of the Gulf experience, whether during the Hajj pilgrimage season or during the normal tourist season. Today, the Gulf Cooperation Council members, in adopting a five percent tariff level towards the rest of the world, have signaled a welcome to all trading nations to send their wares to Gulf ports, whether for local consumption or for re-export.

Trade, whether local or long distance, has always been part and parcel of the Gulf vocation. It has brought jobs and riches to local workers and expatriates alike. With the flow of oil half a century ago, many people from different corners of the earth found a home and a trading partner in the Gulf region. The Arabian Gulf, in short, is part and parcel of the modern world trading system.

With shrinking international frontiers, the Gulf is positioning itself once again through WTO and other bilateral and multilateral trade agreements to take advantage of the world mercantile system. However, the GCC states in particular, have always been wary of conflicts and war. The military might of its member states is miniscule, particularly in number of troops, vis-à-vis the huge size of neighboring regional armies. Perhaps the GCC states may have some advantage in terms of airpower with its neighbors. The Gulf states have never viewed themselves as military bullies and therefore, had always adopted a reactive defense doctrine, not a proactive one.

In this sense, the elites and public in the Gulf region have always been averse towards the militaristic posturing of other powers either within or without the region. The small size of the population and the limited resources meant that even if some resources were directed towards those armies, more was actually spent on building cities, supporting and educating nomadic tribes in modern living and techniques. Gulf states viewed their armies as job creating and training institutions rather than a fighting force or a war-like institution. In short, given the size and nature of their military, Gulf nationals see themselves as part of the world trading system, but not part of the contentious and militaristic world system.

During the past three decades the reality of the situation was the co-existence of the two systems—the mercantile and the militaristic. Each had its own logic, and each found some sort of equilibrium in which people welcomed and participated in the first system and shunned the second system. What the world is witnessing today is a massive rise of the second bellicose system with injurious repercussions to the Gulf region. The Gulf states, and the GCC in particular, must consider new ways and chart an escape route from the scenarios created by the prophets of war, who visit their region with conflict, occupation and destruction. Failure to escape from this predicament would mean that everything that has been built during the past thirty or forty years is thrown into the inferno of belligerent history, and coming generations will lose hope in the future.

Reinventing an Equilibrium

If the Gulf states yearn for a mercantilist world system, they must work very hard to achieve it by balancing their trade structure between local, regional and long distance or international trade. For a trading culture, autarchy in the region would amount to economic suicide. The challenge facing the elites of our region is to find a nexus between those three spheres of trade in order to create a thriving economy for an expanding population.

Another aspect of what I call defeating the war system is to seek allies among Western elites in the United States and Western Europe who believe in human dignity and dialogue among civilizations. The ideologues of confrontation among civilizations constitute part of the rhetorical ideologues advocating the war system. They must be shunned and marginalized, for if they succeed, they will turn the world into battlefields in which human beings of all races, colors and religions are sacrificed in order to realize their grandiose designs. Another aspect of the disequilibrium that has created the malaise is cultural domination or the public perception that such domination is occurring. The religions and cultures of the world cannot afford such marginalization or even conscious political manipulation of religion. Europe and the world suffered for a long time from religious contentions and/or cultural alienation.

The re-emergence of the "harmonious" world trade system requires the re-introduction of trust among all people and states. Unfortunately, for the past three or four years, trust has been certainly absent in regional politics—in a way it is an abandoned value or commodity. Its reintroduction can well serve the interests of regional states as well as the international community of nations. Mistrust and precepts of aggrandizement and accusation can only lead to further unrest and instability with concomitant costs and grief.

The Gulf Cooperation Council in the Middle

The Gulf Cooperation Council, since its inception in 1981, saw itself as a collective bulwark to stop regional powers expanding their influence and power over the region. This was the logic of its cooperation in the seventies in the face of the Iranian Revolution's tendencies to export itself or some of its ideological variations to the Gulf region as a whole. The eighties witnessed urgency on the part of GCC planners to contain the chaos of the Iran-Iraq war to the two protagonists. In the nineties, the GCC with the support of the United States and other allies succeeded in containing the menace of Saddam Hussein and forced the withdrawal of Iraqi troops from Kuwait. Following the events of September 11, 2001 and the ensuing regional troubles whether in Iraq or in the war on terrorism, the GCC elites are confronted with a totally new paradigm.

First, the challenge posed is not external and not purely political or military. Second, the external protector is now assuming a new function, rather like the old Swiss mercenaries who guarded the Italian city-states. The old partner or ally is changing its identity—sometimes taking on the role of a protector and at other times the role of an occupier. In addition, it is no longer satisfied with the status quo and views itself as an agent of total change—in politics, in the military domain, and in the religious-cultural domain as well. This tendency, which is an aberration of the dominant power's long-established policies and political regime, could present a huge challenge to GCC elites.

Those elites do understand the need to modernize their political system, but their *modus operandi* is that of slow and cumulative change. Now, they are being pushed and shoved by outside powers to accelerate the pace of change. The major power is not ready to address its own shortcomings in the region. Instead, it is shifting the blame for the malaise in a truly Freudian manner on Gulf societies that supposedly nurtured those outcasts.

The blame is not merely limited, as we have shown, to the cultural and religious sphere. Suddenly regional monetary institutions and the whole *hawalah* or money-transfer system, which served as a means of wealth distribution between the Gulf region, and adjoining regions in Asia, is being blamed. Al-Rajhi and Al-Mutawa indigenous financial houses are no longer in vogue in official Washington circles. Instead, those *hawalahs* have to be entrusted to "Western Union" and similar financial institutions. The preferences of outside powers are given precedence over local and regional financial needs and requisites. To maintain their independence, GCC countries would have to intensify their cooperation—culturally, politically and economically. This is part of the rationale behind the GCC's move to integrate their member states through the Free Trade Region and Customs Union formula enacted in 2002. Economic cooperation is envisaged and moves are being made to integrate the member states through the establishment of bank branches in each other's capitals and cities. The existence of these banks would enhance competition in this sector as well as allow consumers and entrepreneurs more competitive lending rates, which is regionally higher than other countries of the world, particularly for small businesses.

Another daunting task for GCC planners is their ability to reduce their skewed dependence on the long-distance trade of a single commodity in favor of a balanced regional and international trade based on multiple commodities. This has posed a challenge to development planners since the mid-seventies. Although the Gulf has established a world-class petrochemical industry, which accounts for between eight and nine percent of the world petrochemical production, Gulf states continue to compete with each other instead of integrating their operations in this field.

Some have suggested that inter-GCC trade could be enhanced through product specialization. Saudi Arabia has the largest territory and population within the bloc. To a certain extent, it has specialized in providing agricultural products, dairies and other land crops. Due to low return on the prices of those commodities, it would be difficult for the country to continue playing this role. However, for political and social reasons, Saudi agriculture, which contributes eight percent of GDP and employs a significant segment of the population, would continue to find a market for its products in the GCC countries. The establishment of a customs union, as we have seen in Europe, entails the free movement of people as well as goods between member states. Although a visa is no longer needed and most GCC nationals move between member countries using national identification cards and not passports, there is no true freedom of movement for the factors of production. From time to time, a Saudi lady broadcaster may work in a UAE radio station or a Bahraini banker may work in an neighboring country, but we have yet to witness the free movement of GCC workers between member states. And although most GCC states suffer from varying levels of unemployment among its youth, the free movement of would-be workers would not only boost economic integration but would also help to further social integration through mixed marriages, as their forefathers have done who worked in pearl diving and other vocations related to the sea and in the process built a web of interlocking social and family relations spanning the whole breadth of the Gulf region.[3]

The GCC after the Occupation of Iraq

The traumatic invasion of Iraq in March 2003 by the United States and Britain revived the whole nineteenth century experience on the doorsteps of the Gulf. Saddam was an adversary with whom nobody could sympathize and his military might was a source of unease rather than comfort. Nevertheless, his removal by force followed by the occupation of Iraq was both an injury and an insult to every Arab, regardless of his feelings towards the dictator.

The occupation of Iraq by an outside power and the forceful dismantling of the modern Iraqi state was an unprecedented event in modern Arab history. It is not that outside powers were respectful of internal Arab politics, but their attempts to change the status quo throughout the twentieth century were conducted through *coup d'états* against uncooperative regional regimes. This was the case with the 1936 coup in Iraq and subsequent coups in Iraq, Syria and other regional states. Military leaders during the second half of the twentieth century became the preferred pawns of external powers, be it the United States or the Soviet Union. Once the political establishments in those Levantine countries proved uncooperative, the military was ready to topple even democratically elected regimes. This was the *modus operandi* in most military dominated countries and societies in the Near East, in most Arab states, and in Turkey.

Luckily for the Gulf states, they escaped the military coup phenomenon that swept most of their Arab and Middle Eastern neighbors. This resulted in good measure from the legitimate social base that the leaders of these countries enjoyed, as well as the increasing inflow of wealth, which allowed these regional states to extend the benefits of a welfare state to many classes and areas within their boundaries.

The ship of state in most GCC countries, with the sole exception of Oman, was able to navigate through the rough sea of regional politics in the 1950s and 1960s without major civil wars, coups or social unrest. From the 1970s, the GCC states were blessed with an increasing flow of wealth that benefited their countrymen and expatriates alike. Even though the regional system was not always conducive to political development, the combination of oil wealth and an internationally agreeable milieu produced a tangible leap in economic development in those societies.

With the current military and political threats emanating from coercive international powers, it would be very difficult for regional states to conduct their policies as they had done earlier. The movement of people from the pasturelands of the countryside to the cities has also brought with it a decline in tribal attachment. Even today when the first migrant generation tends to live together in certain quarters of major cities, the second generation tends to move away from their parents both physically and emotionally and cling ideologically to new symbols and values that perhaps differ from their forefathers. While there is a real need for sociologists to conduct research in this field, one finds that the new political groups emerging on the scene are no longer confined to a certain tribal or even regional base. This development has both a positive and a negative dimension.

On the positive side, there is an emerging national identity being formed among the young generation, and future political parties can find an intra-national identity base. On the negative side, the old tribal value system is fading among the new generation and there is a need to introduce new thoughts and ideologies to fill the vacuum. In this regard, the Islamists in most GCC countries have been most successful in guiding such a transformation, perhaps, even by modifying the old value system.

What is lacking in the Gulf states is an inspiring Gulf-wide emotional attachment that would complement, and at a future point, replace local attachments. For twenty years, the GCC had emphasized collaboration among the elites and technocrats. With an increasing flow of tourism among the six states, a new awareness of each other's society is taking place. What is needed is a major interlocking web of interaction among the youth. Annual football tournaments tend to intensify feelings of nationalism among the competing national teams. However, there have been no GCC-wide teams in different sports that could represent the GCC as a whole at international sports competitions.

In the political sphere, every major power attempting to expand its domain of influence since the dawn of history has been keen to develop an appropriate ideology to convince folks at home, to disarm ideological opponents abroad and to win potential supporters. The US utilized the so-called mission of "keeping the world safe for democracy" in the 1950s and the 1960s. The USSR, likewise, during the Cold War era rallied its supporters around the call of "universal socialist solidarity."

Today with the collapse of the peace paradigm, we are back to point zero. "You are with us or against us. Even if you are with us, we will change you or reform you." Thus even traditional allies of the US will find this new cry of the country's ultra political right extremely menacing because it would mean that such change is based on a continuum of policies ranging from political and economic pressures to outright military invasions. Of course, this zeal is sometimes couched in attempts to expand the democratic process around the world. Yet, this Napoleonic contention cannot escape the attention of peace groups in the US and abroad as such rhetoric is nothing but a veil for outright expansion of a big power's influence. Even such a contention is not completely accurate because the major power elites know fully that any democratic process in these lands will bring adversarial Islamic movements to the fore.

In the aftermath of the Iraq invasion, GCC leaders will find it uneasy to work with a forceful and prevailing ally. They are not only caught between external pressures and internal dissent against the existing international regime, but their ability for maneuvering will also be extremely limited because the external power, acting in haste, could even humiliate its own regional allies.

The only option open for those allies, including GCC states, is to work within the US political system, seeking internal partners in the US who would influence proposals before they become an outright foreign policy. Such a task would require a dynamic foreign policy mechanism that not only monitors developments on the American political scene, but also a new approach to public opinion and the US media. GCC interaction with the governing elites in the US has historically been limited to the State Department and other leaders of the executive branch of government. The support of some US business groups has sometimes been sought, particularly those groups with business interests in the Gulf region. The GCC *modus operandi* sprang from earlier leaders' experience with British and European governments, whose executive branches have full control over their countries' foreign policies.

On the other hand, the Israeli lobby in Washington has historically been very influential at the grassroots and political party levels. Once candidates assume the political seat of power in Congress, they would be very familiar with the issues advocated by local political leaders and tycoons of the media, banking and industry. If those leaders also support continuous US backing of Israel, they would compel their representatives in Congress to follow suit. The convergence of elements of the religious right in the US, particularly in the south, in the last twenty years towards the Zionist agenda has also empowered the Israeli lobby with such momentum and force at the Congressional level that no other country can match the Israeli influence and power in those political circles.

To their detriment, the Arab states in general and the GCC states in particular, have historically ignored the US Congress and as a result the Israeli lobby reins supreme in the political body. Henceforth, if the Gulf states were to adopt a more vigorous approach, they would pay more attention to US Congressional debates because any laws or regulations enacted in that body will impact on the global political situation. The Congress in Washington for many years to come will become the Congress of the World, passing laws pertaining to peace and war that will touch the life of every human being on this earth.

Three Mindsets of the Future

The challenges facing the Gulf states have resulted in three different reactions. The first, as alluded to earlier, is what I call the fear among some elites of leaving "the mother country's womb." They have lived for a long time in a sort of emotional and psychological dependence on long-established ties that compel them to wooing one big power or another. I think this approach is detrimental to the future growth and cooperation among the six Gulf states. Once such a vision is adopted, each group of national elites would be competing with the others for the admiration of the superpower, even if such an embrace would harm the region-wide goals of economic and political development. The bargaining power of the GCC states, both individually and collectively, will also suffer in the process, because intrinsic national goals will be confused and perhaps, even in conflict with the GCC goals. Joint policies would have to be sacrificed or delayed as individual priorities of member states are reordered.

A healthier approach would find its way in trying to adapt national goals into a region-wide policy. Once adopted, then the whole bloc can stand up and defend those policies. Even if forced to change such policies, the bloc as a whole would negotiate adjustments that would gain a reciprocal price for such change. And the interests of the bloc would not, therefore, be damaged by escape clauses from certain members of the bloc.

In due time, the psychological dependence on outside powers would be replaced in a healthy manner by a collective mechanism of protecting the interests of the regional entity as a whole, even if it means a diminution of some marginal advantages enjoyed by the member states.

The second mindset already in place is the mass rejection of the influence and manifestations of outside powers in the region. This is a natural outcome of solidarity with other Arabs and Muslims who are killed and humiliated in Iraq and Palestine. For too long, the world has neglected the fate of these people and there is an emerging feeling of outright anathema to Islamic customs and values in various western countries similar to the political milieu of the 1930s in Europe. For many people living in this part of the world, the racism of Israeli Prime Minister Ariel Sharon has been contagious, touching policies in the United States and Europe that are affecting Muslims within their own societies and in the world at large.

The reaction in the Gulf towards these developments has been threefold. The first reaction has been to cling fervently to Islamic values in an unnatural and unhealthy fashion. Islam has always been a religion of tolerance and justice, regardless of race and color. It had also tended to adjust its teaching to existing ethnic cultures and values. By the same token, Muslim societies when they felt cornered, marginalized and confronted by crises have unfortunately exhibited certain intolerant tendencies, particularly by small ideologically fervent groups. However, the mass of Muslim public opinion has, by and large, clung to *wasatieh* or centrist and non-radical tendencies.

The second reaction among certain segments of the public is to advocate detaching oneself wholly from western influence and rule. Their argument is based on the contention that certain governments have malevolent policies toward Arabs and Muslims, and there is neither logic nor room for engaging with those who wish ill for Arab and Muslim societies. The logic of what I earlier termed the "world conflict and war paradigm" is met locally by people who seek divorce from this world of occupiers and conquerors, as they view them.

I think this reaction is perhaps too cautious and pessimistic, for the world conflict paradigm cannot thrive forever. The prophets of conflict can only survive in the short run, not in the long run. People all over the globe cannot afford the wages of war, since their very existence relies on economic exchange and commerce. A major cornerstone of this economic exchange is the creation and maintenance of peace, not the propagation of chaos and war. Peaceful people and groups hold the key to the future, for it is they who will succeed in the long run. In advocating retraction from the world mercantilist system, such groups in the GCC are only being influenced by passing emotions and not by any real comprehension of intrinsic interests. The GCC states and societies cannot afford to marginalize themselves once they have become a central part of the modern capitalist system.

The third and more realistic reaction in the Gulf has been the accommodative approach. It neither fears the future nor mourns the past. It is more confident of itself and its role in the world. The fact that the Gulf states hold some forty five percent of world oil reserves, and constitute the largest economic power in the region ought to induce any major world power to befriend the GCC states and enhance relations with the GCC governments and people. The control of the GCC states over such huge reserves has been exercised to serve the interests of oil consumers the world over just as it serves the interests of the producers. This is so, especially in times of wars and crises that befall our region. For example, the Saudi Ministry of Finance has estimated that increased oil production during the 1990 Gulf War has saved western oil consumers approximately forty billion dollars.

Such statistics pertain only to one Gulf oil producer and not to other Gulf oil producers, nor does it cover other regional crises during which increased oil production in excess of normal Gulf capacity, has saved the world economy billions of dollars. The extent of such savings has yet to be accurately calculated by economists.

The accommodative approach tends to face the facts and not avoid them; it tends to bring to the fore the latent sympathy of billions of people with the Islamic holy lands and to utilize such emotions and rationale in defeating unfavorable designs in the future. The war-like paradigm with its concurrent modalities of racial and ethnic aggrandizement cannot hold the people hostage to such a mentality. It has to be logically shown to be defective and injurious to all people and cultures and harmful to world exchange and commerce. The lessons of the past are clear—states that prioritize aggrandizement and force-projection tend to defeat their own purposes in the long run.

Gulf proponents of an accommodative open system in the region tend to be more realistic about current challenges and future opportunities. They base their policies on not-too-wishful thinking—the abhorrence of war and the gains of trade. They try to build, negotiate and interact. Locally, they bring to fore the centrist values that in due time would defeat some of the radical approaches. Regionally, they seek to cement similar positions and advocate a region-wide emphasis on the regeneration of culture, learning and the modalities of ordered change. The old tribal modalities are no longer capable of carrying the day and regenerating the engines of development.

A new modernization agenda that emphasizes reform of the educational system is the cornerstone of any meaningful advancement. This explains why Saudi leaders projected these ideas at the Kuwait Gulf Summit in December 2003. In this approach, the agencies of a region-wide mechanism are employed to advance an essentially national modernization program. Such an approach is tantamount to enmeshing national goals into a region-wide agenda. Like-minded regional allies are also mobilized to advance essentially national and local problems.

Low Politics Vs High Politics

For a long time, the European experience of regional integration has always focused its energies on harmonizing low political issues in a cumulative way as a means of expanding the extent of cooperation among their member states. This *modus operandi* has worked generally well in the European collective experience. Once engaged in high politics, such as the finalization of the European constitution, the process exhibited certain limits and flaws. Nonetheless, the process worked beautifully in bringing together the European economies into a large bloc that engages the world in a way that is most beneficial to its member countries. Europeans today can travel within their boundaries freely. They can study, work and invest in any of the member states, and live under the same laws of work and taxation in any member country.

Likewise, the Gulf states can and do follow the European model of integration. The added local dimension in our experience is the national dimension. For some GCC countries, identification with a nation state transcends any other identification. In other member countries, the reverse is true, and identification with one's own national feelings, at least among the young is at low ebb. The realigning of educational systems would bring values of socialization at an early stage to some common standard.

Low politics such as emphasis on economic cooperation would bring tangible results that can be felt by the ordinary man. Cooperation in education, investment, trade and employment can encourage a certain positive outcome, which would enlarge the public base for the idea of building a united form of government, or similar policies or parallel processes in the Gulf arena.

Low politics may also have the positive element of avoiding direct interference from outside powers. Some of the European states may see the GCC as another potential EU-like model. Others will view it as merely an economic, not a power-based alliance. Even local political elites will not fear the rise of a competing power base for their established bases of authority. The existence, perhaps even the success of the regional development enterprise would have built a strong substructure, which even if it does not reach the desired unification, would at least keep the organization intact amidst turbulent regional and international challenges.

Back to Basics

Existing challenges will push the GCC nation states into depending more on their power base. Just like vines clinging to the branches of a tree when buffeted by a storm, the current turbulence will force leaders to work closely with the public to chart a new course for the future of their nations.

Challenges directed towards GCC countries and societies are likely to initiate a healthy process of open debate and the reordering of priorities. What Saudi Arabia is witnessing today, for example, is intensive debate among thinkers and leaders of different schools of thought, which is a very healthy process. Hopefully, this would not only open up a wider dialogue for interchange of ideas and thoughts among intellectuals and religious leaders but would also be part of a new generative model of modernization in culture, education and good governance.

Given a strong base, the GCC states will regain confidence and strength that would make them deal with internal and external challenges on a stronger footing. The greater the external pressure, the greater the need to re-engage new segments of the population and new sections of the intelligentsia. The regional domain can provide an arena in which ideas, thoughts and experiences can be shared between and within societies. The success of the regional enterprise would depend on its ability to recognize those new strata and bring them into its fold at the regional level.

By exchanging modernization programs and ideas at the regional and national levels, we can realize and bring to fruition the process of national social differentiation and regional sense of identity.

Conclusion

The Gulf region is in a peculiar situation as a producer of a single commodity that is traded mostly via a long-range exchange. This kind of exchange is likely to lead to instability and war, as it brings to the region external powers wanting to acquire control over a strategically valuable commodity to fuel their economies and means of transport. The imbalance between long-distance trade and regional trade may lead to a certain degree of dependence that lessens the ability of regional leaders to negotiate with outside powers on a meaningful footing.

This chapter has advanced the notion that elites in the GCC are caught between the external pressure of "the empire builders," who are not only interested in usurping natural resources but also in imposing their cultural and religious vision, and the local population of the Gulf societies, who are proud of their own culture, religion and values. The GCC, as a regional organization, is caught in the middle.

Furthermore, the Arabian Gulf is largely wedged between the trading system and the war-like system. By its nature and traditions, the Gulf region is close to the proponents of trade and commerce and not to those who seek conflict and war. Pacifism has been an ethos for both the states and the people of the Gulf region, who have recognized for many centuries that peace has a dividend, while the wages of war are too prohibitive.

In this context, the role of the GCC is not merely to function as a trading bloc but to emerge as a forum offering a middle ground, where ideas of modernization and change are pursued in order to create a local and a national consensus for introducing new modalities of change and political development.

19

The Role of the GCC in Promoting Policies of Regional Coexistence

Abdul-Ridha Assiri

The events witnessed by the Gulf region in the late 1970s and early 1980s, as manifested in the Iranian Revolution and the outbreak of the Iran-Iraq war, had important effects that expedited the formation of a new regional system—the Gulf Cooperation Council (GCC). This stemmed from the fact that these two events represented both an internal threat and an external challenge to the traditional regimes in the region. The GCC, which is now nearly a quarter of a century old, emerged as an organization seeking to enhance coordination and integration between states with similar systems in regard to different fields: politics, economics, military and security affairs. The GCC is also meant to function as a shield to ward off external threats.

Needless to say, regional groupings have become imperative for small countries in the aftermath of the disintegration of the Soviet Union, and the emergence of a unipolar international order for the first time in history. Regional systems consist of two or more countries located in one geographical region. These systems are usually permanent institutions invested with a kind of will and an independent legal identity and their aim is to realize common goals. Some of these systems enjoy supra-national authorities and powers that transcend constitutions and national institutions.

It is worth noting that the concept of forming regional systems is not a product of the moment, nor a current historical phase. Neither is it a novelty in the world of economics and politics. Those who study its emergence find difficulty in tracing its roots back to past centuries, especially after the age of industrial development in Europe. However, these regional systems generally take the form of military alliances, which go far beyond mere commercial or economic groupings. Regional systems are built on the basis that working in a unitary, integrative or coordinated manner at the regional level is more effective than doing so at the global level. Countries of the same region have common interests and close ties, which enable them to contribute effectively to solving the problems of the area or region.

The interwoven interests of communities make local problems a source of tension and a fountainhead of danger to neighboring countries. This necessitates a collective inclination to find effective solutions to such problems. In the light of rapid changes in communications and transportation, it has become difficult to contain the focal points of tension within their national context, or even the regional milieu. For instance, issues of desertification and water scarcity require regional or even international cooperation to find solutions. Failure to adopt this approach means ongoing conflicts and disputes over water sources. Thus crises exacerbate and problems become more complex, with final effects reflected at the international level. Therefore, regional systems are considered the first stage and a basic building block, in the process of forming a global vision of the future of the earth and humankind. They represent a step toward developing, enhancing and integrating the desired world order.[1]

Similarly, it was natural for Arab states on the western coast of the Gulf to proceed to establish a regional organization comprising states with similar political, social and economic systems. The foundation of the GCC can be viewed as a process of creating a local defense mechanism rendered inexorable by the state of tension and alertness following the Iranian Revolution in the late 1970s, and the subsequent outbreak of the Iran-Iraq war. Owing to the intensity of that war, the GCC states sought to establish a military system (the Jazeera Shield Force) with the task of confronting the potential dangers of war threatening these states.

The Iran-Iraq war constituted a central challenge to the ability of the GCC system to confront external dangers. However, the greatest challenge to the efficacy of the system centered on the ability of member states to dispel skepticism about their mutual intentions. This was so because the Qatari-Bahraini, Saudi-Qatari and Emirati-Omani and other border conflicts represented a setback to the principles of cooperation, coordination and integration to the extent that the majority of the resolutions taken at the GCC summit meetings had become unenforceable. The most significant of these is approving a common Gulf currency, applying the common security accord and agreements on facilitating the movement of citizens and financial resources. The crisis of confidence between regimes has led to the loss of confidence on the part of citizens in the importance of the Gulf system and its efficiency. This loss of confidence extends even to the possibility of developing the political and economic systems of the GCC states.

At another level, the Iraqi threat to Kuwait, followed by the Iraqi invasion and occupation of Kuwait in 1990, represented one of the greatest challenges to the system. This challenge centered on the ability of the Gulf system and its member states to meet the politico-military consequences of the invasion and occupation. That crisis attested to the inability of the GCC states to confront the military forces of the former Iraqi regime. This led to assistance being sought from foreign forces under the umbrella of international legitimacy, which was the only means of dealing with the situation resulting from the invasion and occupation of Kuwait. The incident revealed the limited capabilities of the Council in dealing with such threats in the absence of foreign forces acting under an international legal cover. Also, the "liberation" of Iraq and the toppling of the regime of the former Iraqi president Saddam Hussein in 2003 created new and serious security threats to the GCC states. In this context, a crucial question arises as to whether the GCC can assign itself the task of rebuilding confidence between countries of the region, establishing a security system comprising neighboring states and constraining the likelihood of confrontation, a prospect that has plagued the region for more than two decades.

Within the historical context of the GCC and the development of its mechanisms and institutions, different ideas, various proposals, scientific models and several policies have been presented to remedy its political deficiency and overcome the difficulties arising from lack of confidence among the GCC states, and between these states on the one hand, and their neighbors on the other. Undoubtedly, the relations between the GCC states and their neighbors in this sensitive part of the world are as important as the relations among the GCC states themselves. This chapter presents a thesis for rapprochement between the nations and states in the Gulf region. It calls for developing relations between them via cooperation among governments in the form of common organizations and collaboration between professional, economic and cultural elites. This model consists in analyzing and comprehending the interests sought by various governments. It also blends idealism with realism to provide an integral political formula that has proven its feasibility in building confidence between different nations and regimes.

The Concept of Regional Coexistence and Realization Strategies

According to Winston Churchill's evaluation, the Second World War was supposed to be the last of all wars, whether global or regional. It was hoped that the world, after the war had ended, would direct its efforts towards building welfare states and distancing itself from armed conflicts. However, beautiful dreams seldom materialize in the stony ground of reality. In a few years after the end of the Second World War, the conflict between former allies emerged after the world had been divided into two blocs: the Eastern and Western blocs. Churchill, who dreamt of an end to all conflicts, called the Soviet Union and its satellite states the "Iron Curtain states." Thus a new form of conflict, known as the Cold

War, came to the fore. However, it would be more accurate to call it war by proxy because the two superpowers engaged in rivalries and armed conflicts via allies and agents, contenting themselves with the provision of financial and propagandist support. Sometimes one of the two superpowers would engage in an armed conflict with an ally of the other without the two confronting each other directly (such as the wars in the Korean Peninsula, Vietnam and Afghanistan).

However, relations between the East and West took another course with the appearance of the concept of "peaceful coexistence." This concept was first formally presented at the nineteenth conference of the Soviet Communist Party in 1952. It became a general policy of the Soviet Union after the twentieth conference of the party. Peaceful coexistence was supposed to mean, in the context of the Cold War, the possibility of having simultaneously two social systems in the world, both socialist and capitalist, and building proper socio-economic relations based on avoiding outbreaks of war between the two systems. This did not exclude rivalry between the two in the context of specific rules. Though peaceful coexistence basically meant accepting the other as a legitimate party to the international system, it acknowledged at the same time the existence of domains of rivalry and cooperation between the two parties. The application of this policy succeeded in solving many problems between the East and West, and in avoiding the outbreak of a Third World War.

Nikita Khrushchev, credited with promoting the concept of peaceful coexistence when he held power in the Soviet Union, was expelled from the Soviet Communist Party and from power in 1964 for various reasons—including the policy of peaceful coexistence. Moreover, adopting this policy led to the outbreak of the Sino-Soviet conflict. In spite of this, Khrushchev's successors also continued the same policy. Apart from this, they developed the policy into a higher level of coexistence via the concept of "*détente*," which was ceremoniously instituted in the American-Soviet agreements signed in Moscow in 1972. *Détente* was not merely a Soviet policy at that time, but a general policy of the two superpowers. It did not refer to a conformity between the two superpowers as regards the course taken by the international system, but to "alleviating" the causes of conflicts between them, and creating controls to rein in the possibility of escalating conflicts between the two, within the context of agreements on arms control and direct contacts. The policy of *détente* paved the way for rapprochement between the East and the West within the framework of the Conference on Security and Cooperation in Europe (CSCE), the basic principles of which were laid down in the Helsinki Accords signed in 1975. These accords laid down the principles of coexistence between the East and the West till the end of the Cold War. The accords referred to confidence-building measures as a tool to guarantee the stability of the regional agreements reached, and also proposed three areas for cooperation between the East and the West – political, economic and cultural – thus confirming the interconnections between these fields. The accords succeeded in realizing coexistence between the East and the West, and contributed to the transformations that led eventually to the collapse of the communist regimes.

The disintegration of the Soviet Union neither ended the struggle between the East and the West nor realized the dreams of British Prime Minister Winston Churchill nor those of US President Woodrow Wilson before him. The twenty eighth President of the USA, Wilson was contemporaneous with the First World War and his "Fourteen Points" formed the basis of the League of Nations. In the aftermath of the collapse of the Soviet Union, ethnic conflicts erupted in the states that were heirs to the communist regimes (the Balkans and the Caucasus) and in the Third World (Africa). However, it was clear that the intensity of the new conflicts, together with the opportunities created by the end of the Cold War, accorded greater importance to the concept of coexistence. In this context, the concept of "Common Security" came to prominence. This concept was first used by Michael Gorbachev, the last President of the Soviet Union, in the context of what was known as "New Thinking," which is supposed to mean that security is not a conflict value, but a common value according to which all states can coexist.

The concept of common security spread after the collapse of the Soviet Union because it gained a new significance: namely, the endeavor on the part of conflicting parties to achieve security "together" and not via confrontation. This concept is similar to the concept of dialogue instead of conflict between civilizations, which was posed by the Iranian President Mohammad Khatami. The concept was also associated with the new concept of "peace-building" proposed by Boutros Boutros-Ghali, the former UN Secretary-General, in his report entitled *An Agenda for Peace*, which he presented in 1992. "Peace-building" means those operations seeking to end the causes of violent conflicts, and to ensure that they will not recur. These operations include measures at the national, regional and global levels that guarantee building and enhancing peace. Such measures lead to coexistence between parties that have been involved in conflicts, and guarantee the continuity of such coexistence between them.

Regional coexistence can be defined as an agreement between a group of countries in a particular region, which have no alternative save the acceptance of a "common existence" within this region in the context of certain agreed principles. These rules ensure common security for all, so that each party gains some benefits and costs are equally shared in pursuit of preserving the security and wealth of the region. In fact, "regional coexistence," or rather coexistence between countries on a regional level, is one link in the chain of coexistence processes between countries on a world level. This pattern of co-existence requires the adoption of a number of interconnected strategies which guarantee sustaining the concept of coexistence. There is consensus in coexistence literature on the multiplicity of realization strategies. At the national level, there is reference to building political systems on the basis of effective public participation in the process of decision-making. This is based on the assumption that coexistence between democratic states is easier and more enduring than coexistence between undemocratic states. This view has been proved in the literature of international relations. Economic openness is another strategy, together with broadening the private sector role in the process of development without shedding off the guiding role of the state. At the regional level, there are a number of strategies, the most important of these being as follows:

- Resolving regional conflicts via peaceful means as prescribed in international charters.
- Establishing arms control regimes that guarantee a balanced regional system in which all parties feel secure and include the surrender of weapons of mass destruction and the control of arms imports and their domestic production.
- Developing a number of "confidence-building measures" among the parties, which comprise a series of measures such as assuring each party that it will not be attacked suddenly by the others, informing other parties of arms imports and planned military exercises, and establishing direct contact lines between all of them.
- Establishing a "regional system" for general security, comprising cooperation in the fields of economics and environment, such as founding joint ventures for investing in national and common wealth, and working collectively to protect regional environment against deterioration, especially with regard to water scarcity, desertification and biodiversity.

As for the supra-regional level, coexistence includes links with trans-regional and global institutions, which sustain coexistence between regional states, such as the links between non-European states and the Conference on Security and Cooperation in Europe.[2]

Developments in the Strategic Environment of the GCC States after 9/11

Examining the GCC strategies to achieve regional coexistence requires an understanding of recent developments in its strategic environment. This strategic context is the stage on which the Council's strategies for achieving regional coexistence are to be implemented. The strategic environment of the Gulf region has witnessed radical transformations after 9/11. Most of the concepts formulated prior to this date have become history, and that there is now a need to plan new strategies that conform to new contextual realities. The changes in the strategic environment of the Gulf region after that date will be summarized in the following section.[3]

First, the events of 9/11 have proven to decision-makers in the United States that US territories are not isolated spatially from global problems. Immunity from such problems was the concept adopted in varying degrees by American politicians prior to these events. Based on this new premise, the United States has reformulated its national security priorities at both the internal and external levels. At the internal level, it has modified the laws governing immigration, residence and visits. The margin for personal liberties has shrunk and the authorities governing security and intelligence agencies have been strengthened. This orientation has culminated in the creation of the Department of Homeland Security, which was officially instituted in late 2002.

At the external level, the United States administration has placed "counter-terrorism" at the top of its priorities. It has invented the concept of "pre-emptive or preventive strike" as a basic strategy for dealing with the sources of threat and danger before they grow and flourish. Moreover, the United States has adopted the concept of dividing the world into two camps—allies and enemies. This division is based on a classification of world states according to their degree of support for the United States in its long war against terrorism. As a major complementary aspect of the new strategy, the concept of "non-proliferation" has been modified to become "dispossession of weapons of mass destruction," with reference to the states classified as enemies, irrespective of whether or not they are signatories to the international conventions banning weapons proliferation. Additionally, the United States has extended the list of weapons of mass destruction to include several materials and devices capable of dual usage. What has helped the United States to follow this strategy, which was announced by President Bush in his address of September 2002, is its unilateral wielding of military and economic power under a unipolar world order in which it enjoys hegemony. This has led to a tendency on the part of many major powers towards conforming to the new US strategy in the hope of avoiding its dangers or gaining benefits.[4]

Second, the events of 9/11 led to the strengthening the trans-Atlantic Euro-American coalition, and to its being substituted for the United Nations (UN) in the domain of preserving international security. The North Atlantic Treaty Organization (NATO) started to play new roles beyond its borders, such as the task it assumed in Afghanistan, where it has taken command of the International Security Assistance Force (ISAF) since August 2003, and its emerging role in the Mediterranean region within the framework of the Atlantic-Mediterranean dialogue which began in 1996. However, as a result of escalating US unilateralism in contemporary international relations, and the adherence by the United States to its own idea of how to deal with international terrorism, some European countries began to criticize the new US orientation. Perhaps this explains the Franco-German criticism of the US policy towards Iraq. The Euro-American divide was exacerbated owing to the fact that a number of East European countries, which recently joined NATO, adopted a position more supportive of the United States than the original member states of the alliance.

Third, East Asia has witnessed in the aftermath of those events, a strengthening of the traditional coalition between the United States and Japan. This is considerably stronger than the ties that prevailed in the darkest stages of the Cold War, particularly after the escalation of the US-North Korean conflict over the latter's nuclear installations. This escalation has heightened Japan's threat perception regarding North Korea. As a direct result of these developments, Japan has sought to strengthen its coalition with the United States on the East Asia front by providing financial aid to the Iraq reconstruction fund. At the Madrid Conference for donor states, Japan undertook to contribute 5 billion dollars towards the reconstruction of Iraq (1.5 billion as grants and the remainder as loans). Moreover, the US-Japanese coalition gained prominence in July 2003, when Japan agreed to send a paramilitary unit to Iraq to participate in maintaining security and providing humanitarian aid. This move by Japan is unprecedented since the end of the Second World War. Many Japanese consider this a violation of their constitutional principles which ban the involvement of Japanese troops in areas of military conflicts unless carried out under the umbrella of international law. Although the Japanese contribution has been rather symbolic and oriented more towards humanitarian aid, it is viewed as a turning point in the transformation of Japanese foreign policy.

Fourth, the most significant international changes in the aftermath of the 9/11 events have possibly occurred in south and central Asia, where the relations between the United States and the major rival powers in south Asia – India and Pakistan – have been strengthened. These two antagonistic states, which fought three bloody wars during the last century, have found it imperative to support the United States directly in the war against the Taliban regime in Afghanistan and Al Qaeda organization. The attitude of the then ruling Hindu nationalist party (the Bharatiya Janata Party, BJP) towards events in Afghanistan is understandable and justifiable since the party represents the more fundamentalist factions of conservative Hindus. Moreover, India aspires, via its contribution to the war against the Taliban, to broaden the concept of terrorism to include fighters seeking the independence of Kashmir. However, the stand of the Pakistani government has undermined the legitimacy of the regime of President Pervez Musharraf. This stand is exemplified by its strong support of the United States and the fact that it has waged military campaigns against members of the Taliban and Al Qaeda organization on the Pakistan-Afghan border, apart from escalating the campaign against extremist religious parties inside Pakistan.

Viewed from another angle, the Indo-US alliance is one of the most important developments in the Indian sub-continent since the foundation of Pakistan. For over half a century, India has been a bastion against the spread of American influence in the region and a major and active party in the Non-Aligned Movement (NAM). It is worth mentioning that the Indo-US rapprochement had started prior to the events of 9/11. However, those events expedited these rapprochement attempts and facilitated India's orientation towards developing normal relations with Israel, the strategic ally of the United States. This orientation reached its zenith when the Israeli

Prime Minister Ariel Sharon visited India in September 2003. The visit was the first by an Israeli Premier to India since diplomatic relations between the two countries were established in 1992. Following that visit, an agreement was signed between the two countries, according to which Israel is supposed to sell Phalcon advanced radar systems to India. The sale of this radar system will add to the military imbalance between India and Pakistan.[5] This new situation has confronted Pakistan with a dilemma, forcing it to strengthen its relations with the United States, defuse tensions with its bitter neighbor in their bloody struggle over Kashmir and intensify its confrontation with the internal forces of political Islam. Moreover, the disintegration of the former Soviet Union has allowed the United States a military presence in Kazakhstan and Uzbekistan, bringing both East and South Asia within the circle of American influence.

Fifth, the US-Israeli alliance is a unique one. It is not only an alliance built on mutual interests, but one that derives its continuity and development from religious Talmudic teachings. It also draws on the general perception among US politicians of their constant need for Jewish political and financial support, especially since the Jewish lobby is the most active one in the domains of US politics and media. However, the events of 9/11 have strengthened this conventional alliance in an unprecedented manner, since the United States has supported the liquidation campaign waged by Sharon against Palestinians in the name of countering terrorism. Paradoxically, this support comes from a Republican administration, although Republicans have been traditionally less supportive of Israel than the Democrats. In his address on June 22, 2002, George Bush claimed that the source of the Palestinian-Israeli conflict is "Palestinian terrorism," emphasizing Israel's right to liquidate terrorists in self-defense. Moreover, despite the historic concessions made by Arabs in the Arab Peace Initiative, which was approved unilaterally in an Arab Summit Conference held in Beirut in March 2002, Israel has escalated its liquidation campaign against Palestinian people with overt support from the United States. The United States has actually protected Israel against any condemnation by the UN Security Council. The US support and bias towards

Israel have been clearly embodied in the articles of the "Road Map" plan of 2003. The first article in this Palestinian-Israeli peace plan permits moves to liquidate the Palestinian resistance and disarm Palestinians without disarming Israelis living in the occupied lands.[6]

The United States has also set itself the task of dispossessing Arabs and Iranians of weapons of mass destruction without referring directly or indirectly to Israel's possession of such weapons. This is part of the strategy of sustaining the strategic imbalance in the Middle East in favor of Israel.

Generally, we can refer to the genesis of a new regional reality in the Middle East, which is manifest in the following two elements. First, American power is now a regional reality as the United States has become capable of direct moves in the Middle East, and of engaging in major wars in the region on short notice. This is attributable to the fact that the temporary military facilities which were the mainstay of the US military presence in the region have changed into "military bases." Also, the United States now threatens to occupy some of the regional states so as to change their political regimes either in the same way as it did in Iraq, or in some other form. The most notable examples of these threats are those directed at Syria and Iran. What is more important is the fact that the United States is now capable of moving its military force unilaterally. Second, the Arab world is disintegrating and is clearly unable to influence developments in the region. This is evidenced by differences among the Arab states with regard to the attitude towards the US campaign on Iraq and the failure of attempts to reform the Arab League. Thus a new reality in the Middle East is gradually taking shape, which is characterized by direct US military intervention and growing Israeli dominance with a corresponding disintegration of the Arab world.[7]

Sixth, a number of regional and supra-regional institutions have sprung up in the wider strategic sphere bordering the Gulf region. Perhaps the most important of these in the economic domain is the Indian Ocean Rim Association for Regional Cooperation [IOR-ARC], which was established in 1997 with its headquarters in Mauritius. It includes three Arab states, Yemen, the Sultanate of

Oman and the United Arab Emirates, and sixteen other states, the most prominent of which are India, South Africa and Australia. This association was established with the aim of liberalizing trade and facilitating the movement of investment and technology between the countries of the region. Moreover, the Conference on Interaction and Confidence-Building Measures in Asia (CICA) was established in 2002 with its headquarters in Kazakhstan. It is intended as an Asian framework for confronting new security threats, such as terrorism, trafficking drugs and small arms. This conference comprises Egypt and Palestine, in addition to fourteen other states. The most prominent of these states are India, China and Russia, in addition to Israel. None of the Gulf states are members in this conference except Iran. This forum is expected to develop into a model similar to the Organization for Security and Cooperation in Europe (OECD).[8]

Seventh, the strategic environment in the Gulf region has changed in a radical way, whether at the level of alliances or political processes. There are two viewpoints with respect to the strategic importance of the Gulf region after the 9/11 events.[9] The first viewpoint centers round the fact that the Gulf region has become the focal point of regional interactions in the Middle East. Developments in the Gulf region have assumed greater importance since the beginning of 2002 as a result of the escalating US campaign against Iraq compared with the declining relative weight of the Arab-Israeli interactions. According to the vision of the *Arab Strategic Report for 2002/2003,* the shifting of the focus of regional interactions to the Gulf is not related to crucial strategic developments within this region. Rather, it relates to particular political orientations adopted by the US conservative right as part of a trend to move away from the Arab-Israeli conflict and focus more on the Gulf region.[10]

The second viewpoint centers round the fact that the repercussions of 9/11 events have undermined the strategic role of the Gulf region. There has been a gradual erosion of the Gulf's geopolitical importance due to the campaign to undermine its strategic role as the predominant provider of oil to the world. According to this viewpoint, as expressed by Vahan Zanoyan, increasing concern over internal instability in the Gulf region – that is, its inability to continue playing the role of a reliable provider of oil – has led western powers to seek alternative energy suppliers outside the region. These moves have included unraveling the conventional, strategic ties between the United States and the Kingdom of Saudi Arabia. The latter is viewed by the conservative US right as supporting the kind of religious extremism that bred the terrorists who carried out the New York and Washington bombings. Thus the shift towards Russia and Iraq as two other oil suppliers has begun.[11] In my judgment, this viewpoint does not reflect reality because the Gulf region has been at the heart of the US strategy towards Iraq. Also, the present campaign in Iraq shows that the United States is pursuing greater dependence on Gulf oil and not on alternative sources in other regions.

Internal pressures within the Arab Gulf states have increased, which are aimed at more domestic political reforms. This is due to the existence of a growing middle class that calls for political participation. Also, it is no longer possible to ignore external western pressures towards political reform as a means of combating "terrorism." This is especially true since the United States has announced the "Greater Middle East Initiative" project and France and Germany have presented the initiative for a Strategic Partnership for a Common Future with the Middle East in March 2004. Both projects are geared towards democratic transformation and introducing political reforms in Arab countries and some Islamic states. Moreover, the outlines of an US-Iranian confrontation over the Iranian nuclear issue have begun to emerge. In addition, the situation in Iraq remains unstable. As a matter of fact, Iraq is projected to suffer more eruptions and splinter along sectarian and ethnic lines as a result of the US tendency to perpetuate the destructive situation there.

The Strategies of the GCC to Achieve Regional Coexistence

The foregoing review of the strategic environment of the GCC states in the aftermath of the 9/11 events indicates that there have been radical changes. Pressures for internal transformation have become too great to be ignored by the countries of the region. Moreover, the nature of international alliances has radically changed. Thus, the Gulf region is heading towards unprecedented changes that call for the formulation of creative strategies to deal with them. It must be acknowledged that the GCC states face major hurdles in their attempt to shape such strategies. The most significant hurdle is that these strategies need to be formulated amid a state of disarray and under the shadow of militarization in the region ushered in by escalating internal terrorism, external military intervention, or a potential Iranian-US confrontation stemming from discord over the Iranian nuclear program. Dealing with this situation requires searching for possible strategies, which are applicable within available capabilities and under current international constraints. This also requires focusing on strategies directly related to regional coexistence. In this regard, one may distinguish between four levels of strategies—national, regional, trans-regional and global levels.

The National Level

Internal reform is a major approach and an effective tool to promote regional coexistence. It has become fashionable in the literature of international relations to maintain the view that the ability of democratic regimes to peacefully coexist is far greater than the ability of totalitarian and authoritarian regimes to resolve conflicts with their neighbors. Democratic regimes draw their continuity and popular support from the peaceful settlement of their conflicts.[12] Undoubtedly, the events of contemporary history amply attest to the validity of this argument, although this is not universally true. Several armed conflicts have occurred despite the existence of democratically elected regimes. This is evidenced by the Indo-Pakistani conflict, for instance. However, we can say that democratic systems of government are generally hesitant to resolve conflicts with their neighbors by military means in comparison with autocratic, totalitarian regimes. Within this framework, calls for internal political reform, as an approach to promoting regional coexistence, are intrinsically correct, especially in the case of the Gulf region where regimes and political structures are very similar, and major regional conflicts are non-existent.

Achieving desired political reform presupposes broadening the base of political popular participation in the process of making political decisions. This can be achieved by building the organizations of civil society, holding fair legislative and executive elections which involve all social groups, integrating all groups and factions in the political process, emphasizing transparency, fairness and independence of the judiciary and ensuring that the legislative authorities monitor the work of the executive ones.

Broadening and supporting the base of popular participation enhances the ability of governments to take external initiatives that effectively achieve peaceful coexistence. If the political authority depends on electoral political legitimacy, this will strengthen its position with respect to external forces and reduce the ceiling of concessions to be made to these forces. It will also contribute to strengthening the demands of executive authorities and their adherence to nationalist agendas. To avoid repeating mistakes, we must absorb the lessons of Iraq. The collapse of the Iraqi regime could not have occurred so easily if most of the Iraqi political forces were not alienated from the regime. Consequently, there was a negative attitude on the part of the political forces towards the demise of the regime. Such a situation is inconceivable under a democratic regime. Finally, a democratic transformation in the region will rob Israel of a point it employs against Arabs in global forums—that it is the only democratic state in the Middle East. This is a trump card it uses to liquidate Palestinians and to gain the support of western powers.

It is worth mentioning here that political reform in this context does not mean radical and abrupt internal transformations similar to Gorbachev's reforms in the former Soviet Union. It also does not mean very slow internal changes such as those taking place currently in the People's Republic of China, where gradual transformation has become a camouflage for suppressing desired reforms. Political reform requires determination on the part of political regimes with respect to transformation and they must select pivotal sectors for this process. They must also introduce it gradually, in an organized manner and with a clear objective. Gulf society today comprises a broad middle class base. This middle class has enjoyed three decades of higher education. It is in a position not only to participate in the transformation process, but to lead it as well. The middle class necessarily includes the forces of civil society which have crystallized over the past three decades. As long as the Gulf states subject civil society to strict control, this society can set itself tasks that serve the state and complement its role.[13]

A conventional question usually raised in this context pertains to the relation between political reform and economic reform, and which one has precedence over the other. Arab regimes have traditionally adhered to the argument of postponing political transformation until the process of building free economic institutions in society has been completed. However, historical facts indicate that building a real capitalist economy is difficult in the absence of democratic reform. Gulf systems are basically *rentier* capitalist institutions, which depend on the state for their creation, continuity and growth. Hence they are governed by the state. Democratic reform must mean setting free Gulf capitalism in such a way that it becomes real, productive capitalism. Also, without a democratic transformation it will be difficult to establish an economic system where the possibility of resources being squandered recedes and forms of economic and political corruption are minimized. Such forms of corruption have often adversely affected the reputation of Gulf states. Moreover, ruling elites in the Arab Gulf states have envisaged their societies as having certain unique characteristics that prohibit the adoption of the liberal democratic model. At best these ruling elites have created a "special" democracy under the slogan of inherited values and adherence to norms. This has been done without reference to the reality of this inheritance, or to the applicability of these values to contemporary Gulf societies. The situation has been exacerbated by the fact that the educated class has accepted the idea of Gulf uniqueness and the need for a tailored democracy that fits the heritage model and is therefore prepared to propagate this kind of democracy.

The Regional Level

We cannot envisage a vital GCC role in achieving regional coexistence without consensus among its member states on a common strategy to achieve this coexistence—first, among themselves and second, between them and their neighbors. Achieving this goal requires a progressive approach, bilateral work and collective behavior, both formal and informal. What is required is a regional model by which states support the spirit of cooperation between political leaderships, ruling elites and nations.[14] This process involves the completion of Gulf institutions dedicated to the settlement of potential conflicts that might break out between member states, changing the Jazeera Shield Force into an effective, deterrent force and boosting measures for economic integration between member states through the neo-functional approach, which builds on interaction between professional and economic elites across countries.[15] It also involves implementing the approved program of establishing a monetary union and having a common currency.[16]

However, the greatest task confronting the GCC in achieving regional coexistence is addressing relations with neighboring countries – Yemen, Iran and Iraq – in addition to addressing the Palestinian Question and the threats facing Syria. As for Yemen, it is imperative to integrate it gradually in the institutions of the Council. Yemen is a neighbor to both Saudi Arabia and the Sultanate of Oman. It has acquired a special status after settling its border conflicts with these two countries. Yemen has also adopted an external policy that conforms with the general, external policy of the GCC States. Yemen, on its part, seeks forcefully to join the Council. This will lead to increased coexistence and to its being integrated in the regional Gulf system.

It is also important for friendly relations to be maintained between the GCC states and Iran. This will enable the GCC states to continue their goodwill efforts and to deal wisely with the possibilities of escalating the US-Iranian conflict over Iranian nuclear activity. Iran has adopted a realistic policy towards US intervention in Afghanistan and Iraq. This policy has achieved interests that are vital to the GCC states. Iran has also shown flexibility in dealing with the nuclear issue, which has included signing the additional protocol to the Non-Proliferation Treaty (NPT) in December 2003. It is important that the GCC states explain to the United States that they do not consider Iran to be an antagonistic neighbor, and that they will not provide the United States with military facilities in any possible campaign against Iran. The US-Iranian problem can be resolved peacefully as part of a broader regional effort to halt all nuclear programs in the Gulf region and the Middle East generally, including the Israeli nuclear program.

It must be admitted that the Iraqi issue remains highly complex based on the way the United States perceives this country in the long term. The United States view of Iraq can have serious implications for the stability of this country and it can even lead to its disintegration. The main means of achieving regional coexistence with the new Iraq begins with referring the whole Iraqi issue to the United Nations in order to draw up a timetable for the withdrawal of the coalition forces and the actual handing over of full authority to the Iraqis. If the GCC states concentrate on this strategy, this will yield rapprochement with Iraq. They must also contribute to the reconstruction of Iraqi economy, especially Iraqi debts and the reparations it is due to pay.

Just beyond the Gulf region, an armed struggle is blazing in Palestine. Regional coexistence cannot be achieved without addressing this question. In this regard the GCC states, represented by Saudi Arabia, launched an initiative to settle the Arab-Israeli conflict at the Arab Summit meeting in Beirut in March 2002. The initiative is based on total Israeli withdrawal from the occupied Arab lands in exchange for a comprehensive Arab-Israeli peace. However, Israel has ignored this initiative and responded by committing the massacre in Jenin in April 2002. The failure of this initiative is linked to the grave disparity in the power balance, and total US support of Israel in an attempt to restructure the region in a way that results in absolute Israeli dominance. According to this perspective, the GCC states must direct their efforts towards supporting the Palestinian Authority by virtue of its being the elected legitimate representative of the Palestinian people. The GCC must also support the steadfastness of the Palestinian people in the occupied lands, especially since Israel is planning to wage war in the region and use it as a reason to expel the Palestinian people from the West Bank, or at least from East Jerusalem.

Another point of view is that the GCC must rein in the zeal of some of its members to normalize relations with Israel on the pretext that this will positively affect the track of negotiations between Israel and the Palestinians. Arab-Israeli normalization generally, and Gulf-Israeli normalization in particular, will not lead to softening the extremist position of Israel with respect to the major issues. It will not have positive results with respect to the absolute US support of Israel. Historical events clearly indicate that Israel has exploited the goodwill of the Arabs by imposing more sieges on the Palestinian people and in raising the ceiling of its demands for achieving a settlement.

Regarding the Syrian issue, the signals from the United States, as represented in the statements of its officials, show that Syria is confronting US threats to overthrow the regime and dispossess Syria of weapons of mass destruction. These US threats emanate from Israeli instigation. It seems that the approval of the Syria Accountability Act of 2003 by the Congress in November 2003 and the imposition of some actual sanctions on Syria in May 2004 are not going to be the last of these signals. The GCC role in this issue is to prevent any probable US-Israeli aggression on Syria. This can be achieved by exploiting the distinguished relations between the GCC states and the United States. It must be acknowledged that the GCC faces real problems in its pursuit of regional coexistence, given the Iranian, Syrian and Palestinian issues. However, it seems that the Council has no alternative apart from incessant efforts in this direction.

At the regional level, the GCC states must also accord attention to the issue of achieving Arab-Arab coexistence by activating the Arab League. The Arab League is the only formal political framework in which all Arab states are represented in order to achieve their common interests. Though all the criticisms leveled against the Arab League are acceptable, the Arab League has set itself important tasks in the field of Arab economic cooperation, and in resolving some conflicts between Arab states. It has also succeeded in achieving some forms of Arab security, as evidenced by its management of the Lebanese Civil War and the Iraqi-Kuwaiti crisis in 1961. Currently, we are witnessing an attack on the Arab League and calls for its dissolution on the grounds that it has failed to achieve its objectives. In the final analysis, such calls provide avenues for other competing projects to replace the Arab League. Among these are the Euro-Mediterranean Project (led by the European Union), the Greater Middle East Project (under the patronage of the United States) and the Atlantic-Mediterranean project (supported by the NATO). Each of these projects has its own agendas, which do not conform much with Arab agendas. Those who adopt these projects covertly support calls for the dissolution of the Arab League to provide legitimacy to their alternative projects.

Therefore, the GCC states are called upon to work together with other Arab states to preserve the Arab League, and seek persistently to activate it by paying financial contributions. More importantly, they must show commitment to the resolutions issued by the league and approved by its member states. These resolutions are seldom put into effect. By the same token, we cannot ignore the increasing calls to establish a new and more effective political entity to replace the Arab League. However, this endeavor should take into consideration the importance of not dissolving the League before developing mature ideas about the form and content of the proposed new entity, and the extent to which it can contribute to achieving better Arab cooperation. Such ideas should be discussed widely among the Arab states. In this context, we should remember that when the major powers decided to dissolve the League of Nations and establish the United Nations during the Second World War, they set up the latter first. Once the UN institutions were formed and it started to function in 1945, it was decided thereafter to dissolve the League of Nations in 1946. This overlap between the two organizations reflected concern among the founders about ensuring the feasibility and efficacy of their project if only to a partial extent, before plunging into the uncertainties involved in the process of dissolving a beneficial entity. However, calls to dissolve the Arab League first abound in the Arab World. Those who advocate this are doing so without suggesting an alternative and more efficient Arab project. It has been noted that while some GCC states like Kuwait have clearly made such calls, others like Saudi Arabia have proposed an initiative to activate the Arab League.

The Trans-Regional Level

A number of variables have developed in the wider context of the GCC. These variables have created new opportunities for the GCC states to achieve peaceful coexistence. The most important of these variables is the foundation of new economic and political institutions in the wider context of the GCC states. These institutions include the following:

1-The Indian Ocean Rim Association for Regional Cooperation [IOR-ARC]

This association was founded in March 1997. It now includes 19 countries, three of these being Arab states –Yemen, the Sultanate of Oman and the United Arab Emirates – in addition to Egypt, a dialogue partner. The aim of this association is to liberalize trade and develop investment among member states. The association consists of a ministerial, governmental track and two non-governmental tracks for businessmen and academics. As already noted, only two of the GCC states have joined this association. If other GCC states become part of this association, it could be an important step towards strengthening a strategy for coexistence between the association members, especially since some, such as India and Iran, are important economic partners of the GCC states. In this way the economic and scientific approach based on trade, investment and academic cooperation becomes a means of achieving peaceful regional coexistence with neighboring countries, and peaceful trans-regional coexistence with countries in the broader regional context. It is paradoxical that the remaining GCC states are still very slow to join this association despite the great opportunities it provides at both commercial and technological levels. Moreover, this association is invested with political importance in matters concerning support for key Arab issues.

2-Conference on Interaction and Confidence-Building Measures in Asia

This conference was founded in June 2002. It currently has 16 member countries, two of these being Arab states—Egypt and Palestine. There is no Gulf state in the conference though it provides a framework for security and cooperation in Asia. The conference aims at dealing with "new security threats" like terrorism, trafficking of small arms and drugs, and the like. The GCC states are called upon to join this conference so that they can influence its orientation towards security and cooperation in the Gulf region and central Asia. It is worth noting that Israel is a member in this conference and participates effectively in its deliberations. Cooperating in security affairs with Asian countries on those issues which affect the GCC states is an important approach to achieving coexistence with those countries located in the region surrounding the Arab Gulf states.

3-The Euro-Mediterranean Partnership

This partnership was founded in November 1995 at the Barcelona Conference which issued the Barcelona Declaration on the Euro-Mediterranean Partnership (EMP). This new system includes the countries of the European Union and 12 Mediterranean countries, eight of which are Arab states. The partnership comprises a free trade zone, laying down rules for politico-security cooperation and a dialogue on culture and civilization. The European Union has started an independent dialogue with the GCC states, but this dialogue has been hampered by the EU's insistence on certain provisions for developing it further. On the other hand, there are important links between Gulf security and the security of the eastern Mediterranean. Part of the Gulf oil trade passes through the

Mediterranean; there are also close economic links between the GCC states and the countries of the eastern Middle East and southern Europe. On the basis of this, it is important to launch a Gulf initiative to transform the Euro-Mediterranean Partnership into a comprehensive Euro-Arab partnership in such a way that Israel is expelled from the Euro-Mediterranean equation, thus leading to strengthened relations between Arabs and the European Union.[17]

The Global Level

The Gulf region is one of the areas most closely connected with the interests of major powers because of its significance to the oil trade and its strategic importance. Therefore, it is difficult to imagine ever establishing a security system in the region without this being linked in one form or another to the major powers. However, this does not mean that the major powers will necessarily be part of the system or dominate its strategies. Rather, the primary goal is for the GCC states to concentrate on establishing a Gulf security system that achieves peaceful coexistence in the regional context, free from the pressures exercised by the major powers. Although this is difficult to realize in a unipolar world, the efficacy of the Gulf system and its importance to the regional nations are at stake, especially after the failure of this system following the occupation of Kuwait in 1990, and its inability until today to transcend the crisis caused by lack of confidence and to take unifying steps in spheres such as economy and trade.

As a matter of fact, the question of establishing a Gulf security system is the most complex issue for the GCC because it involves conflicting viewpoints. Whereas Iran opposes any role for the major powers in the Gulf region, some members of the GCC maintain that historical experience renders the presence of major powers inevitable at least in the present stage. These viewpoints can be reconciled through an invitation to an international conference under the aegis of the United Nations to be attended by the major powers that have a military presence in the Gulf region. This conference can be a preparatory stage for a collective declaration on mutual non-aggression, i.e. signatory states refrain from the use of force against other parties, while gradually pursuing reduction of foreign military presence in the region. Confidence-building measures can play an important role in this process through agreements on the part of these states to inform one another (as well as foreign states in the region) of military exercises and moves.

Also, the GCC states can play a central role in facilitating a settlement of the US-Iranian conflict over the issue of the Iranian nuclear program. This role centers on emphasizing the role of the International Atomic Energy Agency (IAEA) in dealing with this issue, the importance of applying the Non-Proliferation Treaty strictly with respect to the right of Iran to possess nuclear technology for peaceful purposes, and stressing the right of the international agency to verify that Iran does not possess any nuclear weapons.

Conclusion

The GCC states can move to build a regional Gulf security system by following several internal, national, trans-regional and global tracks. The point of departure is developing new functional cooperation between economic, academic and cultural elites in the GCC states. This is to be done in such a way that allows for the emergence of Gulf cross-border groupings in the GCC states. The most important of these groupings are the organizations of civil society in their various forms. This will make the Gulf system move out of a framework based on family-oriented ruling elites and into a framework based upon the concept of popular rule. It will also enable it to overcome rigidity and duplication and seek new horizons of advancement and development.

There is a second track which centers round applying the strategy of confidence-building measures between the GCC states and their neighboring countries. In this track, it might be useful to establish a new institutional framework to bring together the GCC states and their neighboring countries. It is worth mentioning that there are important signals in the Gulf region indicating a drive towards peaceful coexistence. One of these is the decision made by the GCC at the Muscat summit held in late December 2001 to accept Yemen in the non-political institutions of the GCC. What is required within the framework of confidence-building is genuine will in addition to bilateral and collective understanding between all regional states in their pursuit of peaceful coexistence.

Peaceful coexistence is not only an urgent need for the Gulf region and its nations, but it is also a major element in the survival and stability of the region. Peaceful regional coexistence is the only guarantee for the states in the region. The nations of this region cannot live under sustainable welfare in its absence. On the other hand, peaceful coexistence will provide all with true opportunities for proper economic development, desired social progress and necessary political participation.

Section 7

SOCIAL CHANGE IN THE GULF

20

The Evolving Role of Media in the Gulf: Privatization, Competition and Censorship

Jamal Kashoggi

The Gulf states have subsidized development in different domains – industry, agriculture, housing and the press. Following the oil boom of the mid-1970s, and within the framework of its ambitious and rapid development plans, the Gulf press witnessed a development leap of several years, bringing it closer in professionalism to countries that had been this field several decades earlier than the Gulf states.

In countries like the UAE and the Kingdom of Bahrain, the state has financed the publication of some newspapers through the Ministry of Information or other state authorities. Other countries like the Kingdom of Saudi Arabia have contented themselves with formulating regulations that allow the Ministry of Information to supervise editorial policies. However, the latter countries have left details pertaining to the establishing, financing and managing of newspapers to the business sector. Hence there is no need for privatization in these countries. As a matter of fact, in legal terms, newspapers are wholly owned by the private sector. However, in the countries of the former group, both the market and businessmen are eager to acquire state newspapers or buy shares in these newspapers because of the investment returns anticipated in this industry.

The Saudi and Gulf markets are both vast and replete in measures that are adequate for financing the publication of newspapers, and for initial publicity and distribution expenses, without any need for governmental subsidies. In this respect, governments should not worry about foreign subsidies trickling down to newspapers in a way that shifts them from the kind of national orientation that conforms with the vision of Information ministers and approved state policies. The limited existing subsidies in Saudi Arabia are now dwindling since they are no longer justifiable. In reality, the budget of the smallest newspaper is several times the amount of subsidy, which does not exceed more than one million Saudi riyals per annum. This subsidy has already been reduced by half and will further be reduced to a quarter of the original amount before being cancelled altogether in 2005. The subsidy is provided in return for government advertisements that the newspapers are obliged to publish gratis. From the end of 2004, newspapers will deal with government advertisements in the same manner in which they deal with commercial advertisements, according to the logic of the market.

Other privileges made available to the newspapers by the Saudi government in the previous stage are also disappearing, such as free air freight facilities. This has been cancelled gradually and newspapers will pay shipping charges in full within a year or two. However, newspapers are currently driven by a pure commercial mentality and will not return past favors. Saudi press institutions are not going to offer preferential rates to Saudia – the national carrier – in advertising and subscription as has been the case for many years. At the same time, newspapers will retain past assets like the lands offered to them free of charge to found their giant institutions.

It could be maintained that newspapers in the Kingdom of Saudi Arabia and most of the Gulf states are independent in terms of financing. Moreover, they are wholly owned by the business sector. However, state bureaucracy does not surrender what it believes to be its right. It withdraws from one aspect of the field only to intrude on another aspect of journalistic work by means of regulations. This is the state of affairs between "independent" press institutions and the Ministry of Information. The appointment of chief editors and the admission of new board members who contribute towards press financing are subject to approval by the Ministry of Information. Even the names of prospective foreign journalists must be approved by the Ministry of Information prior to completing the same entry procedures as required by the relevant authorities with regard to other sectors.

Declaring a state of hundred percent independence of newspapers from the influence of the Ministry of Information (which is always relative to the particular situation) in the Kingdom of Saudi Arabia and other Gulf states does not require making difficult decisions, such as transferring a newspaper from the public to the private sector and halting large state investments in the newspapers. This can be achieved simply by annulling those provisions in the publication laws that confer on the Ministry of Information an essential role in appointing and sacking editors-in-chief and board members, suspending writers and journalists and imposing penalties and fines on them, in addition to the supervisory and directive role of the ministry. All these provisions can be substituted eventually by a well-designed act regulating publication affairs and publication rights, in cases that are referred to the courts.

While there is no formal censor, the relationship between the Ministry of Information and chief editors turns every chief editor into a censor within the domain of his newspaper. Two factors have made daring press coverage, chasing hot news, and constant stretching of the boundaries of press freedom to the furthest limit, avenues for successful newspapers to prove their excellence and win over new readers. These factors are, first, calculations of profit and loss, although these should not take priority over the informational mission of the newspaper, which a professional chief editor would perpetuate and defend with steadfastness. Second, the growing competitive spirit among newspapers over a lucrative market that increasingly attracts press investors because of its incessant growth.

However, wisdom dictates observing the fundamentals of the game of balance between the newspaper and the Ministry—with the

Ministry winning one round, and the newspaper winning the next. This game may result in painfully traumatic situations, the victims often being the chief editor and other journalists. However, this has eventually led to an extension of the permissible boundaries. Newspapers have regained the ground they lost as the Fourth Estate. This loss was inflicted upon them during an era of chief editors who were weak because they either came from outside the profession or because they gave priority to maintaining their positions rather than taking risks in an exciting profession.

Newspapers are almost equally placed in terms of possible advantages to the reader, such as increasing the number of pages, use of colors, speedy distribution, glossy paper and printing via state-of-the-art digital technologies. Such advantages are governed by the availability of money. Money can only be earned by newspapers that are powerful in market circulation—as money is in the hands of advertisers. This kind of position is only attained by newspapers through their daring coverage, expanding interest in public affairs and concern with local issues.

The Shock that Awakened the Saudi Press

The press in the Kingdom of Saudi Arabia experienced shocks that have altered the mode of thinking and management. These shocks have not only spurred development, expanded resistance to constraints but also raised uncertainties regarding a stereotyped, unchanging market.

The first and severest shock experienced by Saudi newspapers occurred in 1986 when the Ministry of Information (paradoxically enough) entered the scene as a fierce competitor by allowing commercials on Saudi television. Within weeks, millions of riyals generated from advertisements in the newspapers evaporated. This happened after a decade during which newspapers enjoyed a monopoly on advertising and received high and unprecedented incomes. It led to severe measures including the retrenchment of tens of employees. I was a new employee in the Saudi Research and Publishing Company in 1987 when 60 employees were sacked at the same time. This happened in August of that year and a cynic, commenting on the incident, described those events as an all-consuming fire. However, this did not devastate the company, which corrected its financial position and continued to lead the publishing market in the Kingdom and the region. Those who were retrenched found jobs elsewhere. This incident manifested the creativity of the market economy and its ability to adjust to very extreme situations.

The Era of Satellite Channels

Following this shock, newspapers had to find a role that could help them regain their importance among advertising companies—a task that could only be achieved by winning over more readers. However, while newspapers were considering competitive ways of attracting readers, they were taken unawares by another shock. This was a real challenge involving a competition for the minds and hearts of the readers. This shock occurred during the Kuwait invasion. Newspapers benefited from increased distribution and from wider boundaries of freedom in that period. However, a new rival appeared on the media stage—satellite channels. The first was CNN, which was followed shortly after by MBC.

Subsequently, the *Al Jazeera* channel appeared. In my view, its establishment was a very important event in the field of the Arab media comparable to the publication of the *Al Ahram* newspaper at the end of the nineteenth century. This imposed a greater responsibility on newspapers, demanding excellent analysis and going beyond merely conveying news—thus ushering in the era of elaborate coverage and expression of strong views.

Successive technological and social changes occurred, such as the Internet, spread of political consciousness, aspirations for reform, and the emergence of an educated class concerned with public affairs. All this conferred on newspapers a greater role which was manifested in covering the tragic event of the fire in the primary school for girls in Mecca four years ago. In that event, newspapers played an unprecedented role, which led to the customary reprimand from the Ministry of Information as well as recognition of their importance. We could say that this important role still continues to ebb and flow in the reform process being witnessed by the Kingdom, especially since May 2003 after the bomb explosions hit the Kingdom and ushered in a new era that would not have been possible under normal conditions.

While media personnel pay lip service to the Ministry of Information by expressing gratitude for the greater margin of media freedom bestowed – it is generally acknowledged, but never voiced explicitly – that in reality, gratitude should go to history, technological developments, the spirit of independent commercialism adopted by some media institutions and the boldness of some chief editors who stretched the limits imposed on them and their fellow journalists. Of course, some chief editors paid the price for such actions.

In my view, there are three landmark institutions in the current history of the Saudi press, which are behind the gains made by newspapers. They are as follows:

- The Saudi Research and Publishing Company, which instituted and developed commercialism and concern over profit and loss in journalism. This spirit will undoubtedly lead to the reliance of the media on itself and the market rather than the state.
- *Al Iqtissadiya* newspaper which, under its previous Chief Editor Mr. Muhammad Al Tunisi, gave priority to local affairs and concerned itself with consumer affairs vis-à-vis the business sector as well as the state, which encouraged other newspapers to follow suit.
- *Al Watan* newspaper, the most recent among Saudi newspapers, which pushed the limits even further and raised modern and progressive issues in a conservative society and within the field of conservative journalism. This is especially true of the pages devoted to expressing viewpoints, which served to enliven the expression of opinions after a long period of repetitiveness and complacency.

Much Remains Ahead for the Gulf Media

There is still a vast, unexplored landscape lying ahead for the Gulf media, particularly in the Kingdom of Saudi Arabia. In a state like

Kuwait, the media might have reached a state of saturation, but in Saudia Arabia, there are still opportunities for greater income. If the scope of journalistic work is doubled, the amount of profit will also be doubled by staggering figures in comparison with other commercial fields. Capital returns in some newspaper corporations reached hundred percent whereas others have maintained a level of fifty percent for several years. Total returns in a newspaper like *Al Riyad* are in the region of 350 million riyals per annum. Half of these returns represent profits—160 million riyals last year.

These are staggering figures. However, per capita expenditure on advertising is no more than 9 riyals per annum, whereas the corresponding figure in Israel is $260 and $70 in Cyprus. What would happen if this amount doubles or triples? It would translate into richer newspapers with additional capacity to spend on writers and reporters. It would also mean a greater political role and more newspapers, more competition and a greater desire for excellence.

Many researchers seek to investigate the impact of satellite media on newspapers. However, it has been proven that the print media remains more influential, especially in the process of change, reform and political development being witnessed by the region. This is so despite the fact that some articles and news items overlap with the content being broadcast by satellite channels. Even with their spread and attractiveness, these channels have not undermined the market for publications, which has doubled during the last decade. Newspaper readers in the Saudi society are estimated to be about 600,000 readers, reading 350,000 copies of newspapers daily.

However, these are still modest figures in a country with a population of 25 million. In Saudi Arabia, the press has achieved a great deal. Yet there remains more scope for growth and development.

21

The Changing Role of Women in the Gulf Region

Munira Ahmed Fakhro

Issues relating to women, particularly their changing role, have been given greater attention at Arab, regional and international levels, especially by the United Nations and human rights organizations. International conferences and programs have been organized and conducted with a view to addressing these issues, enhancing the potential of women and ensuring their equality with men.

In the Gulf region, however, special importance has been accorded to this subject since societies of the region have embarked upon the modernization process rather recently, after the discovery of oil during the 1940s and 1950s. Some trace the beginning of this process to earlier dates, but the pace was not as rapid as it has been after the flow of Gulf oil.

At the start of the last century, women in the Arabian Gulf area used to live a more traditional, routine life where absolute norms and notions prevailed for the majority of the population, which suffered from severe hardships in a so-called "subsistence economy." However, the discovery of oil and the investment of part of its revenues in developmental projects have had broad repercussions on the emerging families and societies.

According to one researcher, the "pre-oil" society in Bahrain, during the first quarter of the twentieth century was divided socially and economically into two groups: first, the elites, including the ruling family, landlords, pearl merchants and leading businessmen; second, the common people including fishermen, peasants and pearl divers.[1]

With the discovery of oil and the growing educational and employment opportunities allowed to both sexes, increasing numbers of women joined the labor force. Consequently, the traditional role of men – as the dominant protectors and supporters of the family – was bound to change to that of partners with equal rights and duties, who would assist their wives in carrying out household chores and rearing children. Nevertheless, women still shoulder greater responsibility than men despite the significant changes that their status has undergone, including among others, their involvement in public functions and new economic activities, working side by side with men.

Therefore, the laws and regulations that govern the relationship between men and women have to be amended with a view to achieving gender equality between the two sexes as a fundamental factor for advancing the process of sustainable development.

This chapter addresses the economic, social and cultural transformations that have influenced the status of women in the Gulf region, as well as issues in relation to the changing role of women in Gulf societies and the measures adopted to remedy the resulting difficulties faced by men and women alike.

Education and Labor: Implications for the Role of Gulf Women

Apparently education and labor are the factors that have played a significant role in changing the position of women in the Arabian Gulf region. As one of the most important elements influencing the growth of any society, education has had an enormous effect on the social and cultural changes that both men and women have undergone in the region. However, its effect on women has been more obvious because it allows them employment opportunities. The more educational opportunities provided for women, the greater are the job opportunities available to them.

No longer is the goal of education simply to teach writing and reading and supplying government departments with clerks and employees. It also aims at providing production and services sectors with well qualified and experienced cadres (such as engineers, doctors, pharmacists, teachers and technicians). This need has prompted the expansion of secondary and higher education facilities. Thus, offering free education and setting up schools and universities to prepare and qualify the increasing numbers of necessary graduates have become important objectives of the development process, and education has become a key channel for mobility, whether it is professional, rural-urban or social in nature.[2]

Education in the Gulf Arab States

The first stage of women's education in the Gulf region was initiated at the beginning of the last century when the American Expedition set up the first girls school in Bahrain in 1909, in which 40 students were enrolled. Several government girls schools were established in 1928, nine years after the first public boys school was opened in 1919.

In 2003, the number of girl-students in the primary, preparatory and secondary stages was 98,931 (50.02% of the total number of students), 15,163 (50.1% of the total) and 13,909 (52.5% of the total), respectively. At the higher education level, girl students represented 62 per cent of the total number of Bahrain University students.

In Kuwait, the first public girls school was opened in 1937. In the academic year 1998/1999, the total number of female students who graduated from the University of Kuwait was 2094 (against 879 male graduates), compared to 1804 female graduates (against 660 male graduates) in 1999/2000.

The 1950s witnessed the beginning of women's education in the United Arab Emirates. In the academic year 1996/1997, the number of female university students was 68% of the total number of students.

This percentage is expected to soar to 80% in the coming years, and the number of female students of the UAE University in Al Ain city is expected to double in 2006, which means that additional seats have to be provided in that year for 13,000 new female students. As for the new Zayed University – established exclusively for women – 400 girl students were admitted to its headquarters in Abu Dhabi, during the academic year 1998/1999, in addition to 1,100 students enrolled in its Dubai branch.[3]

In the State of Qatar, the first boys school was opened in 1952, admitting 240 students. The number of girl students admitted to the University of Qatar has risen from a figure as low as 55 in 1955 to 754 in 1989 (out of 1,085 students of both sexes). The increase in the number of the University's girl students is ascribed to the fact that male students are being sent abroad to complete their higher studies.[4] The number of female teachers in all government schools in Qatar was put at 3,995, representing 84 per cent of the male teachers.[5]

In the Sultanate of Oman, the regular education of women started in early 1970, which is considered late compared to other Arab states of the Gulf region. Before that date, only three boys schools were set up, enrolling no more than 909 students. The proportion of females in the education process has risen from 12.7 per cent in the academic year 1971/1972 to 48.6 per cent in 2002/2003. In 1990, a total of 283 students (160 males and 123 females) graduated from Sultan Qaboos University.

The increase in the numbers of female students started five years later, in 1995, when the numbers of graduates were 307 males and 433 females, while in 2000 these numbers rose to 497 and 574 respectively. Out of the total number of students studying abroad (9,229 students), about 5,453 were female.[6] Most of the female students generally completed their higher studies in humanities, and graduates of these disciplines usually faced many difficulties in getting suitable job opportunities.

In the Kingdom of Saudi Arabia, the women's educational system was initiated in 1960 with the establishment of the General Directorate for Female Education (GDFE) under the supervision of a commission of senior religious scholars. The number of schools has risen from 15 primary schools in 1960 to 38 women's educational administrations and 150 regional educational institutions in 1999.

In the same year, the number of educational institutions set up under the GDFE was around 12,168, of which 170 were teachers training institutes and 72 educational colleges. Private schools and universities, which usually open some of their faculties to girl students, are not included in these statistics.[7]

All the figures mentioned above show that within a few years the number of students has soared to high levels. Nevertheless, the quality of the educational programs still falls short of internationally acknowledged standards, especially since these states have not yet adopted compulsory education systems, resulting in literacy rates that favor males. The UAE and Qatar are the exceptions where illiteracy rates are lower among women, as shown in Table 21.1.

Table 21.1
Rates of Illiteracy in the Arab Gulf States, 2003
(Age Category: 15-24 years)

State	*Females (%)*	*Males (%)*	*Both Sexes (%)*
United Arab Emirates	18.5	24.0	22.2
Kingdom of Bahrain	15.0	8.1	10.9
Sultanate of Oman	32.8	17.0	24.2
State of Qatar	15.0	18.6	17.5
State of Kuwait	18.3	15.5	16.5
Kingdom of Saudi Arabia	29.2	15.4	21.3

Source:ESCWA, *Report by Arab Women's Center* (New York, NY: United Nations, 2003), Table 5, page 14.

In terms of enrollment rates at the university education stage, the available data reveals a wide gap between the two sexes in favor of females, due to the fact that most males either complete their higher studies abroad, or join the military or the police force after the secondary education stage, whereas such opportunities are hardly available for females under the prevailing norms and customs.

Table 21.2 highlights the aspects of educational discrimination concerning the social role that women are expected to play. In general, they are encouraged to join those fields of employment that people feel are more appropriate for women, like school teaching, nursing and some civil service jobs. Female engineers, for example, would find it extremely difficult to obtain jobs in construction sites, and they end up staying at home as mothers or wives.

Table 21.2
Distribution of Students by Fields of Study, 2001

	Education		*Arts and Humanities*		*Business Management, Law, Social Sciences*		*Sciences*		*Engineering*		*Health and Medical Care*		*Others (Cultural + Miscellaneous)*	
State	*F (%)*	*M (%)*	*F (%)*	*M (%)*	*F (%)*	*M (%)*	*F (%)*	*M (%)*	*F (%)*	*M (%)*	*F (%)*	*M (%)*	*F (%)*	*M (%)*
Bahrain	77	23	76	24	58	42	72	28	32	68	75	25	38	62
Kuwait	79	21	79	21	63	37	73	27	49	51	73	27	–	–
Oman	–	–	62	38	46	54	51	49	9	91	51	49	55	45
Qatar	91	9	88	12	77	23	79	21	42	58	–	–	37	63
Saudi Arabia	75	25	34	66	31	69	44	56	1	99	39	61	24	76
UAE	95	5	86	14	57	43	82	18	44	56	65	35	83	17

Note: M = Male; F = Female.

Source: ESCWA, *Report by Arab Women's Center 2003* (New York, NY: United Nations, 2003), Table 11, page 22.

During the period 1995-1999, these states allocated about 15 per cent of their total public expenditures for education. Of these allocations, the UAE spent 16.7% for female education programs and 1.8% for male education; Bahrain spent 12% and 4.4%; Oman spent 16.7% and 4.5%; Kuwait spent 14% and 5%; and Saudi Arabia spent 22.8% and 7.5% respectively.[8]

Gulf Women and the Labor Market

Women in the Gulf region used to carry out their daily household works, such as raising children and housekeeping. A large segment of working women, both in the urban and rural areas, practiced traditional professions such as stitching local dresses and weaving palm-leaf baskets, while shouldering full household responsibility, especially when their husbands went away for month-long pearl diving missions.

Women joined the labor force first as school teachers and nurses before entering the main fields of employment at later stages. The share of Kuwaiti working women has risen from 1.4% in 1975 to 12.8% in 1980, 25.3% in 1993, and to 30.2% in 2001. In 1995, out of the total female labor force, the percentages of graduates from universities, secondary and intermediate schools, and holders of post-secondary school diplomas were around 37%, 23%, 17% and 20%, respectively.[9]

According to one researcher, the eagerness of Kuwaiti women to join the labor force is attributed more to their desire to invest their knowledge and experience, assert themselves and prove them rather than for purely economic or financial benefits.

Nevertheless, the Kuwaiti women's contribution to the development process is still limited due to attempts in confining them to the above-mentioned careers, or perhaps to mere household duties.

Statistics published by the Kuwaiti Civil Service Council in May 1995 reflects some discriminatory abuses against women working in government departments. The number of female civil servants has reached 55,080, of whom 42,508 were Kuwaitis, representing 77.7 per cent of the total number of working women, and 39.9 per cent of government employees of both sexes. Only eight of these women occupy senior posts, that is, 3.5 per cent of total high-ranking officials (236 in the ranks of Deputy Minister, Assistant Deputy Minister, or equivalent posts).

Political Islam movements and organizations have had a considerable effect in curbing the role of women and preventing them from assuming influential positions in which vital decisions are made. Strong calls are being made to enforce early retirement on women with a view to denying them any opportunities to reach high government positions.

According to data published by the General Establishment of Social Security on June 30, 1994, the numbers of female pensioners in the public, private and oil sectors were 40,626 (of whom 88.4% were less than 40 years of age), 296 (96% under 40 years) and 1525 (78% under 40 years), respectively.[10]

As expected, social work and teaching were the first fields of employment for Saudi women since early 1960. The number of female school teachers soared by 12.4 per cent per year to reach more or less 200,000 teachers in 2001 from 5000 in 1970. Female teachers at the university level represented around 38 per cent of the total number of teaching staff, while their shares in sectors such as medical care, nursing, pharmaceutics and administration have risen to 40%, 54%, 29% and 8%, respectively.

Job opportunities for women in the private sector were relatively lower due to inappropriate rehabilitation and training programs designed for women. The percentage of jobs available for women in the banking sector did not exceed 4 per cent.[11]

The share of Bahraini women in the overall labor force was almost the highest compared to other Gulf countries (except for Kuwait), as it has risen from 3% to 6% during the period 1971-1981. In terms of the domestic workforce, this percentage increased during the same period from 5% to 13.7%, rising to 19% in 1999 and to 25.7% in 2001. In 1995, the services sector in the Kingdom of Bahrain was the largest field of employment for the female workforce. The Bahraini Ministries of Education and Health have absorbed around 58 per cent and 22 per cent of the female labor force adding up to 80 per cent of the total female labor force.[12]

The available data has shown that in 1995 the number of female (including foreigners) in Qatar was 181,620 (32.3% of the total population). Out of the total female labor force, working women represent 11% (of whom 25% were Qataris).Generally speaking, the contribution by Qatari women to the total labor force has almost doubled to reach 12% in 1995 from 6.2% in 1986.

In the government sector, the female labor force, the biggest compared to other sectors, constituted 36.6 per cent (8141 employees) and 40.2 per cent (12,397) in 1995 and 1999, respectively.[13] In 1993, the shares of women employed in teaching, administrative and services fields were 57.8%, 19% and 22%, respectively. Government school teachers comprised 78% of the total number of Qatari working women.

According to one study, the most important problems that hinder the integration of Qatari women in the development process are the limited opportunities given to them to acquire education, jobs and training and the tendency to restrict women's roles to maternity and household duties.[14]

In the Sultanate of Oman, the number of female employees in the government departments increased from 1364 in 1980 to 18,641 by end of 2000. The share of working women in the private sector during the Five Year Plan (1996-2000) more than doubled from 4435 in 1996 (15%) to 10,048 (18% of the total number of Omani workers of both sexes, which is estimated at 55671persons).

The rate of economic activities conducted by Omani women in 1998 has reached 18 per cent as a result of the increase in development rates in general, even though it is regarded as lower than that realized in the rest of the Arab countries.[15]

Several fields of activity in the UAE have witnessed rising contributions by women, from 10 per cent in 1991 to 13 per cent in 1998, while their shares in sectors such as agriculture, industry/electricity and basic infrastructure (including construction, transport/communications and trade) stood at 4%, 2% and 19%, respectively.

Despite the considerable increase in numbers of female graduates from the Higher Colleges of Technology, who are equipped with the necessary skills for employment in the private sector, women's share of the labor force in this sector was still low, as their number in 1995 has not exceeded 138 (out of 6463 workers of both sexes) due to social and cultural constraints, and increasing reliance on expatriates and preference for them in the labor force, which in turn impedes wider participation by women in this sector.[16]

Women and Changing Gulf Family Relations

The family represents the cornerstone of society as a key source of stability, communication and development, and a natural framework of emotional and material support for all members of the family. The major duties of the family can be summed up as the creation of sentimental, social and economic ties between the two parents and caring for their children.[17]

The traditional Arab family is viewed as the "extended" patriarchal (fatherly) family type. In its broader definition, paternity (fatherhood) means the domination by the father over the mother and children and the imparting of an institutional trait to this domination which could be extended to include women in society as a whole.

As is the case with Arab societies in general, the family has undergone several changes resulting from the rapid transformations taking place at the cultural and economic levels.[18]

A Gulf researcher has noted that during the past five decades the family in Gulf societies has witnessed almost radical changes both in its structure and traditional functions, though the direction and type of these changes differ from one society to another, depending on the degree of modernization and the role played by the family in shaping the framework of economic, social and perhaps political relationships and alliances.[19]

Most studies assert that the economic and political transformation that the region has witnessed since the discovery of oil have brought about many changes in the form and composition of the Gulf family, and have contributed to the evolution of the smaller "nuclear" family which enjoys independence from its larger, original (or the so-called "root") family in terms of dwelling and economic resources. Thus, by detaching itself from its "root" family, the "nuclear" family is reducing the influence of the extended family on a given social environment.

In most cases, the new "nuclear" family consists of a husband, wife and no more than four children. A survey published in 1980s has indicated that nuclear families in Saudi Arabia represent 76 per cent of the number of families under discussion. A research study conducted in Kuwait in the mid-1970s had shown an increase in the members of nuclear families which represented around 59.1 per cent of the study sample. As for the State of Qatar, a study issued in 1989 has shown that the Qatari family averaged 7 members, as is the case in the UAE where the prevailing type is that of large-member families.

Another study conducted in 1994 has concluded that 44.7 per cent of members of samples surveyed belong to large families of 6-10 members. In Bahrain, according to the 1991 population census, the percentage of Bahraini families of 3-5 members stood at 33.8 per cent of total number of families.[20]

Although the above statistics reflect the prevalence of the type of nuclear family in the Gulf region, the extended family system still has a strong effect on the members of the nuclear families throughout their whole lives. Members of an extended family would often bear the living expenses of their married sons and daughters living in separate dwellings. They would also have an influence on the cultural views and ideas of these families.

According to one researcher, the family constitutes the core social unit of Arab society that ensures the continuity of cultural values and identity and offers support and stability in times of individual and societal distress.

Nevertheless, the changes that Arab society has witnessed in the last decades, and the increasing numbers of women joining the labor force, have had a big effect on the structure of the Arab family, leading it to adopt the concept of partnership between man and woman in taking decisions, caring for children and shouldering household expenditures.[21]

A recent report by the UN Economic and Social Commission for Western Asia (ESCWA) indicates that the family in Arab states remains the dominant unit in the formation of the social order. And in this social/cultural framework, said the report, women are confronting heavy pressures to create a family almost immediately after marriage so that they can lay the foundations of their social status. Hence, the woman's role as a wife and mother takes priority to her social role, thus turning marriage and motherhood into a duty rather than an option.[22]

However, the changes that the family structure has undergone have had no effect on the traditional roles of both men and women, and working women are still performing their roles in shouldering the household and family burdens and caring for their husbands and children.

By undertaking these two responsibilities in the absence of traditional assistance from their extended families, working women are obviously encountering additional psychological and physical pressures. This situation necessitates revisions by Gulf states of their policies towards working women in particular, and the family establishment in general, in order to suit their new family and living conditions and their growing share in the labor market. Indeed, by joining this market, working women are fulfilling their part in bearing a specific portion of living costs, and in curbing the flow of the foreign labor force which is flooding the work markets of the Gulf region and leading to serious political, cultural and social consequences.

Women's Organizations: Effect on the New Female Role

1-Kingdom of Saudi Arabia

The progress achieved in the field of voluntary community service in the Kingdom of Saudi Arabia reflects the historical development of its social and economic conditions and its rapid population increase. Hence, such activities are the biggest among the GCC countries in terms of scale and commitment to social and health care services.

Until the early 1990s, the internal changes that community service has undergone were seemingly slow due to weak interaction with external experiences and the strict adherence to specific activity patterns. Since the beginning of 1990s, such works have witnessed a qualitative stride in assuming a greater role in enhancing the national economy and in pushing for social change.

Meanwhile, these developments were accompanied by a considerable increase in the number of civil and charity organizations and associations, in the 1980s when approximately ten women's societies were established.

Women's societies were the first civil organizations to be officially registered in the Kingdom of Saudi Arabia due to the traditional stance adopted towards women based on the notion that their role in the community can only be line with social work and service. The number of women's organizations has increased during the 1980s as ten of these societies have been authorized, and the women's civil activities have gained greater importance after the adoption of the Seventh Five Year Development Plan. This plan provided for activating the role of women in the field of voluntary community service in order to enhance their contribution to the country's development process and national economy.

Women's social activities can be associated with the three stages that Saudi society has undergone. First, the traditional stage (prior to 1960) during which community service was limited to charitable activities through individual or family efforts and in accordance with the prevailing norms and practices. In this context, women played a significant part in disseminating and enhancing religious guidance and awareness and in providing assistance to needy families, widows and orphans.

Second, the transitional stage (1960-late 1980s) which coincided with the establishment of the state's apparatus and services and the implementation of the first women's education programs (1960). This was followed by the employment of women in the public sector and the setting up of social development departments under the Ministry of Labor and Social Affairs (MLSA), including supervision offices for female associations and women social care institutions.

Until the 1990s, activities of women's organizations concentrated on traditional social care aspects, notably assisting needy families, caring for children and the handicapped, fostering women's causes and raising all kinds of awareness (on health, cultural, religious and social issues).

The third is the current stage (from 1990 to the present) which has witnessed several qualitative developments influencing the community service sector in the Kingdom, in general, and women's organizations, in particular. Among the key factors behind these developments were the greater knowledge gained in this field, increasing openness on external Arab and international experiences, and growing awareness of the new developmental concepts and information/communications revolutions.

However, the shortage of resources required to finance the projects and programs set up by these associations has led to growing awareness of the importance of rationalizing and reorienting the activities of charity associations with a view to advancing the intended goals of sustainable development and addressing some of the major issues faced by Saudi society, such as poverty, unemployment and the declining share of women in the labor market.

On the other hand, the community service sector has assumed a more important role in the Saudi economic sphere since the Gulf War of 1991 due to the decline in resources available to both the state and individuals.

The Seventh Five Year Plan has underlined the necessity of bolstering the community service sector as the third partner in the development process and stimulating the role of women in this sector in order to absorb the increasing numbers of female graduates who failed to find jobs in the public and private sectors.

The MLSA report of November 2003 has indicated that the total number of civil associations was 277 (including 23 female organizations and 38 charities) whose membership consists of 31,262 members (2929 were women), while the number of workers therein was 6430 (out of whom 1897 were females).

Saudi community service associations for women have realized the importance of implementing qualitative leaps forward to improve their range of activity with a view to achieving the desired level of contribution to the broader development process, as well as the required changes in the laws and regulations that govern the overall civil works and activities.

Among the factors that have motivated such a drive were the increasing level of openness to external experiences, broader knowledge of social work principles and the emergence of female leaders (especially the pioneer founding figures) hailing from the Saudi royal family and ruling elites.

The most significant accomplishments recently achieved by women's organizations are the implementation of new projects and programs employing modern developmental concepts, such as setting up training and rehabilitation centers, productive projects run by families, offering small loans and new job opportunities for women, as well as designing rehabilitation programs for the disabled and protecting their social rights.

In addition, these organizations succeeded in creating cooperation and coordination bridges with the elite Saudi female communities in the academic, media and social arenas. This has resulted in wider media coverage of women's causes, demands and accomplishments; broader debates over the desired structural changes; and more in-depth focus on women's civil and legal rights and their participation in economic activities, and issues related to poverty and children.

Saudi women's associations were able to organize a series of annual coordination meetings for three successive years which were attended by many female leaders from all segments of Saudi society. These meetings, which were held in Abha (Southern Province), Al Qaseem (Central Province) and Al Ahsa' (Eastern Province) in 2000, 2001 and 2002, respectively, have addressed the problem of weak collective performance and inter-association coordination without reliance on support from governmental authorities in dealing with urgent issues.

By holding these meetings in smaller cities, the women's community service sector has gained strong momentum enabling it to boldly raise key sensitive questions like developing the regulations that govern the activities of women's organizations and the undertaking by those organizations of a new developmental role in society. Several working papers have been submitted to the last meeting held in Al Ahsa' addressing issues like women's legal rights and domestic violence.

However, these meetings encountered strong protests by both religious and governmental circles, being viewed as an attempt to breach the accepted limits set for the social role assigned to women. Yet, such attitudes have resulted in public opinion harboring more appreciation and sympathy for these gatherings, which have become a means to lobby action to tackle the problems facing the community service sector, particularly those pertaining to the proposed amendments to the statutes of women organizations and the setting up of nationwide communication networks linking these organizations.

As a result, it was not long before the competent government authorities took the initiative to form a supreme commission for civil organizations (including a female one) and to revive the process of developing the regulations governing their activities. After the third women's meeting, a national committee was created to counter acts of domestic violence.

Nevertheless, despite the insistence by women's associations to proceed with holding their annual meetings, they failed to obtain the necessary official permission to organize the 2003 forum.[23]

The foregoing review of Saudi women's associations reveals their limited impact on the making of political and even social decisions, and the fact that there are conservative, mainly religious forces still blocking the progress march of the Saudi women's movement.

Any progress that may be realized by this movement is contingent upon the creation of a social movement encompassing both sexes; a trend that is apparently in the formative stage at present, as various civil forces, including women's groups, have embarked upon a process of dialogue and discussions that would pave the way for the imminent emergence of an embryonic civil society movement.

2-State of Kuwait

In 1962, professional organizations and trade unions began to be established and these were joined by women who sought to improve their work conditions and to advance the country's trade union movement. The origins of women's voluntary services can be traced to the formation in 1962 of the Arab Women's Renaissance Society which endeavored to set up social, sports, health and cultural projects and highlight women's activities in these fields. Its name was changed in 1971 to the Family Renaissance Society, but it was ultimately dissolved in 1980.

The Women's Cultural and Social Society was created on February 3, 1963 to promote cultural, social and sports activities among its members. In December 1964, the Kuwaiti Women's Federation was established encompassing the two above-mentioned societies with a view to assisting women's associations in implementing social and charity projects, coordinating their efforts by holding meetings and conferences and protecting the interests of Kuwaiti women. However, this Federation was dissolved in 1977 following the withdrawal of the Women's Cultural and Social Society from it.

On November 22, 1975, *Al Fatat* club (Young Women's Club) was set up as a public service society to encourage members to practice social, cultural and sports activities, and on November 17, 1981, the *Bayader Al Salam* society (Harvest of Peace society) was formed as a women's Islamic organization aimed at raising women's cultural awareness, setting up nurseries, organizing training courses and handling family problems. On January 3, 1982, the Islamic Care Society was created to disseminate Islamic culture and traditions, improve the methods of teaching the Holy Quran, and to care for needy children and achieve collective social responsibility.

In February 1982, the General Federation of Kuwaiti Labor formed its Working Women's Committee which was entrusted with the role of supporting and advancing women's causes, including their contribution to the country's economic growth process.

This was followed by many purely "male" organizations, including religious, social and charity organizations, setting up their own women's committees, such as the Social Reforms and Heritage Revival Societies. Those newly-formed committees sought to encourage women to conduct charitable activities and to fulfill their traditional duties in accordance with the provisions of Islamic Law. In general, the increase in women's forums, societies and committees has accorded women's activities a broader base of public support.

During the 1980s, the Kuwaiti women's movement had witnessed great momentum at the Arab and regional levels through effective participation in the activities conducted by the broader Gulf women's movement, especially when the Committee for Coordination of Women's Action in the Gulf and Arab Peninsula was established by a Kuwaiti initiative. The formation of this Committee provided the opportunity to boost communications, interaction and coordination processes among women leaders in the Gulf states with regard to national and regional issues.

Following the occupation of Kuwait by the forces of the Iraqi regime, popular committees were formed outside Kuwait prompted by spontaneous patriotism to the homeland and were successful in addressing and handling the problems and difficulties suffered by Kuwaitis living in exile.

Meanwhile, similar committees had been set up inside the country, and on September 15, 1991 (after the liberation of Kuwait) all of these committees were combined together under the Kuwaiti Voluntary Women's Association for Social Services. On June 25, 1994, the Kuwaiti Federation of Women's Societies was established.[24]

Despite the fact that the majority of these organizations were established by elite woman personalities, it can be said that they are still unable to play an effective role in political life or to confront religious and tribal forces that still prevent women from entering institutions such as the Kuwaiti parliament.

3-State of Qatar

The Qatari Society for Rehabilitation of the Handicapped was the first civil society established in 1976 and in March 1978 became a branch of Qatar's Red Crescent Society (QRCS), which was internationally recognized in 1981. The QRCS women's branch was opened in 1982 to organize special programs to protect and provide health care for children, mothers, the handicapped and the sick in general. In early 2000, the name of the women's branch was changed to the Social Development Administration.

In 1992, the Qatar Charity Society was set up, and its women's branch opened in 1993. The Qatar Foundation for Education, Science and Social Development was established in 1996 as a socio-economic, charitable and voluntary organization aimed at utilizing the energies of individuals and enhancing their productivity levels in order to improve the economic status of limited-income families.[25]

Pursuant to an Emiri decree, the Supreme Council for Family Affairs was established as a legal entity by the end of 1998. An independent budget was allocated for the Council, which is chaired by Sheikha Mouza bint Nasser Al Misnad, wife of the Emir. The Council's most important goals are enhancing the role of family in society, implementing the objectives envisaged in the Universal Declaration of the Rights of the Child, the Convention on the Rights of the Child and other international instruments, and improving the conditions of working women.

A careful review of the accomplishments achieved by Qatari women will show that they were not gained as the result of civil or private efforts, but rather the outcome of decrees and decisions taken by the political leadership. It can be said that the Qatari women's movement has witnessed several advances over the past few years: obtaining rights of suffrage, the right to run for municipal council elections (and probably for parliamentary elections later), and the right to education, since they were given the opportunity to join newly opened foreign universities and colleges.

4-The United Arab Emirates

During the past two decades, UAE women have made significant achievements in the fields of education and employment. The proportion of educated (literate) females has soared from 4 per cent in 1960 to over 80 per cent in 1993. For example, the number of female students of the UAE University in Al Ain now represents approximately 75 per cent of the total number of students. During the period 1980-1997, the percentage of illiteracy has dropped from 51 per cent among males and 77.6 per cent among females to around 7 per cent for both sexes. Yet, the share of women in the labor market stood at 6 per cent.

In 1976, in order to strengthen their voluntary social service role, the UAE women embarked upon the process of creating their own associations when a group of educated ladies took the initiative and the first two women's societies were established in the emirates of Ras Al Khaimah and Dubai under the name of the Women's Renaissance Society, with activities concentrated on the eradication of illiteracy, setting up cultural and social programs, and organizing seminars and lectures dealing with the prevailing women's situation. Similar societies were established in Abu Dhabi (1973) and in each of the emirates of Sharjah, Umm Al Quwain and Ajman (1974). Sharjah's *Al Ittihad* (federation) Society and Abu Dhabi's Women's Renaissance Society were both set up pursuant to governmental directives.

The UAE General Women's Union was established in 1975 under the leadership of H.H. Sheikha Fatima Bint Mubarak, wife of the late H.H. Sheikh Zayed Bin Sultan Al Nahyan to bring all the country's six women's societies under one umbrella. The Union's founding members were the First Ladies of the emirates of Abu Dhabi, Sharjah, Umm Al Quwain, Dubai, Ajman and Ras Al Khaimah, and its main responsibilities are to improve women's conditions and expand women's activities throughout the country. At the domestic level, the Union was instrumental in establishing local branches, introducing health and social awareness projects, adult education and literacy programs, handicraft centers, charity fairs and bazaars, and vocational training classes.[26]

An examination of the functions and events organized by women societies during the 1993 cultural season indicated that the traditionally presented religious education programs seemed to be the dominant activity performed by all UAE women's associations without touching upon women's causes and their daily problems and difficulties. The most important achievement so far attained by these organizations is the literacy eradication program that has enabled considerable numbers of female learners to complete their university education.[27]

The total number of women's societies' members was estimated at 1918, with at an average of 170 members per society. Generally speaking, these societies are suffering from inaction and ineffective managements, which have failed to attract many female university graduates. The composition of the boards of the managements have not changed since these societies were first established, and the last meeting of the general assembly of the Dubai's Women's Renaissance Society was held 13 years ago.[28]

5-Sultanate of Oman

The Omani women's movement began in the early 1970s when the Omani Women's Society, the first of its kind, was created in 1972. Other societies with the same name were set up in different provinces raising the total number to 23 societies in 1998. The major activities of these organizations are designing literacy eradication programs, setting up summer and children's clubs, building nurseries and kindergartens, conducting dressmaking classes, caring for the disabled and other functions, most of which are directed to improve the situation of the local women.[29]

The activities of these societies are confined to the voluntary works already mentioned rather than addressing urgent political issues, which is partly due to immature institutions of civil society and the lack of qualified national cadres capable of bringing about and introducing the desired changes.

6-Kingdom of Bahrain

The Bahraini women's societies, whose number presently stands at nine, have become key centers for political and social activities, especially after women were granted their rights to vote and to run for elections. A large number of women have joined different social and professional associations such as for engineers, lawyers as well as sports clubs in which the memberships used to be exclusively limited to males.

The origins of organized female activities in Bahrain can be traced back to the 1950s when a women's social club was set up in 1954 and headed by the wife of the British Chancellor. The Club's main activities concentrated on first-aid training and organizing hospital visits. Two years later the club went out of existence, and the first women's society in the Gulf region, the Bahraini Women's Renaissance Society, was set up in 1955.

The Society, with its present membership estimated at around 500 members, has been involved in building nurseries, kindergartens and organizing dressmaking workshops in poor areas; publishing serious studies that deal with pressing social problems and Bahraini women's conditions in general, like the increase in divorce rate and children's problems; and offering legal advice and psychological counseling services through a highly successful project designed by the Society for this purpose.

In 1960, the Mother and Child Care Society was established, and because the majority of its members belong to economically and politically influential segments of society, it was able to undertake big social and educational projects that other societies have failed to create, such as *Al Amal* (hope) Institute for the Disabled, the Women's Studies Center – the first of its kind in the Gulf area – and other programs relevant to women's economic and social development.

Following the same guidelines adopted by the Bahraini Women's Renaissance Society in terms of goals, functions and liberal trends, the Awal Society was set up in the city of Al Muharaq and both societies advocated radical reforms relating to women's political rights and the issuing of a new personal status law consistent with the era's requirements and realities. The Society's membership stands at about 250 members.

The Al Rifa'a Cultural Society, which was established in 1970 in Al Rifa'a city, was active in the charity and cultural fields. With a gradually declining membership (of about 50 members) and most projects at a standstill, except for one kindergarten in Al Rifa'a city, the Society's board of management resigned, but the new board has failed to attract new members. Therefore, the Ministry of Labor and Social Affairs decided to dissolve the Society. A group of 25 ladies are now considering reopening the Society in accordance with new guidelines and programs.

The International Women's Association, set up in 1975, is the only organization in which the majority of its membership (150 members) belongs to foreign minorities living in Bahrain.

As a result of reform initiatives launched in the Kingdom during the past two years (notably granting women the rights to vote and run for the elections), new nine women's societies have been authorized, raising to 14 the total number of such societies. Four other societies are currently under constitution. These are the *Fatat Al Reef* (rural women) Society; the Bahraini Women's Society (associate of the political leftist Democratic Forum organization); the Future Women's Society (an Islamic Shiite association under the *Al Wifaq* (accord) political movement; and *Al Bahrain* Women's Society (under another female Shiite group).

The Islamic Sunni political groups (such as the Reform and Islamic Societies) have contented themselves with setting up women's branches under their main societies.

Other women's associations have been established in the current year. They include the Hamad City Society; Bahrain Women's Development Society; Women for Al Quds Society; Al Hoor Women's Society; and Al Muharaq Women's Society.

The trend of political openness that the Kingdom of Bahrain has witnessed during the last few years has resulted in the formation of the following three bodies which will have substantial impact on the women's movement within the Kingdom:

1- *The Supreme Council for Women's Affairs:* This is a governmental organization presided over by Sheikha Sabeeka Bint Ibrahim Al Khalifa, wife of His Majesty the King of Bahrain. It comprises 15 members representing prominent female leaders. Among its key tasks is the designing of women and family-related policies, plans and programs. The very composition and duties of the Council allow women the opportunity to perform a greater part in the economic, political and cultural areas. The Council reports directly to His Majesty the King in order to avoid any bureaucratic delay in arriving at resolutions. The Council has its own independent budget, and the appointment of its Secretary-General to the rank of minister represents an advanced step that would pave the way for women to assume ministerial posts.

2- *The Bahraini Women's Federation:* (under constitution): The Federation's founding committee was first formed of representatives of women's and religious associations, women's committees of mixed societies (both sexes), and independent individuals, in order to ensure integration and coordination between the activities of civil and government sectors. The governmental authorities concerned, however, were slow to authorize the creation of the Federation and rejected any representation of women's committees. The women's associations accepted this condition by agreeing to a request made by the Ministry of Labor and Social Affairs to reduce the Federation's membership to include these associations only. In my view, this would weaken the women's movement as a political force expected to play an important part in drawing up women and family-oriented policies.

3- *Women's Petition Committee:* The Committee was created in May 2001 and was chaired by Mrs. Ghada Jamshier. The Committee's membership consists of a group of ladies affected by unjust court rulings (pertaining particularly to cases of divorce, alimony, children's custody) and other female figures who advocate the Committee's goals of reforming the religious judiciary, restructuring the legal system, subjecting elderly judges to the pension law provisions, enacting a modern personal status law, and strengthening the role of the Supreme Council of the Judiciary in monitoring and examining the rulings made by the religious judiciary. In its endeavor to realize these objectives, the Committee, with a membership that has risen from 208 to more or less 500 women during the past two years, has resorted to public demonstrations and sit-ins, issuing press releases and statements, setting up channels of communication with local and foreign mass media, exposing the actions of corrupt judges, and distributing brochures and leaflets aimed at raising awareness of injustice and oppression inflicted upon women and families stemming from the slow implementation of these reforms.

As a result of these activities, the Committee was able to attract the attention of the public and other women's associations. However, a group of judges strongly resisted the Committee's accusations by filing lawsuits based on charges of defamation and insult.

The setting up of the Committee – in my judgment – has come as a response to the impatience of large segments of women resulting from the failure by both civil and government establishments to resolve women's problems and to enact a modern personal status law that would endorse their legitimate rights.

Generally speaking, the women's movement in Bahrain is regarded as the most advanced in the Gulf region, probably due to the relatively mature Bahraini civil society institutions compared to those of other Gulf countries. The movement, nevertheless, like other civil society groups, have been weakened by those laws and regulations that hinder their efforts to attract new members and by the lack of support urgently needed to expand their educational, cultural and social programs.

Applicable Policies to Enhance Women's Role in Gulf Societies

It has been obvious that, during the last few decades, Gulf women have moved from their traditional roles of the wife, mother and household manager to a new and broader role that combines the former together with an economic and social part generated by the educational and employment opportunities allowed to women.

However, the new role has not been associated with the policies and social institutions that should have been designed and created to protect women's gains.

The following chapter discusses the issues that would affect the women's development process and suggests some applicable measures that can be adopted with a view to fostering their new role and augmenting their accomplishments.

1- Enacting Personal Status Laws in Gulf States

Despite the ostensible equality between men and women as envisaged in labor laws in effect in the Gulf states, article 61 of Bahrain's labor law, for example, provides for paid maternity leave of no more than 45 days in addition to an unpaid 15-day leave. Most of the laws passed in these states follow the same pattern. As for divorce and child custody cases, the Islamic law provisions are literally and strictly applied in all countries of the region. None of the Gulf states has enacted a personal status law except for the one passed in Kuwait which itself requires several amendments.

An attempt was made in Bahrain to enact a similar law and a draft thereof was submitted for public debate. However, owing to the strong protests against the law by religious groups, both Sunni and Shiite, it was dismissed and set aside indefinitely. No other attempt was made in this respect in the remaining Gulf states (Qatar, Sultanate of Oman, UAE and Kingdom of Saudi Arabia).

The "personal status law" is meant to govern situations in connected to the status and legal capacity of individuals, as well as those matters related to the family such as, *inter alia,* engagement, marriage, divorce, mutual rights and duties of both wife and husband, and their financial relationship.

In my judgment, such a law will have to provide for certain rights for the wife, such as her right to joint ownership of the place of dwelling and the right of the mother to pass her own nationality to her children in cases where her husband has a different nationality. In addition to Islamic law (sharia) and jurisprudence, the law must derive its provisions from the principles enshrined in the United Nations Charter, human rights conventions and any other international instruments advocating equality between males and females and banning all acts of violence against women.

Several Arab states have legislated new laws that guarantee more or less equality between men and women, such as the Tunisian law which forbids men from having more than one wife at the same time. The Egyptian law was amended to allow the wife the right to seek divorce through the so-called *Khula'* procedure, that is, in return for a monetary compensation to be paid by the wife to the husband.

The law that has recently created a huge uproar in all the Arab countries is Morocco's Personal Status Law (Family *Moudwawwana*) which was revised and all proposed amendments therein were endorsed. The law derives its stipulations from a profound understanding and judicial interpretation of the Islamic law (sharia) and establishes the principle of equality between wife and husband in caring for the family by recognizing the wife as the husband's main partner in terms of their respective rights and duties. It also dropped the condition of "obedience by the wife to her husband," imposing strict and rigid restrictions on polygamy.

Indeed, the law even prohibits polygamy when the husband feels unable to ensure equal and just treatment between his wives. This is in compliance with the following verse from the Holy Quran: "… If you fear that you shall not be able to deal justly (with them), then only one."(Surah Al-Nisa, 4, Ayah 3).

The new amendments made to the Moroccan law endorse the wife's right to seek divorce on account of harm inflicted upon her as a result of violation by the husband of the conditions provided in the marriage contract, such as violent treatment, desertion, long absence and parsimony.

In addition to many provisions that are in favor of the family and women, and to protect children's rights, new stipulations were forged to be in harmony with the requirements of the international conventions and agreements ratified by the Moroccan government. These stipulations address issues such as kinship, children's custody and guardianship, breast-feeding, alimony, religious guidance, education and sponsorship.[30]

Needless to say, the enactment by Arab Gulf states of personal status laws similar to the Moroccan code is seemingly an unattainable goal owing to the strong protests by different segments of Gulf societies. Yet, the only feasible approach – in my point of view – is to instill these concepts in the new generations through the established curriculums.

In addition, government authorities concerned and women's organizations might consider adopting strategies based on passing one provision dealing with each of these urgent issues during a given period of time. Thus, as time passes, perhaps the gradual addition of these provisions would facilitate the enactment of a complete and integrated personal status law.

2-Strengthening Ties with UN Agencies and Implementing Women's International Conventions

Since joining the United Nations, Arab Gulf states have taken part in most of the meetings held by international organizations, especially the four women's summits and the UN General Assembly's Special Session held in New York from June 5-9, 2000, known as the "Beijing + 5" session.

Gulf countries have also signed several international instruments and treaties which advocate the political participation by women and prohibit violence and discriminatory acts against them. Thereby, these states are committed (at least morally) to honor the resolutions adopted by international conferences, particularly the so-called "Beijing Recommendations, 2000."

Questions related to women have been seriously addressed with greater attention and interest by the United Nations at the First World Conference on Women, Mexico, 1975 at the end of which the International Decade for Women (1976-1985) was announced and the objectives, "Equality, Development, Peace," were adopted by the international community.

The goals and achievements attained by women during the Decade were evaluated at the Copenhagen and Nairobi Conferences in 1980 and 1985, respectively. Perhaps the most important of these gatherings was the Fourth World Conference on Women (Beijing, 1995) which has crowned the women's struggle by issuing a world program of action (the Beijing Platform for Action) under which about twelve strategic goals have been approved in various fields: poverty; education and training; health; violence against women; armed conflict; women and economics; power and decision-making; institutional mechanisms to empower women; human rights and women; women and media; and women and environment.

The Conference called upon governments, international community, civil communities and private sectors to take the necessary measures to promote and advance these goals.

Despite reservations expressed by most Arab states with respect to several recommendations of the program of action as inconsistent with the Islamic law and traditional cultural norms (especially those concerning abortion and inheritance), these very states have sought to implement some of the program's commitments. These reservations, however, were assessed at the UN General Assembly's Special Session (New York, 2000), during which hurdles impeding the implementation of the Beijing Conference's recommendations were also examined and identified.

The forthcoming Conference that the UN plans to organize in 2005 ("Beijing+10"), ten years after the 1995 Beijing World Conference on Women, will almost certainly boost women's accomplishments at both international and Arab levels, and those gained by Gulf women in countries like Qatar, Bahrain and Oman – as already mentioned – will be cited as major achievements in the reports to be submitted by those states to the forthcoming Conference.

The ratification by Gulf governments of the Convention on Elimination of all Forms of Discrimination Against Women constitutes a qualitative stride in the context of stances taken by governments towards women's causes and a commitment to implement the Convention's stipulations should yield significant dividends in favor of women.

Furthermore, except for the Kingdom of Saudi Arabia which was represented by purely male delegations, the participation by both governmental and non-governmental delegations comprising a majority of women members, in the above-mentioned conferences had increased the level of openness to the outside world, especially since these delegations have taken part in public debates in their respective countries and sought to advance some of the goals set by the conferences in question. Besides, informal meetings held on the sidelines of these conferences between the Gulf and other delegations, both governmental and non-governmental, have allowed women leaders to exchange information and viewpoints, widened horizons of cooperation, and bolstered their ambitions to create a better future.

On the other hand, research studies conducted by the UN Development Program (UNDP) for the host countries in the Gulf region are meant to identify the nature of the problems that women and societies suffer from, and propose solutions that might help respective governments to overcome any obstacles encountered by women's movements in those countries. Similar contributions are made by different UN organs and specialized agencies, notably the UN Development Fund for Women (UNIFEM), the UN Fund for Population Activities (UNFPA), International Labor Organization (ILO) and the UN Economic and Social Commission for Western Asia (UN ESCWA), with a view to fostering the role of women and helping to achieve the goals determined by the Beijing Conference.

Conclusions

Issues relating to women have been brought to the fore in most nations of the world and accorded greater attention by the United Nations and human rights organizations. As a result of the discovery of oil, increased educational and employment opportunities for women, and substantial rise in living standards as growing numbers have joined the labor market, women's conditions have witnessed significant changes in the Gulf states. Consequently, the nature of the traditional roles played both men and women should also have changed accordingly.

On the other hand, the family in Gulf societies has witnessed radical structural changes, such as moving from the status of an extended family to a nuclear one, though the former has maintained its influence on the latter.

In the Gulf region, as is the case in the Arab world as a whole, the family remains the dominant unit comprising the social fabric. Women, therefore, are still facing heavy pressures to create a family in order to establish their social status.

Nevertheless, the changes that the family structure has undergone have not led to a shift in the traditional roles of both men and women, and working women have been obliged to perform their traditional role in shouldering the household and family burdens and caring for husbands and children while simultaneously fulfilling their responsibilities as working women.

By undertaking both these roles – the traditional and the new – working women are obviously encountering additional psychological and physical pressures, a situation that necessitates revisions by Gulf states of their policies towards working women in particular, and the family establishment in general, in order to suit their new family and living conditions and their growing share in the labor market.

A careful revision of the activities of female associations and societies in the region has revealed the marginal effect that these organizations have had on fostering the new role of women.

In the Kingdom of Saudi Arabia, women's societies have exerted a limited impact on the making of political and social decisions. It is also clear that there are powerful elite political and religious forces, which control the political decision-making process and are still blocking the progress march of the Saudi women's movement.

Similarly, it can also be said that the Kuwaiti women's associations are too powerless to confront religious and tribal forces that are still preventing women from entering institutions such as the Kuwaiti parliament.

Many accomplishments have been achieved recently in the State of Qatar, though not initiated by the civil society institutions but rather as an outcome of actions taken by the political leadership. This includes granting women the right to vote and to run for municipality elections (and perhaps for parliamentary elections in future).

The UAE's women's societies are suffering from inaction and ineffectiveness of their boards of management. The role performed by the General Women's Union is confined to setting up health and social awareness projects and organizing literacy and adult education classes, avoiding any involvement in political issues concerning women.

In the Sultanate of Oman, the activities of female societies are restricted to traditional voluntary service rather than addressing urgent political questions facing Omani women.

The women's movement in Bahrain is regarded as the most advanced in the Gulf region. Nevertheless, this movement has been subdued lately by restricting the membership of the Bahraini Women's Federation to include only women's societies. Hence, the women's movement remains weak, like other civil society organizations which are still governed by laws that hinder their efforts to attract new members.

Women's societies in the countries of the region depend heavily on annual subsidies and aid presented by the respective ministry of labor and social affairs as these societies are not allowed to receive any grants or donations from abroad.

In conclusion, this study calls for the adoption of comprehensive policies that would promote the new role of women in the Gulf states and enhance and boost their potential, energies and accomplishments, such as enacting advanced personal status laws, and strengthening the ties of the Gulf states with UN agencies and implementing conventions adopted by them regarding women's issues and causes.

22

Women in the Gulf and Globalization: Challenges and Opportunities

Badria Abdullah Al-Awadhi

The term "globalization" still invokes fears and skepticism among theorists, economists, politicians and even clergymen particularly in the developing states, including Arab states. As a phenomenon, "globalization" has become the key axis of debates in academic forums and conferences, television programs, and specialized UN committee meetings, as well as a main item on the agendas of special conferences on human and women's rights, with a view to understanding both the positive and negative impact of this phenomenon on such rights.

In general, while some believe that globalization is a beneficial process, a key to a better future for economic progress, and an inevitable reality that is impossible to abrogate or avert, others harbor a hostile and skeptical attitude towards globalization arguing that it heightens inequality among peoples, jeopardizes job opportunities and living standards in developing countries, and impairs the process of social development.[1]

Integration into the global economy has resulted in varying developmental dividends and there are limitations in globalization's role in curbing impoverishment—as the experiences of some Latin-American, African and South Asian developing countries during

1970s and 1980s have shown. This confirms the fact that opportunities available under globalization are not without risks. This conclusion can be derived from the continuing big differences between high and low-income states which the International Monetary Fund (IMF) attributes to the inability of the latter to integrate into the global economy at the same pace as other states, partly because they opt for certain policies and partly due to circumstances beyond their control.[2]

The Glossary of The Globalization Website explains and interprets this term as the "expansion of global linkages, organization of social life on global scale, and growth of global consciousness, hence the consolidation of world society."

Some of its forms are described as follows in the website:

- "The inexorable integration of markets, nation-states, and technologies to a degree never witnessed before—in a way that is enabling individuals, corporations and nation-states to reach around the world, farther, faster, deeper and cheaper than ever before...and the spread of free-market capitalism to virtually every country in the world."
- "A social process in which the constraints of geography on social and cultural arrangements recede, and in which people become increasingly aware that they are receding."
- "As experienced from below, the dominant form of globalization means a historical transformation: in the economy, of livelihoods and modes of existence; in politics, a loss in the degree of control exercised locally... and in culture, a devaluation of a collectivity's achievements...Globalization is emerging as a political response to the expansion of market power."[3]

I-Anti-Globalization Attitudes

Some Arab intellectuals regard globalization as a colonial phenomenon intended to dominate people's destinies and natural resources, stemming from the belief that the poorer nations of the world possess what humanity needs of these resources but are incapable of manufacturing them. Hence, they conclude that globalization is the new face of colonialism, and that whoever wins the economic markets will gain influence all over the world.[4]

In justifying his objection to globalization, the Islamic intellectual Dr. Yousuf Al Qardhawi asserts that globalization is in the interest of the stronger powers because they will be definitely and ultimately be the winners in the race with the poorer powers. He raises the question: Why have westerners – the Americans and Europeans themselves – in Seattle and elsewhere, protested against globalization? Even though economic globalization will benefit the rich North at the expense of the poor South, peoples of the developed countries still resist it, which proves that globalization is widely and increasingly considered an unacceptable practice throughout the world.

In a broader sense, it may be said that globalization seeks to create a linkage between its economic, cultural and political elements, integrate financial markets with the international trade and direct investments, and incorporate capital and labor force into the global economy. This will lead to the infiltration of frontiers and nationalities by the multinational corporations and those employing massive capital and ultimately result in the total abolishment of borders.

On the other hand, both opponents and advocates of globalization agree that it cannot be seen as an accidental phenomenon, rather a reality that has undergone a set of world-wide changes during the past two centuries ending up with the globalization that we have today.

> Therefore we have to learn how to accommodate to this reality and understand its dimensions, repercussions and consequences, and admit that a revolution is erupting in which values, morals and conceptions are changing. This understanding will allow us to deal with it through both interaction and resistance; namely, by resisting the negative aspects and interacting with the positive ones, such as those pertaining to information technology advancements.[5]

Those who accept this stance argue that globalization gives expression to the technological and economic development prevailing in the world since 1980. Therefore, we will have to make every effort towards scientific and economic progress, in order to deal with globalization in a positive, rational manner stemming from knowledge and awareness so as to avoid its adverse impact on the culture and Islamic identity of the Arab people. In this connection, Dr. Al Qardhawi reaffirms:

> Nations must maintain their own traits and specialties, since diversity will lure humanity to advance forward… It is needed because each party will benefit from the other… Scientific globalization is desirable but cultural globalization – that is, the abolition of individual traits and attributes – is unacceptable.[6]

Such a dispute over the concept and content of globalization poses the real challenge confronting people in the Arab region in general, and Gulf women in particular, especially in an era of globalized ideologies, cultures, economies and politics. This is also apparent in the ways and means of tackling this phenomenon with all its cultural, social and religious dimensions, regardless of the views depicting globalization as a new form of colonialism or a form of American or European domination over the fate of Arabs. Indeed we must employ the mechanisms and elements of globalization to disseminate Arab cultural values and tolerant Islamic ideology among peoples of the world.[7]

This brief review of the stances of Arab scholars towards globalization and its forms and dimensions helps us to reflect on the economic, social and political status of women in the Gulf region. Thus, it is important to identify the main challenges confronting Gulf women and the opportunities available to enhance their human, political and economic rights in the era of globalization, taking into account the diminishing chances of women's empowerment within the Arab social structure, as the first Arab Human Development Report 2002 (AHDR) has shown.

The Report has listed this issue within the three imperatives, the absence of which has impeded the advancement of human development in the Arab world: absolute respect for human rights and freedoms; empowerment of Arab women and the unleashing of their energies in an environment of equity, justice and impartiality; and encouragement for the acquisition of knowledge in all social activities.[8]

II-Political and Legal Challenges Confronting Gulf Women under Globalization

Under international law and human rights conventions, women are categorized as being among the segments most vulnerable to world economic and political developments since they are less skilled and less educated than men, particularly in the developing countries. As a result, in the era of globalization and informatics, women are viewed by states, societies and big corporations as an unskilled and relatively cheaper labor force.

In these circumstances, some intellectuals fear that from an economic point of view, globalization will not foster women's interests, and that the balance will be in favor of working men. Work opportunities for women are expected to be limited to unpaid or low-wage jobs because of insufficient income generation in their respective countries. Politically, women could be sidelined from domestic affairs, while culturally they would lose their individual identity and independence in favor of global culture.

At the same time, advocates of globalization suggest that its impact could vary and yield considerable benefits depending on the nature of women's groups and movements. To gain these benefits, new standards for treating women could be established to assist women's social organizations, and to improve the conditions for women in societies where they are oppressed or treated as inferior to men. This can be achieved through asserting the importance of equality without gender discrimination in order to attain economic growth and sustainable development, since investments in women's development projects and plans will have positive and multi-faceted effects.

The UN Secretary-General's Report on sustainable development, global economy and women's role in the development process, presented to the 58th Session of the General Assembly in 2003, admits that although globalization has created more jobs for women in many countries, it has been the source of many problems and difficulties confronting women in developing and less developed countries as a result of unstable economic activities.[9]

At the regional level, the participation by Gulf women in the political, economic, social and cultural development process in the majority of the Gulf states dates back to the early 1950s when they were offered educational opportunities that paved the way for them to acquire jobs as school-teachers and civil servants. Although there are variations from one country to another, it was only in the beginning of the 1960s and 1970s that women's participation in economic and social life became an economic necessity for them and a national need for the development of society. Education has played a key role in promoting women's capabilities to assume positions in spheres that were earlier considered exclusive to men, such as the diplomatic corps and the police force and to take up high ranking posts such as ministers and deputy ministers.[10]

The following are some of the main challenges that Gulf women confront in the era of globalization, and the possibilities of overcoming the political and legal challenges in particular:

1-Political Challenges (The Right to Political Participation)

The right to political participation is one of the fundamental human rights according to the principle of the equality of rights and duties as enshrined in the UN Charter and international human rights conventions. Theoretically, national constitutions of the Arab Gulf states endorse this principle, but the major obstacle lies in the ways and means of activating women's right to political participation. Despite the participation by women in various aspects of public life and exercising their rights on equal terms with men in some of these states, this right has long been the subject matter of major political, religious and social debates.[11]

In comparison, women in some Asian, Latin-American and African countries have been granted this right much earlier, such as Albania (1920), Mongolia (1924), Ecuador (1929), Turkey (1930) and Sri Lanka (1931).

Generally speaking, the women's struggle in some of the Gulf states against those opposing their political rights was only successful by the end of the 1980s due to the lack of awareness among most women themselves of the importance of these rights and the neutral stances taken by several governments. This includes the stance of the Kuwaiti government in dealing with parliamentary initiatives that called for the abolition of the legal ban imposed on the right of Kuwaiti women to political participation. Such stances have allowed the conservative and religious groups the opportunity to thwart all the above-mentioned initiatives by democratic means.[12]

In the 1990s, this challenge did not constitute an obstacle for women in three Gulf states – Sultanate of Oman (1997), State of Qatar (1997) and Kingdom of Bahrain (2002) – by virtue of direct initiatives by the political leaderships, which enabled women to take part in the public life of Gulf societies by practicing political activities on an equal footing with men. This radical transformation has been promoted by international and regional developments advocating the abolition of all forms of discrimination against women.

In addition, this development indicates the need to reexamine the concept of citizenship for women, who are still regarded in some Gulf countries as ineligible for full citizenship rights. Three other Gulf states have not yet recognized this basic right of women – State of Kuwait, the United Arab Emirates, and the Kingdom of Saudi Arabia – despite the distinguished scientific, social and economic status that women have gained in these states.[13]

The Sultanate of Oman was the first Gulf state to grant women the right to political participation. In 1997, for the first time four women were appointed as Members of Oman's State Council. The Council at the present time includes six female members. The State of Qatar followed suit in the same year, when it granted Qatari women the right to vote and stand for elections to the municipal councils at which the percentages of female voters and candidates were 44.9 percent and 3 percent, respectively.[14]

In February 2001, under the Charter of National Action which endorses the principle of equality of rights and duties of all citizens – both men and women – before the law, the Kingdom of Bahrain recognized the right of women to political participation. Although Bahraini women's participation in the parliamentary and National Assembly elections held in May and October 2002 were effective and highly organized, none of the female candidates achieved the required percentage to win the membership of the Council. Four women were later appointed as members of the Shura (consultative) Council to underline the role of women in the political process.[15]

The rapid development witnessed in the political conditions for Gulf women has been spurred by the direct initiative of political leaderships despite the fact that some Gulf states that have activated the right of political participation are not yet parties to the International Convention on the Elimination of all Forms of Discrimination Against Women, 1979 (CEDAW).[16]

Many international human rights organizations view this Convention as an international charter for women's rights in the political, social and economic fields. It is even considered by some as an integral part of the universal world order that obliges states to observe the basic rights of women. Hence, the political leaderships of some states have been keen to promote women's political rights by adopting the so-called "quota system." Under this system, some seats in the Shura (consultative) councils are assigned for women to allow them the opportunity to participate in the decision-making process in societies where restrictive, outdated norms and traditions hinder women from entering parliaments through direct elections, as the elections held in Qatar (1997), Bahrain (2002), and – to some extent – in Oman (2003) have shown. Hence the initiatives have been taken by some Gulf political leaderships to appoint women as members in their national representative councils. In 2003, for example, four women were appointed to the Bahraini Shura Council, and six to Oman's State Council.[17]

At the international level, the parliamentary membership ratio of women to men is still low. This is attributed to political or economic reasons obstructing the equal exercise by women of their political rights alongside men, as envisaged in the recommendation of the UN General Assembly pertaining to the implementation of decisions adopted by the 1995 Fourth World Conference on Women. By 2002, the percentage of female members in the parliaments of all states did not exceed 14%.[18]

As already mentioned, the most important challenge facing Gulf women lies in the promotion of public awareness by women of the usefulness of this right, which is in harmony with the Islamic law

(*Sharia'a*). Indeed, it is the opinion of most Islamic scholars that it is a religious duty no less important than other family-oriented duties that women are required to carry out. They assert that Islam treats men and women equally in terms of their rights and duties.[19]

Accordingly, it has become important to highlight the positive effects that exercising the political right can have in advancing other women's rights, promoting human development, consolidating and codifying the initiatives of political leaderships to activate this right, and benefiting from the ratification by Gulf states of international human rights instruments (particularly the International Convention on Elimination of all Forms of Discrimination Against Women of 1979) by taking the appropriate steps to modify or abolish all discriminatory laws, regulations and practices against women.[20]

As for women in those Gulf countries that have not yet activated women's right to political participation, they will have to embark on an active endeavor – through women's and civil society organizations – to gain this right, the denial of which is no longer acceptable in the era of globalization nor suitable considering the status that women have acquired, academically and socially, and the role they play in the process of social development.

Also, women will have to create a supportive popular base to assist them in conducting their political activities, since any changes in the prevailing social conditions of Gulf societies cannot be accomplished in the absence of prior social, ideological and psychological shifts rendering women's right to political participation acceptable.

The Kuwaiti women's struggle for this right has shown the extreme complexities involved in the process of modifying discriminatory laws against women (such as the Kuwaiti Election Law No. 35 of 1962) as long as the legislative power is determined by the conflicting interests of parliamentary representatives, the judicial system is involved in enforcing merely the letter of the law, and the executive power is unable to meet the constitutional procedures for enacting or amending the laws pursuant to the principle of separation of powers as stipulated by Article 50.[21]

For all these reasons, the initiative of the Emir of Kuwait dated May 16, 1999 to remedy the incorrect, unconstitutional trends was rejected by the overwhelming majority of the Kuwaiti National Assembly on November 22, 1999 on constitutional, tribal and partisan grounds. On November 30 of the same year, a draft law to amend the above-mentioned election law with a view to recognize women's right to political participation was tabled by a number of legislators and supported by the government. The National Assembly vetoed this by a margin of only two votes.[22]

The conditions of women in the UAE and Kingdom of Saudi Arabia are similar to those of Kuwaiti women regarding the absence of political participation even though the political and constitutional situations in these two states are somewhat different from the existing political system in the State of Kuwait.

It is expected that new developments in Saudi Arabia could boost women's role in the political participation and decision-making processes, as one can conclude from the recommendations of the Second National Saudi Forum for Ideological Dialogue, which concluded its work in Mecca on January 4, 2004. Ten female academics and intellectuals took part in the Forum together with 50 male scholars, which reflects the intention of the Kingdom's political leadership to further advance the status of Saudi women.[23]

This shift was reiterated by the twelfth recommendation of the Forum, which calls for the "...promotion of women's role in all fields and the establishment of specialized national organizations to provide care for children, women and families." In addition, the eighteenth recommendation by the participants was that the theme of the Third Forum could be one of the following topics: the relationship between the Ruler and the subjects; women's rights and duties; popular political participation; and education.

In assessing this historical experience in which Saudi women have officially and publicly taken part in a national forum, Dr. Hind Bint Majid Al Khthaila observed that the presence of women was effective and amounted to a remarkable achievement. "Though many think that our participation was not official, it was effective and substantial and performed at the desired scientific and academic levels as was reflected in the Forum's recommendations," she added.

Dr. Amal Al Tua'mi, a Saudi writer and academic, noted that "this participation is an important event for Saudi women since they represent the female half of Saudi society and are qualified to take part in such a dialogue and make the right decisions."

Commenting on this event, Prince Saud Al Faisal, the Saudi Foreign Minister, underlined the importance of women's participation in the Forum. By focusing on women's conditions throughout the Forum discussions, the State is recognizing the role that Saudi women play in the national development process.[24]

In order to promote Saudi women's rights at yet another level, the Kingdom of Saudi Arabia in 2002 ratified the International Convention on Elimination of all Forms of Discrimination against Women of 1979, under the condition that its provisions do not contradict the Islamic Law (*Sharia'a*), the Constitution or the national law.[25]

Given the new developments that have taken place in Gulf countries, the right of women to political participation on an equal footing with men has become one of globalization's merits aimed at the expansion of popular participation in the decision-making process.

Hence, women in these countries will have to resist all attempts to marginalize or deny their citizenship rights acquired by birth. Also, they are called upon to make use of the mechanisms and outcomes of globalization that oblige states to endorse and implement women's right to take part in the political and decision-making processes through democratic institutions and civil society organizations (provided that this does not contradict with their obligations under international human rights instruments), with a view to narrowing the existing gap between the legal and constitutional provisions and the ground realities and practices applied in the Gulf states at all levels.

2-Legal Challenges at the National and International Levels

This section examines the provisions of some national laws concerning women's rights and the suggested amendments to certain discriminatory laws and regulations against women that hinder the attainment by Gulf women of the standards set by globalization, which considers respect for human rights and fundamental freedoms as an integral part of international conventions and charters on human rights, including women's rights.

Such a review will reveal the gravity of legal and political challenges that Gulf women are confronting at the national and international levels which are preventing their integration in an era of globalization. It must be recalled that legal challenges are particularly difficult to deal with, being founded on inherited Arab beliefs, traditions and legacies. In addition, the nature and complexities of the legislative process make it extremely difficult to amend the discriminatory local laws against women.

A) At the National Level

The national laws still constitute a major impediment for the human aspirations of Gulf women in the era of globalization, thus decreasing their opportunities to take part in the social development process owing to the legal constraints that dishonor the general rights of citizenship. The most notable of these laws are:

Nationality Law: The majority of Arab Gulf states deny the children of a Gulf woman married to a foreigner the right to acquire their mother's nationality, and they are treated as foreigners once they are 18 years old. Therefore, this law should be amended to allow the Gulf mother the right to grant her own nationality to the children born in her homeland, or allow the children the right to chose between the nationality of their mother or father.[26] The children must be also allowed the right to ownership of real-estate passed to them by inheritance, since the law that prohibits foreigners from the acquisition of real estate is still applicable to them.

Those children should also be accorded the same benefits of free education and health care services as provided to the children of male nationals who are married to foreign women.

Personal Status Law: Until the end of 2003, only two of the Gulf states (Kuwait in 1984, and Oman in 1996) had issued a personal status law. Steps are currently being undertaken by the United Arab

Emirates and the State of Qatar to enact similar laws in the near future.

It is worth mentioning that the Kuwaiti personal status law of 1984 contains many stipulations that deny women the freedom to choose their spouses or to unilaterally terminate their marriage contracts by their own will without having to impair their legal rights.[27] Besides, these laws vary in terms of the right of divorced women to custody of children and other women-related issues. Such legal diversity and variations are common not only in Gulf states but also in the legislation of those Arab countries which are guided by a single Islamic school of thought (*mazhab*). Indeed, legislators in these countries tend to adopt the most stringent of these schools regarding women's rights, with a view to ruling out any possibility of gaining from the fairness and flexibility of Islamic Law provisions in tackling issues concerning the fundamental rights of women.

For example, the unjust articles of the Kuwaiti law concerning women's rights have not been revised since the law was passed two decades ago, as the legislators (who are mostly men) have shown no interest in amending the law in order to be in harmony with the new developments witnessed in the status of women. Laws dealing with women and children's rights are now nicknamed as the "law of the poor."[28]

Criminal Law and Procedures: This is one of the laws that have a direct effect on women's rights in Arab Gulf states. Despite the amendments made to this law, it remains unfit for the new position of working women and the kinds of crimes they are exposed to, such as sexual harassment, kidnapping and rape. In addition, criminal laws in the Gulf countries usually tend to mitigate the punishments inflicted upon males who commit so-called "crimes of honor" (husbands, fathers, sons and brothers) despite the fact that such crimes deprive women of their fundamental right to life due to mere suspicion or social reasons. With regard to the crime of adultery, the law discriminates between adulterer and adulteress in terms of the punishment it imposes upon each of them. The existing codes lack any stipulations that can tackle crimes such as abortion in rape cases or the use of women in commercial promotions, in such a way as to protect their rights and dignity.[29]

On the other hand, the current legislation in Gulf states also lacks the applicable criminal terms needed to protect women from crimes of domestic violence which are treated as ordinary offenses. This situation necessitates adopting new laws under which severe punishments should be applied to those who perpetrate such repulsive crimes by misinterpreting the right of husbands under the Islamic Law (*Sharia'a*) to discipline their wives.

Labor and Social Codes: The increasing numbers of working women have made it imperative for labor codes in both public and private sectors to honor the rights of those women according to international labor standards. New laws need to be enacted to govern work relations in the new spheres that Gulf working women have joined recently. Likewise, the existing labor legislation should also be amended to delete any gender-discriminatory clauses and to meet working women's needs for vocational training and promotion. Any violations of these laws must be pinpointed and monitored by specialized governmental committees.

For their part, Gulf women should realize that any reforms aimed at promoting women's rights is an extremely difficult task in societies where males impose their domination on the legislative branch and where women, as a result, are unable to exercise any effective pressure as long as they are deprived of their political rights. Hence, their influence will have no impact on the process of issuing new laws related to women's rights or amending the existing discriminatory provisions.

B) At the International Level

The international conventions already signed by Gulf states should contribute to the consolidation of women's rights and the development of the relevant national legislation. The Gulf countries are parties to many international instruments, especially the International Convention on Elimination of all Forms of Discrimination against Women (1979) and other International Labor Organization's agreements. However, the way in which international labor standards are implemented by these states reveals the inconsistency of local standards concerning Gulf working women with the international norms. This necessitates the enhancement of national standards to ensure conformity with ILO agreements.[30]

Working women in the Gulf will have to make use of communication mechanisms, systems and facilities that globalization has produced to get acquainted with international labor standards in order to improve their conditions in collaboration with the ILO's specialized agencies, as well as to effectively participate in international forums in order to benefit from the successful experiences of other Arab and Islamic countries.

Gulf states are also invited to bridge the existing gap between the rights of women as envisaged in relevant binding agreements and the practical implementation of these rights at the internal level. This can be achieved by creating national mechanisms to boost women's capabilities to practice their own rights as effectively as possible.

Another challenge for women in the Gulf area is the obvious contradiction between the terms of national legislation and those of international law and conventions in terms of human rights, the lack of true equity between men and women, the under-representation of women in the legal branch, and the improper implementation of laws dealing with issues related to family and women's rights, and fundamental freedoms.

Therefore, Gulf states have to keep pace with the changes in women's conditions to be in harmony with the advantages of globalization, taking into account Arab identity and the very special traditional traits of Gulf societies. These states are called upon to design and put into effect plans and programs for the eradication of legal illiteracy to help Gulf women grasp the links between their rights and other aspects of life, irrespective of the social and educational status they have attained.

Gulf women will also have to persuade decision makers in those states which have not yet joined the International Convention on the Elimination of all Forms of Discrimination against Women (1979) to sign and implement the stipulations of the Convention, thereby enabling women to identify with the era of globalization by improving their conditions and promoting their human rights.

III-Opportunities Provided for Gulf Women Under Globalization

The 2003 United Nations Development Program (UNDP) Report – jointly prepared with the Arab League – on the "Arab World Development Goals in the New Millennium: Achievements and Aspirations," indicates that despite the general decrease in illiteracy rates, 50 per cent of women are still illiterate, and the number of females among all the Arab parliamentarians represents less than 5 per cent.[31]

In the light of these circumstances and the political and legal challenges that Gulf women are encountering, it could be suggested that globalization will have beneficial effects on them. It creates new standards to deal with women's human and legal rights by advancing their rights to political participation, and encouraging social organizations in these countries to play a more significant part in promoting the social and legal status of women in Gulf societies.

On the other hand, Gulf states are obliged to honor their international commitments under human rights agreements and the International Convention on the Elimination of all Forms of Discrimination Against Women (1979), in order to improve women's political and economic conditions, achieve equality between males and females in the political, economic, social, cultural and civil fields by enacting national laws prohibiting discrimination against women, and taking the necessary measures to change the social and cultural norms and practices that make discrimination against women an acceptable *de facto* reality.

In this connection, the Kuwaiti *Al Qabas* daily published the findings of a field study on some educational and developmental issues, including school curriculums, with the intention of reforming the current educational situation in Kuwait.

The study affirmed the continuation of gender-discrimination against women in the preparation of curriculums and textbooks that stress the inferiority of women in Kuwaiti society. The researcher noted that similar surveys conducted in Egypt, Lebanon, Algeria, Morocco, Kuwait and Bahrain have demonstrated that women in general are unfairly treated in these curriculums and that improving their image would involve substantial reforms in the concepts and contents of the Arab educational systems.[32]

Moreover, the study also revealed that this trend is discernible in the contents of the Arabic-language textbooks of the secondary stages (pre-university), and that Kuwaiti curriculums in general are authored in accordance with the instructions from the GCC's specialized educational committees. It can also be said that women all over Arab Gulf states are not excluded from this tendency. Hence the importance of the international initiatives calling for the revision of the Arab teaching curriculums, despite concerns by some Arab scholars and intellectuals about the political, social and cultural implications and dimensions of such initiatives being seen as the negative effects of globalization.

The same study highlights the disproportionate role that Gulf women are playing in reality, compared to the status they have achieved in Kuwait and Arab societies. This situation represents further evidence regarding the impact of ideological factors, social values and traditions which are seemingly playing a more exceptional and decisive part in addressing women's issues than economic factors.[33]

In the light of the foregoing discussion, one might wonder about the state's commitments to abolish all forms of discrimination against women in the educational field. These commitments were based on realizing equity between men and women by eliminating any stereotyped concepts of the roles of men and women in this field and by revising the existing textbooks and school curricula and adapting teaching methods.[34]

The above-mentioned field study may provide the answer. In terms of writing Arabic language textbooks, the study concluded that women's share (23%, 27 authors) represents around the quarter of that of men (77%, 91 authors), bearing in mind that the number of female students and teachers is no less than that of their male counterparts.

However, opportunities are still available for Gulf women in the era of globalization if they prove capable of strengthening their knowledge and educational potential and skills, bringing about meaningful shifts in the prevailing social conditions, and promoting the role of civil society and human rights organizations, which remains limited due to restrictions and constraints imposed on these organizations under the terms of their statute.

Women's organizations are still governed by the law which regulates the establishment of public service associations and determines the nature of permissible functions conducted by these organizations. The majority of laws in force in GCC states prohibit all political activities by civil society groups whose functions are limited to social and charitable works.

The creation of women's societies in the Gulf area dates back to mid-1950s with the establishment in 1950 of the Bahraini Women's Renaissance Society. By 2002, the number had increased to around 69 women's associations in the GCC countries, mostly in the Sultanate of Oman (29), the Kingdom of Saudi Arabia (19), and the Kingdom of Bahrain (10).[35]

It is worth mentioning that the constraints imposed on public service associations in general, and female organizations in particular, have weakened both the influence they might exercise to achieve the desired transformations and the pressures they need to exert on decision-making circles, especially in this era of globalization where civil society organizations are the instruments for strengthening democratic values.

Since most current national laws stress the stereotyped role of women in society, Gulf women are invited to exert serious and strenuous efforts to correct the existing traditional social patterns of conduct. They should also work hard to root out pre-Islamic practices, which still constitute the major obstacle that hinders women from taking advantage of the positive, favorable aspects of globalization.[36] In this effort, they ought to be guided by the following verse from the Holy Quran: "Verily, Allah will not change the condition of a people until they change that which is in their hearts." (Surah Al Ra'd, Ayah 11).

H.H. SHEIKH HAMED BIN ZAYED AL NAHYAN is a Member of the Abu Dhabi Executive Council and Chairman of the Abu Dhabi Department of Economy. He is also the Deputy Chairman of the Bani Yas Sports Club.

H.H. Sheikh Hamed is the Chairman of the Board of Directors of the Abu Dhabi General Authority for Health Services, and Chairman of the Executive Committee of the Union Water and Electricity Company. He is the Head of the Higher Corporation for Specialized Economic Zones and Head of the General Holding Company.

H.H. Sheikh Hamed Bin Zayed Al Nahyan obtained his Master Degree (Honors with Distinction) in Economics from the University of Wales in the United Kingdom in 2000.

H.H. SHEIKH ABDULLAH BIN ZAYED AL NAHYAN has been Minister of Information and Culture of the United Arab Emirates (UAE) since March 1997. H.H. Sheikh Abdullah has also been the Chairman of the Board of the Emirates Media since January 1999, and is the Honorary President of the Abu Dhabi Committee for Classical Music.

H.H. Sheikh Abdullah was the Under Secretary of the UAE Ministry of Information and Culture from 1995 to 1997, the Chairman of the Annual Conference of Arab Ministers of Culture, the Chairman of the Annual Conference of the GCC Countries from 1998 to 1999 and the Chairman of the UAE Football Association from 1993 to 2001.

H.H. Sheikh Abdullah Bin Zayed Al Nahyan was awarded a Bachelor of Arts degree in Political Science from the UAE University at Al Ain in 1995.

H.R.H. PRINCE TURKI AL-FAISAL BIN ABDUL AZIZ is the Ambassador of the Kingdom of Saudi Arabia (KSA) to the United Kingdom. Prince Turki is one of the founders of the King Faisal

Foundation and Chairman of the King Faisal Center for Research and Islamic Studies in the KSA.

Prince Turki served as an Advisor in the Saudi Royal Court and became the Director General of the Saudi Intelligence Department.

H.R.H Prince Turki Al-Faisal Bin Abdul Aziz pursued his higher education in Princeton and Georgetown Universities in the United States of America.

H.E. SHEIKH HAMAD BIN JASSIM BIN JABR AL THANI has served as Qatari Minister of Foreign Affairs since September 1, 1992. While being the incumbent of this office, he was also appointed First Deputy Prime Minister.

From 1982 to 1989, he served as Director of the Office of the Minister for Municipal Affairs and Agriculture. On July 18, 1989, he became Minister of Municipal Affairs and Agriculture, and on May 14, 1990, he was concurrently appointed Acting Minister of Electricity and Water, a position he held for two years. Other important positions held by H.E. Sheikh Hamad include: Member of the Higher Council of Defense, which was established in 1996; Chairman of the Qatari Permanent Committee for Supporting Jerusalem, established in 1998; and Member of the Committee in charge of preparing the Permanent Constitution, which was established in 1999. He is also a Member of the Council of the Ruling Family which was established in 2000, and a Member of the Higher Council of Investment of State Reserves, which was established in 2000.

H.E. Sheikh Hamad Bin Jassim has supervised many successful projects in the agricultural sector, and is largely responsible for the significant progress made in this sector. In addition to the important posts mentioned above, H.E. Sheikh Hamad Bin Jassim was also Chairman of the Board of the Qatari General Electricity and Water Corporation, Chairman of the Central Municipal Council, Director of the Office of Private Emiri Projects, Member of the Board of the Qatar Petroleum Company and Member of the Higher Council of Planning.

H.E. MR. MOHAMMED ALI ABTAHI studied theology at the Seminary for Theological Studies in Mashhad, Iran. He began his career with the Islamic Republic of Iran Broadcasting (IRIB), where he was the Programming Director General for IRIB's Mashhad branch in 1980. In 1983, he was appointed Director General of IRIB for the Bushehr and Shiraz branches. Later he moved to Tehran as Director General of IRI Radio, a position held until 1987. A year later, he became Assistant Director General for Overseas Broadcasting Services of IRIB.

At the end of 1988, Seyyed Mohammed Khatami, then Minister of Culture and Islamic Guidance, appointed Mr. Abtahi as his Deputy Minister for International Affairs. As Iran's chief cultural liaison, he traveled widely. When Mr. Khatami resigned his position in July 1992, Mr. Abtahi also resigned and rejoined the IRIB. Following this, the IRIB Chief appointed him as IRIB representative in Beirut, where he moved in 1994. Upon his return from Lebanon, he established a center for dialogue between different religions, currently headed by his wife, Mrs. Fahimeh Moussavi-Nejad.

When Seyyed Mohammed Khatami announced his presidential candidacy, Mr. Abtahi relinquished his IRIB job and joined Mr. Khatami's election campaign. President Khatami appointed Mr. Abtahi as the Chief Secretary of the presidential office in 1996 to conduct the affairs of the Executive Branch. In Mr. Khatami's second tenure as President, he appointed Mr. Abtahi as Vice President for Legal and Parliamentary Affairs, a position intended to strengthen relations between the Executive and Legislative branches, which he held until his resignation in October 2004.

H.E DR. MUHAMMAD ABDUL GHAFFAR has been Minister of State for Foreign Affairs in the Kingdom of Bahrain since April 2001. He has held various positions since joining the diplomatic corps within the Bahraini Ministry of Foreign Affairs in 1975. He served as Ambassador of Bahrain to the United States of America (1994-2001), then as Non-resident Ambassador Plenipotentiary of Bahrain to Canada (1996-2001), and later to the Argentine Republic (1998-2001). He was the Ambassador and Permanent Representative of Bahrain to the United Nations headquarters in New York (1990-

1994), and had been a member of Bahrain's Permanent Mission to the UN (1979-1984). In 1977, he served as a junior diplomat at his country's embassy in the Hashemite Kingdom of Jordan.

H.E Dr. Muhammad was a member of his country's delegation to the UN General Assembly sessions from 1979 to 1984. Additionally, as a member of the Bahraini delegation, he participated in successive summits of the Gulf Cooperation Council (GCC), meetings of the Arab League, and summits of the Non-Aligned Movement (NAM).

H.E. Dr. Muhammad holds a Ph.D. in Political Science from the State University of New York at Binghamton in 1991, a Master's degree in Political Science from New School University in New York in 1981 and a Bachelor's degree in Political Science from the University of Poona (now Pune) India, in 1974.

H.E. MR. KAMEL M. AL-KILANI held the position of Minister of Finance in the cabinet of the Iraqi Interim Administration.

Born in Baghdad in 1958, he graduated from the Administration and Economics Faculty of Al Mustansiriyah University in 1988.

H.E. Mr. Al-Kilani worked earlier for Thomas Cook, a private banking establishment for approximately one year. Thereafter he worked in the private sector as a businessman specializing in trade and industry, especially the contracting sector, until his appointment as the Minister of Finance.

THE RT. HON. JOHN MAJOR, CH is former Prime Minister of the United Kingdom and a leading authority on the changing global landscape. An astute economist, Mr. Major was elected to the British Parliament in 1979, joined the Cabinet as Chief Secretary of the Treasury in 1987 and went on to serve as Foreign Secretary and Chancellor of the Exchequer before becoming Prime Minister.

During his seven years as Prime Minister, he instituted public sector reforms and left behind a strong economy. Mr. Major initiated an unprecedented effort to secure lasting peace in Northern Ireland and continues to lend his support to the British government on that issue. In 1999, he was awarded one of the United

Kingdom's greatest honors – The Companion of Honour – bestowed on him by HM Queen Elizabeth, in recognition of his initiation of the Northern Ireland peace process.

Since leaving the British Parliament in 2001, Mr. Major has taken up various business interests including Chairman of the European Board of the Carlyle Group, Washington, DC; Special Advisor of Credit Suisse First Boston; Chairman of the European Advisory Council of the Emerson Electric Company, St. Louis; and Member of the European Board of Siebel Systems, California. He is also the author of *John Major: The Autobiography*. On the death of Diana, Princess of Wales, Mr. Major was appointed legal guardian to Their Royal Highnesses Princes William and Harry. He serves as President of the National Asthma Campaign and Patron of the Child of Achievement Awards, also working with the Consortium for Street Children, Mercy Ships and other charitable organizations.

HON. GARETH EVANS AO QC has been President and Chief Executive of the Brussels-based International Crisis Group (ICG) since January 2000. A Member of the Australian Parliament for 21 years, Gareth Evans was Senator for Victoria (1978-1996) Minister for Resources and Energy (1984-1987), Minister for Transport and Communications (1987-1988) and Foreign Minister (1988-1996). He holds BA and LLB (Hons) degrees from Melbourne University and an MA degree in Politics, Philosophy and Economics from Oxford University.

Mr. Evans was one of Australia's longest serving Foreign Ministers. He was recognized as the Australian Humanist of the Year in 1990, won the ANZAC Peace Prize in 1994 for his work on Cambodia, was made an Officer of the Order of Australia (AO) in 2001. He was awarded an Honorary Doctorate of Laws by Melbourne University in 2002. In 2000-2001, the Government of Canada appointed him Co-Chair (with Mohamed Sahnoun) of the International Commission on Intervention and State Sovereignty (ICISS). He had previously served as a member of the Carnegie Commission on Preventing Deadly Conflict, co-chaired by Cyrus Vance and David Hamburg (1994-1997).

Among other current positions, Mr. Evans is Chair of the World Economic Forum Global Governance Initiative's Peace and Security Expert Group; a member of the International Task Force on Global Public Goods; a Fellow of the Foreign Policy Association; a member of the International Council of the Asia Society; and a member of the International Advisory Board of the Pew Global Attitudes Survey.

GENERAL ANTHONY C. ZINNI, USMC (RET.) joined the United States Marine Corps in 1961 and was commissioned as Infantry Second Lieutenant in 1965 upon graduation from Villanova University. He has held a number of command and staff assignments. A tactics and operations instructor at several Marine Corps schools, he was selected as a Fellow on the Chief of Naval Operations Strategic Studies Group.

His operational experiences include two tours in Vietnam, Operation Provide Comfort, Operation Desert Thunder, and Operation Desert Fox, and the Maritime Intercept Operations in the Gulf. He was involved in the planning and execution of Operation Proven Force and Operation Patriot Defender in support of the Gulf War. He has attended several military schools including the National War College. He holds a Bachelor's degree in Economics, Master's degrees in International Relations, and in Management and Supervision, and honorary doctorates from William and Mary College and the Maine Maritime Academy.

He currently serves on the boards of major US companies. He has held several academic positions: the Stanley Chair in Ethics at the Virginia Military Institute; the Nimitz Chair at the University of California-Berkeley; the Hofheimer Chair at the Joint Forces Staff College; the Harriman Professor of Government appointment; and membership on the board of the Reves Center for International Studies at the College of William and Mary. He has worked with the University of California's Institute on Global Conflict and Cooperation and the Henry Dunant Centre for Humanitarian Dialogue in Geneva. He is also a Distinguished Advisor at the Center for Strategic and International Studies and a member of the Council on Foreign Relations.

DR. SHAMLAN YOUSEF AL-ISSA is Chairman of The Center for Strategic and Future Studies at Kuwait University. He was Chairman of the Department of Political Science, Kuwait University (2002-2003) having held several academic positions over the years. A personal consultant to the Kuwaiti Minister of Information (1991-1992), he also worked as a consultant for the Higher Council of Planning for the Cabinet's "Study of the New Population Policy, 1992," and for the Studies and Research Agency at the Emeri Diwan (1982-1985). A consultant for the Ministry of Foreign Affairs (1979-1981) he played a major role in the consultations that led to Kuwait's preparation paper for the establishment of the GCC. He worked at the Kuwait Ministry of Information from 1970-1978. He has been a member of the Social Sciences Faculty Council since 1998.

Dr. Al-Issa gained his M.A and Ph.D from the Fletcher School of Law and Diplomacy and his B.A from California University. He has published research papers in coordination with other academic institutions. These include "Attitudes of the Students of the United Arab Emirates University on the Gulf Crisis," co-researched with Dr. Jamal S. Al-Suwaidi, Director General of ECSSR (Social Sciences Magazine, 1991) and another co-researched paper "Expatriate Labor in the Arabian Gulf: Problems, Prospect and Potential Instability" (Social Sciences Magazine, 1978).

Dr. Al-Issa is a founding member of the Child Learning and Evaluation Center and the Kuwaiti Friendship Committee; a member of the Board of Directors of the Kuwaiti Olympic Committee; a member of the National Committee for Fighting Drugs of the Kuwaiti Ministry of Interior; a Member of the Kuwait Union's Board of Directors for Tennis (1988-1998); Chairman of the Kuwaiti Popular Committee in Al-Ain, during the Iraqi invasion of Kuwait; and a member of the Board of Directors of the Popular Committee in Abu Dhabi (September 1990-April 1991).

DR. BENJAMIN R. BARBER is the Gershon and Carol Kekst Professor of Civil Society at the University of Maryland and a principal of the Democracy Collaborative with offices in New York, Washington and the University of Maryland. He consults regularly with prominent political and civic leaders in the United States and Europe, as well as with institutions such as the Corporation for National Service, the United States Information Agency, the National Endowment for the Humanities; and in Europe, UNESCO, the European Parliament, the Swedish Parliamentary Commission on Democracy and "Mission 2000" (the French Millennial Commission). Dr. Barber is also Chairman and Chief Strategic Vision Officer of Bodies Electric, a democracy software company co-founded with Dr. Beth Noveck and Peter Dolch.

Dr. Barber holds a certificate from the London School of Economics and Political Science and an M.A. and Doctorate from Harvard University. Among the books he has published are *Strong Democracy* (1984); *Jihad Vs McWorld* (1995 and 2001); *A Passion for Democracy* (Princeton University Press, 1999) and *The Truth of Power: Intellectual Affairs in the Clinton White House* (W.W. Norton & Company, 2001). His latest book is *Fear's Empire: Terrorism, War and Democracy* (W.W. Norton & Company, 2003).

Dr. Barber's honors include the Palmes Academiques (Chevalier) from the French Government (2001), the Berlin Prize of the American Academy of Berlin (2001) and the John Dewey Award (2003). He has also been awarded Guggenheim, Fulbright and Social Science Research Fellowships, honorary doctorates from Grinnell College and Connecticut College, and he held the Chair of American Civilization at the Ecole des Hautes Etudes in Paris in 1991-92. He was a founding editor and Editor-in-Chief of the international quarterly *Political Theory* for ten years, and writes frequently for leading magazines, newspapers and other scholarly and popular publications in America and Europe.

DR. MAHMOOD SARIOLGHALAM is Associate Professor of International Relations at the School of Economics and Political Science at National (Shahid Beheshti) University in Tehran, Iran. He is also Research Director at the Center for Scientific Research and Middle East Strategic Studies; a member of the Council of 100 Leaders at the World Economic Forum; Senior Advisor at the Center for Strategic Research; and Editor-in-Chief for the *Middle East Quarterly* and *Discourse*.

The focus of Professor Sariolghalam's teaching is Third World Political Economy, International Relations Theory, International Relations Methodology, and International Politics of the Middle East. He completed his B.A. in Political Science/Management at California State University in 1980, M.A. in International Relations at the University of Southern California in 1982, Ph. D in International Relations at the University of Southern California in 1987, and a Post-Doctorate in International Relations at the Ohio State University in 1997.

Professor Sariolghalam's books include *The International Dimensions of the Western Conflict; Development, The Third World and the International System*; *The Evolution of Method and Research in International Relations*; *Rationality and Development*; *Foreign Policy of the Islamic Republic of Iran: Theoretical Renewal and the Paradigm for Coalition*; *Research Methodology in Political Science and International Relations*; and *Rationality and the Future of Iran's Development*. In addition, Dr. Sariolghalam has published a number of articles in prominent journals such as the *Journal of Political Economy*, the *Journal of Policy Science*, *Security Dialogue*, the *Brown Journal of International Affairs*, *Energy Focus*, and the *Washington Quarterly*.

DR. JERROLD D. GREEN is currently Executive Vice President for International Operations and Partner at Best Associates, a merchant banking firm in Dallas, Texas. Prior to this he spent ten years at the RAND Corporation in Santa Monica, California, where he served as Director of International Programs and Development, Director of the Center for Middle East Public Policy, Professor of International Studies at the RAND Graduate School (RGS), and a Senior Political Scientist. He was Associate Chairman of the Research Staff (1996-1998) and Head of the International Policy Department and Corporate Research Manager (1994-96). Dr. Green's career began at the University of Michigan but he later joined the University of Arizona and became Director of the Center for Middle Eastern Studies. He was Visiting Professor at the University of Southern California and at UCLA.

Dr. Green joined RAND in 1994 where he wrote on Middle East politics and he has conducted extended research in Egypt, Iran and Israel. He has been a Visiting Fellow at the Chinese Academy of Social Science's West Asian Studies Center, Beijing; a visiting lecturer at the Center for African and Middle East Studies (CEAMO), Havana, and delivered papers at conferences sponsored by the Iranian Institute of International Affairs, Tehran. Dr. Green is a member of the Council on Foreign Relations; the International Institute of Strategic Studies; and the Advisory Committee of The Asia Society of Southern California.

Dr. Green holds a B.A. degree from the University of Massachusetts as well as M.A. and Ph.D. degrees in Political Science from the University of Chicago. His work has appeared in *World Politics, Comparative Politics, Ethics and International Affairs, Survival, Middle East Insight, Politique Étrangère, The World Today, The RAND Review, The Harvard Journal of World Affairs, and The Iranian Journal of International Relations.*

DR. GEOFFREY KEMP is Director of Regional Strategic Programs at the Nixon Center. He served in the White House during the first Reagan administration and was Special Assistant to the President for National Security Affairs and Senior Director for Near East and South Asian Affairs on the National Security Council staff.

Prior to his current position, he was a Senior Associate at the Carnegie Endowment for International Peace where he was Director of the Middle East Arms Control Project. In the 1970s he worked in the Defense Department in the Policy Planning and Program Analysis and Evaluation Offices and made major contributions to studies on US security policy and options for southwest Asia. In 1976, while working for the Senate Committee on Foreign Relations, he prepared a widely publicized report on US military sales to Iran. In the 1970s, he was also a tenured faculty member at the Fletcher School of Law and Diplomacy at Tufts University.

Dr. Kemp received his Ph.D in Political Science at the Massachusetts Institute of Technology and his MA and BA degrees from the University of Oxford. His books include: *Forever Enemies? American Policy and the Islamic Republic of Iran*; *The*

Control of the Middle East Arms Race and Strategic Geography and the Changing Middle East (co-author, 1994); *Energy Superbowl: Strategic Politics and the Persian Gulf and Caspian Basin* (1997); *America and Iran: Road Map and Realism* (The Nixon Center, 1998); and *Iran's Nuclear Weapons Options: Issues and Analysis* (The Nixon Center, 2001). His most recent publication is *US and Iran, The Nuclear Crisis: Next Steps* (April 2004).

DR. BADRIA ABDULLAH AL-AWADHI has been Director of the Arab Regional Center for Environmental Law (ARCEL) at the University of Kuwait since 2001. A practicing lawyer with her own law firm since 1994, Dr. Al-Awadhi has served as the Dean of the Faculty of Law and Sharia as well as Lecturer of International Law at the University of Kuwait. Her more recent publications include: *Toward Equality and Equal Opportunity* (1999), *Women's Legal Guideline on Kuwaiti Family Law* (1997), *Environmental Laws in the Gulf Co-operation Council Countries* (1996), *Iraqi Aggression and Trampling of Humanitarian and Legal Principles* (1992), *Women and the Law: A Comparative Study* (1990).

Dr. Al-Awadhi has been a member of many international organizations: the International Federation of Women Lawyers (FIDA) and International Law Association (USA); International Commission of Jurists, and Committee of Experts on the Application of the Conventions and Recommendations of International Labor Organization (ILO) from 1983 to 1996 (Switzerland); International Council of Environmental Law (ICEL) and Vice Chairperson for West Asia Commission on Environmental Law at The World Conservation Union (Germany); International Institute of Humanitarian Law (Italy); Arab Thought Forum (Jordan); Arab Association for International Arbitration (France); the GCC Commercial Arbitration Center (Kingdom of Bahrain); The Euro-Arab Arbitration System and The International Bar Association (IBA) in the UK.

Dr. Al-Awadhi was awarded a Bachelor of Law (LL.B) degree and a Master of Law (LL.M) degree from Cairo University and completed her Ph.D. in International Law in 1975 from the University College at London University.

DR. FALEH A. JABAR is a sociologist and Senior Fellow in residence for the year 2003-2004 at the United States Institute of Peace (USIP). He was a lecturer at the Department of Law, Governance and International Relations, London Metropolitan University and is Research Fellow at the School of Politics and Sociology, Birkbeck College, University of London. He is also Director of the Iraq Cultural Forum (ICF) a research group working with the School of Politics and Sociology, Birkbeck College and the School of African and Oriental Studies (SOAS) in London. This Forum has developed into the Iraqi Institute of Strategic Studies.

Dr. Jabar is an expert on Iraq and Middle Eastern politics and sociology. He has attended conferences organized by the International Institute for Strategic Studies (IISS), Royal Institute of International Affairs (RIIA), United Nations Educational, Scientific and Cultural Organization (UNESCO), United Nations Economic and Social Commission for Western Asia (UN-ESCWA), The Emirates Center for Strategic Studies and Research (ECSSR), and Review of International Social Questions (RISQ) in The Netherlands. He has delivered lectures at universities and forums all over the world.

Dr. Jabar's more recent English publications include: *The Shi'ite Movement in Iraq* (Saqi, 2003); *Tribes and Power in the Middle East* (Saqi, 2002) and *Ayatollahs, Sufis and Ideologues* (Saqi, 2002). An author for the International Crisis Group (ICG) his contributions include: *Iraq in War*, April 2003, *Governing Iraq*, August 2003, and *The Shi'ite of Iraq*, September 2003. His recent paper is *Post-Conflict Iraq* (USIP, May 2004) and his forthcoming book is entitled *Post-Conflict Iraq: A Nation in Making* (USIP). He has published several papers and contributes to *Le Monde Diplomatique*, The Middle East Research and Information Project (MERIP), *The Financial Times*, *The Times* and other publications.

MR. FREDERICK D. BARTON has been a Senior Advisor at the Center for Strategic and International Studies (CSIS) since 2002 and Co-Director of the Post-Conflict Reconstruction Project. He has co-authored two reports on Iraq: *A Wiser Peace: An Action Strategy for a Post-Conflict Iraq* (released in January 2003) and *Iraq's Post-Conflict*

Reconstruction, A Field Review and Recommendations (released in July 2003).

A graduate of Harvard College in 1971, Mr. Barton earned his Masters in Business Administration with an emphasis on Public Management, from Boston University in 1982. He received an Honorary Doctor of Humane Letters from Wheaton College, Massachusetts in 2001. Mr. Barton was the Frederick H. Schultz Professor of Economic Policy and Lecturer of Public and International Affairs at the Woodrow Wilson School of Princeton University and continues there as a Visiting Lecturer. His courses address durable solutions for refugee situations and practical political initiatives that can redirect a citizenry away from war.

As the United Nations Deputy High Commissioner for Refugees in Geneva (UNHCR, 1999-2001), Mr. Barton worked with five thousand colleagues, hundreds of partner organizations and a budget of $850 million to protect twenty two million uprooted people in 130 countries. He emphasized getting closer to the refugees, creating a more trusting internal culture and addressing endemic structural challenges. Mr. Barton was the first Director of the Office of Transition Initiatives in Washington (USAID, 1994-1999) where he helped to start $150 million of political development programs in over 20 war-torn regions, from the Philippines to Rwanda, and from Bosnia to Haiti.

PROFESSOR BASSAM TIBI has been Georgia Augusta Professor of International Relations at the University of Göttingen since 1973 and is currently A.D. White Professor-at-Large at Cornell University. Over the last 25 years he has been visiting professor at more than 30 universities in four continents including Harvard, Berkeley and Princeton. Professor Tibi hails from the Banu Al Tibi family based in Damascus, where he completed his schooling. He holds Ph.D degrees from both Frankfurt and Hamburg universities. He is widely recognized for his expertise in Middle Eastern Affairs, Islam and the West. Professor Tibi has undertaken research in most of the Arab countries. He holds guest lectures at several US universities. He has given lectures in Asia and Africa, taught at the University of

Khartoum in Sudan and the University of Yaoundé in Cameroon and also worked at the Al-Ahram Center in Cairo.

Professor Tibi is the joint founder of the Arab Organization for Human Rights. His regular articles and comments have been featured in international newspapers and magazines. He has published articles in the German magazine *Der Spiegel* and been a guest writer at the German newspaper *Frankfurter Allgemeine Zeitung* since 1987. He has published 25 books translated into 16 languages. Among his recent books are: *Islam: Between Culture and Politics* (Palgrave Press, 2001) and the updated edition of *The Challenge of Fundamentalism: Political Islam and the New World Disorder* (University of California Press, 2002).

Professor Tibi was awarded the Medal of the State—First Class by the German President Roman Herzog in 1995 and was selected by the American Biographical Institute in 1997 as Man of the Year. During the Gulf War, he became well known as a Middle Eastern expert, making frequent TV appearances on ZDF, one of Germany's main television channels.

DR. SALEH ABDULREHMAN AL-MANI is a Professor of Political Science and Dean of the College of Administrative Sciences, King Saud University (KSU), in Riyadh, Saudi Arabia. From 1983-1985 he was Assistant Dean of the College of Administrative Sciences at King Saud University and was also Chairman of the Department of Politics at the university from 1992-1994 and 1996-1998. Additionally, he has been a member of the King Saud University Academic Council since 2000 and was the Editor of the *King Saud University Journal* from1999 to 2003. Dr. Al-Mani has also been a consultant to some government ministries in Riyadh, and to the Gulf Cooperation Council Secretariat.

Dr. Al-Mani has authored several books including *The Euro-Arab Dialogue, A Study in Associative Diplomacy (1984); and The World Challenge, Europe, the Arab World and Japan (1981).* He has written a number of research papers. He has contributed chapters in eleven books in Arabic and English, and writes a weekly column in *Okaz* newspaper, Jeddah and in *Al Ittihad* newspaper, Abu Dhabi.

Dr. Al-Mani has lectured widely at universities and forums in the Gulf region and abroad. He completed his BA in Political Science in 1973 and his MA in 1976 from California State University. He received his Ph.D in International Relations in 1981 from the University of Southern California, where he worked as teaching assistant the previous year. In 1991, Dr. Al-Mani received a research grant from The British Council, which he spent at the International Institute for Strategic Studies in London. He also received a grant from the International Association of University Presidents, Connecticut, USA, in 1996-1997. Dr. Al-Mani also attended an executive program on Gulf security at the Kennedy School of Government, Harvard University in June 2001.

DR. ABDUL-RIDHA ASSIRI is currently Professor of Politics and Chair of the Department of Political Science at the University of Kuwait. He served as Political Advisor to the Kuwait National Assembly from 1993 to 1996. From 1990 to 1991, he was Visiting Professor at the United Arab Emirates University, having served earlier in a similar capacity at the University of Colorado from 1987 to 1989. From 1983 to 1987, he was Associate Dean of the College of Economics, Commerce and Political Science at the University of Kuwait, where he also served as Head of the Department of Political Science from 1982 to 1984.

Dr. Assiri has authored many books and contributed to many specialized studies and research papers, both in Arabic and English including "Kuwait in Contemporary International Politics: Kuwait's Foreign Policy Achievements, Failures and Challenges" (1993). He has authored critical and analytical political articles published in Kuwaiti and foreign newspapers including "Kuwait's Foreign Policy: City-State in World Politics" (1990), in English, and "Government and Kuwait Policy: Principles and Practices" (1996).

He received his Bachelor of Arts degree cum laude from East Tennessee State University, Johnson City, USA, in 1974, and his M.A. and Ph.D. degrees from the University of California, Riverside, USA, in 1977 and 1980 respectively.

MR. JAMAL A. KHASHOGGI is Media Advisor to the Ambassador of the Kingdom of Saudi Arabia in the United Kingdom. Formerly,

Mr. Khashoggi served as Editor-in-Chief of *Al-Watan* (Arabic daily, KSA) and Deputy Editor-in-Chief of *Arab News* (English daily, KSA). As KSA correspondent for the London-based *Al-Hayat* Arabic daily, he also covered Pakistan, India, Afghanistan, Turkey, Algeria, Kuwait, Sudan and the rest of the Middle East. He was Managing Editor and Acting Editor-in-Chief of *Al-Madinah* (Arabic daily, KSA). As a correspondent for *Al-Sharq Al-Awsat* (Arabic daily) and *Al-Majallah* and *Al-Muslimoon* (Arabic weeklies) in the KSA, Mr. Khashoggi covered the Second Gulf War, the Afghan War, the Algerian Crisis, and Sudan. He was also a reporter for the *Saudi Gazette* (English daily, KSA); Assistant Manager of the Information Center at *Okaz* (Arabic daily, KSA); and Manager of the Tihama Bookstores in the Jeddah region.

Mr. Khashoggi is a regular political commentator on Saudi TV and radio channels as well as leading Arab and Western media channels. Mr. Khashoggi has produced and presented political programs for MBC TV. He writes a weekly column in several newspapers and magazines including *Al-Watan* (Arabic daily, KSA), *Al-Ittihad* (Arabic daily, Abu Dhabi), The *Daily Star* (English daily, Lebanon), *Al-Rai Al-Aam* (Arabic daily, Kuwait), and *Al-Sharq* (Arabic daily, Qatar).

Mr. Khashoggi's first publication in Arabic, entitled *Critical Relations: Saudi Arabia after September 11* was published in October 2003. He was awarded his Bachelor's degree in Business Administration from Indiana State University, United States in 1982.

DR. MUNIRA AHMED FAKHRO is an Associate Professor of Social Sciences at the University of Bahrain where she began her teaching appointment as an Assistant Professor in 1987. In 1997-1998, she joined the Middle East Institute at Columbia University as a visiting scholar. The following year she was also a visiting scholar at the Center for Middle Eastern Studies at Harvard University, conducting a research study on "Gender, Citizenship and Civil Society in the Gulf States." She carried out research for UN agencies such as the United Nations Fund for Population Activities (UNFPA), the United Nations Development Program (UNDP) and the International Labor Organization (ILO).

Dr. Fakhro has published three books and many articles focusing on working women and civil society in the Gulf region. The books are entitled: *Women at Work in the Gulf: A Case Study of Bahrain* (Kegan Paul Publishers, London, 1990); *Basics of Sociology* (A text book in Arabic for high school students, co-authored in 1992); *Civil Society and the Democratization Process in Bahrain* (published in Arabic by the Ibn Khaldoun Centre, Cairo, 1995).

Dr. Fakhro has attended several conferences and seminars including international conferences on women in Copenhagen (1980), Nairobi (1985), Beijing (1995), and Beijing + 5 (New York, 2000). She is a member of the board of the High Council for Women in Bahrain, and a member of the National Democratic Action Society. Dr. Fakhro completed her Master's degree from Bryn Mawr College, Pennsylvania, in program planning and administration, and earned her Doctoral degree from Columbia University in 1987 in social policy, planning and administration.

NOTES

Chapter 1

1 All the trends referred to here will be fully documented in the first annual *Human Security Report*, due to be published by Oxford University Press in October 2004, produced by the University of British Columbia Human Security Centre, directed by Andrew Mack (Director of the UN Secretary-General's Strategic Planning Unit 1998-2001, and formerly Professor of International Relations at the Australian National University).

2 The International Crisis Group (ICG) is an independent, non-profit, multinational organization, with over 100 staff members on five continents, working through field-based analysis and high-level advocacy to prevent and resolve deadly conflict. ICG's co-Chairs are (as of November 12, 2004) Former European Commissioner for External Relations Christopher Patten and President Emeritus of the Council on Foreign Relations Leslie Gelb, and its 56 Board members include Zbigniew Brzezinski, Wesley Clark, Inder Kumar Gujral, Fidel V. Ramos, Salim A. Salim, Surin Pitsuwan and Grigory Yavlinsky. For a full account of ICG's work in over 40 conflict or conflict-potential areas see www.icg.org.

3 This analysis is further developed in the Global Governance Initiative, Annual Report 2004, World Economic Forum (www.weforum.org). The Global Governance Initiative Peace and Security Expert Group comprised Gareth Evans (Chair), Ellen Laipson, Andrew Mack, Jane Nelson, Mohamed Sahnoun and Ramesh Thakur.

4 Moses Naim, YaleGlobal Online, December 29, 2003 (http://yaleglobal.yale.edu)

5 Samuel R. Berger, "Power and Authority: America's Path Ahead," Brookings Institution Leadership Forum, June 17, 2003.

6 The Commission's members were Gareth Evans and Mohamed Sahnoun (Co-Chairs), Gisele Cote-Harper, Lee Hamilton, Michael Ignatieff, Vladimir Lukin, Klaus Naumann, Cyril Ramaphosa, Fidel Ramos, Cornelio Sommaruga, Eduardo Stein and Ramesh Thakur. It held comprehensive consultations over a full year, meeting in Asia and Africa as well as North America and Europe, and holding roundtables and other consultations in Latin America, the Middle East,

Russia and China. The ICISS report, with its large supplementary research volume, is available on www.iciss-ciise.gc.ca. On the six criteria for military intervention, the specific context of our report was the responsibility to protect against internal threats; but the language – with just a little generalization – is equally applicable to external ones.

7 Michael Ignatieff, "Why are we in Iraq? (And Liberia? And Afghanistan?)," *New York Times,* September 7, 2003.

8 Efforts have already been made within the Security Council and the General Assembly, led by the Canadian government (who initiated the ICISS commission) and supported by the Secretary-General, to win at least informal acceptance of these criteria, and these efforts will continue in forthcoming months. The argument is that with perseverance and application and declaratory resolutions, guidelines can become norms, and norms can become accepted principles of customary international law, even if they never see the light of day as treaty or Charter provisions.

9 Implementation of the United Nations Millennium Declaration, Report of the Secretary-General, September 8, 2003.

10 The High Level Panel on Threats, Challenges and Change, due to report to the Secretary General in September 2004, was appointed in November 2003. Its members are Anand Panyarachun (Thailand, Chair); Robert Badinter (France); João Clemente Baena Soares (Brazil); Gro Harlem Brundtland (Norway); David Hannay (United Kingdom); Mary Chinery-Hesse (Ghana); Gareth Evans (Australia); Enrique Iglesias (Uruguay); Amr Moussa (Egypt); Satish Nambiar (India); Sadako Ogata (Japan); Yevgeny Primakov (Russian Federation); Qian Qichen (China); Nafis Sadik (Pakistan); Salim Ahmed Salim (Tanzania); and Brent Scowcroft (United States).See www.un.org/apps/news.

11 The Brahimi Panel on UN Peace Operations recommended in 2000 the creation of a new Information and Strategic Analysis Secretariat (EISAS), with a staff of over 50, to bring a strong new focus and professional competence to this task. But essentially because of member states' anxieties about the secretariat becoming *too* competent in its receipt and handling of sensitive intelligence, the proposal was drastically watered down, and has still not been implemented even in its diluted form.

12 See ICG Middle East Reports Nos 2-4: *Middle East Endgame I: Getting To A Comprehensive Arab-Israeli Peace Settlement; Middle East Endgame II: How A Comprehensive Israeli-Palestinian Peace*

Settlement Would Look; and Middle East Endgame III: Israel, Syria and Lebanon – How Comprehensive Peace Settlements Would Look, July16, 2002.

13 See www.geneva-accord.org.

14 See ICG Middle East Report No.17, *Governing Iraq*, August 25, 2003, 12.

15 See ICG Middle East Briefings on *The Challenge of Political Reform in Egypt*, September 30, 2003, and *The Challenge of Political Reform: Jordanian Democratisation and Regional Instability*, October 8, 2003.

Chapter 6

1 *Gulf Strategic Report 2003-2004* [in Arabic] (Sharjah: Research Unit, Dar al Khaleej for Press, Printing and Publication, 2004), 215.

2 Ahmed Shihab, "Openness to Build a Contemporary State" [in Arabic]. See: http://www.alhalem.net/magalat/almosaraha.htm. See also Abdallah Bin Muhammad Al-Ghilani, "Elements of Stability in the Gulf" [in Arabic], at http://www.alwatan.com/graphics/2001/mar/11.3/heads/ot6.htm.

3 Yusuf Hamad Al-Ibrahim, "Reforming Production Flaws in the GCC States: An Economic-Political Approach" [in Arabic], in *Radical Reform: A Vision from Within*, Development Forum, Twenty Fifth Annual Meeting, Bahrain, January 4-6, 2004, 2.

4 Quoted from Torki Al-Hamad, "Globalization and the Search for a Definition" in *Gulf States and Globalization* [in Arabic], Development Forum, Twenty-First Annual Meeting, Dubai, February 3-4, 2000, 32.

5 Ahmed Yousuf, 'Globalization and the Arab Regional Order,' in Hassan Naf'i'a et. al., *Globalization: Issues and Concepts* [in Arabic] (Cairo: Department of Political Science, Cairo University, 2000), 32.

6 Ibid., 33.

7 Ibid., 33.

8 Ahmed Al-Rashidi, "The Phenomenon of Globalization and the Principle of National Sovereignty" [in Arabic], in *Globalization: Issues and Concepts*, op cit., 74.

9 Ibid., 74-75.

10 See Abdul Aziz Al-Dakhil, "Economic Globalization and the GCC States," in *Gulf States and Globalization*, op cit., 55-78.

11 See Goda Abdul Khaliq, "Globalization and Political Economy for the Nation-State," *Globalization: Issues and Concepts*, op. cit., 162.

12 See Al-Dakhil, op. cit., 95.
13 Ibid., 96.
14 Ibid., 107.
15 Mona Al-Hadidi, "Satellite Channels and the Arab World under Globalization," in *Globalization: Towards a Different Vision* [in Arabic], (Cairo: Department of Political Science, Cairo University, 2001), 30-31.
16 Muhammad Ghubash, "The Gulf State: A More Than Absolute Authority and a Less Than Capable Society," in *Radical Reform: A Vision from Within*, op. cit., 8-9.
17 Quoted from Muhammad Ghubash, "The Democratic March in the Arab Gulf and the Arab World and Future Horizons," debate on *Current International Changes and Their Impact on the Future of the Gulf* [in Arabic], The National Council for Culture, Art and Literature, Kuwait, January 5-7, 2003, 7-8.
18 Abdullah Bishara, "Globalization and State Sovereignty: The Case of the Gulf," in *Globalization and Its Impact on Society and State* [in Arabic] (Abu Dhabi: ECSSR, 2002), 72-73.
19 Richard Higgott, "Globalization and Regionalization: Two New Trends in World Politics," *Emirates Lecture Series*, no. 25 [in Arabic] (Abu Dhabi: ECSSR, 1995), 74.
20 As'ad Al-Khafaji, "Islamic Globalization and American Change" [in Arabic], an article published in Elaph website on January 23, 2004: http://www.elaph.com9090/elaph/arabic/frontendprocessArtical.
21 Amin Al-Mushaghiba and Shamlan Al-Issa, "Political Reform in the GCC States," in *Political Reform in the Arab World* [in Arabic], Center for Developing Countries Studies and Research, Cairo University, May 3-4, 2004, 14-15, 18-19.
22 Ali Al-Tarah, "The Social Dimensions of Globalization and Their Impact on the Role of Gulf Women" [in Arabic], *Journal of the Faculty of Arts*, vol. 60, no. 4 (Cairo: Faculty of Arts, Cairo University, October 2000), 332.
23 See Amin Al-Mushaghiba and Shamlan Al-Issa, op.cit., 13-14, 17-18.
24 For more information see Chapter 22 by Badria Al 'Awadi in this book.
25 "The Status of Population and Labor Force in Kuwait in the Years 1994-2002" [in Arabic], Ministry of Planning, Kuwait, May 2003, 12.
26 Isabelle Jacques, "Political, Economic and Social Reform in the Arab World," in *Post-Iraq: What Implications for Middle East Arab Security?* Conference Report on Wilton Park Conferences, March 31-April 4, 2003, 2.

27 Rima Al-Sabban, "Issues and Concerns of the Civil Society in the GCC States: Institutions, Regulations, Minorities" [in Arabic], in *Issues and Concerns of the Civil Society in the GCC States*, Development Forum, Nineteenth Annual Meeting, Dubai, February 19-20, 1998, 3.
28 Fouad Ibrahim, "Civil Society in the Gulf: Hopes and Roles" [in Arabic], unpublished research paper, 5.
29 "National Assembly Election Results, 2003," General Department of Legal Affairs, Ministry of Interior, Kuwait, July 2003.
30 Kazim Habib, "Globalization and Arab World Fears," published in the website of Ibn-Rushd Forum for Free Thinking, no. 2, Summer 2001, 8. See http://www.ibn-rushd.org/forum/GlobalA.html.
31 For a review of the Greater Middle East Project, refer to the web site http://www.arabrenewal.com/index.php?rd=AI&AI0=4266
32 Saad Muhyo, "Greater Middle East or a Grand Maneuver? *Al Wasat*, no. 630, February 23, 2004, 7.
33 Suleiman Nimir, "Saudi Arabia and Egypt Refuse," *Al Hayat*, no. 14943, February 25, 2004.
34 Vahan Zanoyan, *Time for Making Historic Decisions in the Middle East* (Kuwait Center for Strategic and Future Studies, Kuwait University, 2002), 30-38.
35 Ibid., 31.

Chapter 7

1 See Lawrence Wright, "The Kingdom of Silence," *The New Yorker*, January 5, 2004.
2 This is the theme of my study of American foreign policy after 9/11 called *Fear's Empire: War, Terrorism and Democracy* (New York, NY: W.W. Norton and Company, 2003).

Chapter 9

1 See: Kenneth M. Pollack, "Securing the Gulf," *Foreign Affairs*, vol. 82 no. 4, July/August (2003), 8.
2 Pollack, op. cit.

Chapter 10

1 Yadollah Shokri, *A History of Safavids*, (Tehran: Ettellat Publications, second edition, in Farsi, 1985), 3–17.
2 Homa Katouzian, *A Political Economy of Modern Iran* (London: I. B. Tauris, 1985) 24–65.
3 Seyed Javad Tabatabaee, *An Introduction to a History of Iran's Deterioration* (Tehran: Zaman Moaser Publications, 2002), 112–146.
4 Abolhassan Ebtehaj, *Memoirs* (Tehran: Elmi Publications, 1993, second volume), 525–526.
5 See Said Amir Arjomand, *From Nationalism to Revolutionary Iran* (New York, NY: State University of New York, 1984) and Nikki Keddi, *Iran and the Muslim World: Resistance and Revolution* (New York, NY: New York University Press, 1995).
6 Mahmood Sariolghalam, "Arab-Iranian Rapprochement: Regional and International Impediments," in *Arab-Iranian Relations*, edited by Khair-din Haseeb (Beirut: Centre for Arab Unity Studies, 1998), 425–427.
7 Mahmood Sariolghalam, "The Future of the Middle East: The Impact of the Northern Tier," *Security Dialogue* vol. 27, no. 3 (September 1996): 312–315.
8 See Peter Katzenstein, *Between Power and Plenty* (Wisconsin, WI: Wisconsin University Press, 1978).
9 Mahmood Sariolghalam, *Globalization and the Issue of Sovereignty of the Islamic Republic of Iran* (Tehran: Center for Strategic Studies, 2004), 63–75.
10 Jim Rohwer, *Asia Rising: Why America Will Prosper as Asia's Economies Prosper* (New York, NY: Simon and Schuster, 1995), 207–278.
11 Mahmood Sariolghalam, "Understanding Iran: Getting Past Stereotypes and Mythology," *Washington Quarterly* vol. 26, no. 4 (Autumn 2003): 73–78.
12 Mahmood Sariolghalam, *Rationality and the Future of Iran's Development* (Tehran: Center for Scientific Research and Middle East Strategic Studies, 2004, third edition, in Farsi), 9–29.
13 One can also refer to the conformist and multi-layered social and political behavior of Iranians. It is rather difficult to find a harmonious, internally integrated and logically consistent set of behavior in this culture. Japanese, Germans and South Koreans enjoy such harmony. Perhaps, one can even broaden this analysis to encompass the whole region of the Middle East. On this issue see

Graham Fuller, *The Center of the Universe* (Boulder, CO: Westview Press, 1991).

14 Sariolghalam, "Understanding Iran," op. cit., 69–73. Donald Rumsfeld has also expressed a hope that during his lifetime, the Iranian government will change. Former President Rafsanjani rebutted by saying that the American Secretary of Defense will have to take this wish to his grave because the Islamic Republic of Iran is here to stay.

15 Massod Nili et al, *Iran's Industrial Strategy* (Tehran: Center for Advanced Research and Planning of Development, Farsi edition, 2003), 65–121.

Chapter 11

1 For an informative discussion of this dynamic, see Mahmood Sariolghalam, "Understanding Iran: Getting Past Stereotypes and Mythology," *Washington Quarterly*, vol. 24, no. 4 (Autumn 2003).

2 For a particularly insightful portrait of these factors at work, see Ali M. Ansari, "Iran: Continuous Regime Change from Within," *Washington Quarterly*, vol. 24, no. 4 (Autumn 2003).

3 For an analysis of the implications of this sense of isolation, see Farhad Khosrokhavar, "Iran: The Fear of Encirclement" (unpublished, 2003).

4 See Jahangir Amuzegar, "Iran's Crumbling Revolution," *Foreign Affairs*, volume 82, no.1 (January-February 2003): 44-57.

5 One perplexing complication in this history of Pakistani-Iranian enmity emerges in periodic reports that Pakistan has provided assistance to Tehran for Iran's nuclear program. Although the sources of these assertions have generally been credible, this assertion is at odds with political ties between Islamabad and Tehran that are characterized by significant tension. Recent reports now suggest that Pakistani support for Iran's nuclear program may have been unsanctioned and unofficial and pursued by well-placed Pakistanis "seeking personal gain" through secret and illegal technology transfers. The government of President Musharraf is now said to be conducting investigations into such leaks, which may have been motivated by simple personal greed. See for example, "Pakistan Admits Possible Nuke Ties with Iran," *Los Angeles Times*, December 24, 2003, A2.

6 For thoughts on the possible future of this troubled relationship, see Anoushiravan Ehteshami, "Iran-Iraq Relations after Saddam," *Washington Quarterly*, vol. 26, No. 4 (Autumn 2003): 115-129.

7 For an overview of Iran's perspectives of the United States, see Farhad Khosrokhavar, "La politique étrangere en Iran: de la révolution à l'"axe du Mal," *Politique Etrangere*, 1/2003, 77-91.

8 Despite information provided above in note 5, Iranians are still concerned about Pakistan's nuclear capability even though Tehran may have been able to breach Pakistani defenses to its questionably secure nuclear secrets.

9 For a discussion of Iranian foreign policy formulation, see Shireen Hunter, "Iran's Pragmatic Regional Policy," *Journal of International Affairs*, vol. 56, no. 2 (Spring 2003):133-147.

10 See for example Ray Takeyh, "Iranian Options: Pragmatic Mullahs and America's Interests," *The National Interest* (Fall 2003): 49-56.

11 For a fascinating new history of what was for Iran and the Iranian people, if not for the United States, a significant national trauma, see Stephen Kinzer, *All The Shah's Men: An American Coup and the Roots of Middle East Terror* (Hoboken, NJ: John Wiley & Sons, 2003).

12 These Russian recriminations became apparent to the author in personal interviews and meetings with a number of senior Russian officials responsible for ties with Iran. Personal discussions, Moscow, October 2003.

13 There has been substantial analysis of Iran's nuclear program, with some being extremely critical of it and others taking a more benign view. For samples of these writings, see "Iran's Nuclear Calculations," W*orld Policy Journal*, vol 20, no. 2 (Summer 2003); and Shahram Chubin and Robert S. Litwack, "Debating Iran's Nuclear Aspirations," *Washington Quarterly*, vol. 24, no. 4 (Autumn 2003).

14 For discussion of Iran's concern with terrorism, see Gary Sick, "Iran: Confronting Terrorism," *Washington Quarterly*, vol. 26, no. 4 (Autumn 2003): 83-98.

15 For an expanded discussion of the concept of a grand Western bargain in the Middle East, see David C. Gompert, Jerrold D. Green, F. Stephen Larrabee, "How an Atlantic Partnership could Stabilize the Middle East," *RAND Review* (Spring 1999).

Chapter 12

1 Report by the Director General of the IAEA Board of Governors, "Implementation of NPT Safeguards Agreement in the Islamic Republic of Iran," GOV/2003/75, November 10, 2003.
2 *The Military Balance, 2003-2004,* International Institute for Strategic Studies, (Oxford: Oxford University Press, 2003), 121.
3 http://usinfo.state.gov/topical/pol/terror/texts/03061804.htm.
4 http://www.usea.be/Categories/GlobalAffairs/June1903BushIAEAIran.html.
5 http://www.cbsnews.com/stories/2003/11/20/world/main584634.shtm.
6 http://www.whitehouse.gov/news/releases/2004/02/20040211-4.html.
7 http://www.dawn.com/2003/10/28/int3.htm.
8 http://www.daneshjoo.org/generalnews/article/publish/printer_3789.shtml.
9 http://www.frontlineonnet.com/fl2101/stories/20040116000506200.htm.

Chapter 15

1 "The Bush National Security Strategy," International Institute for Strategic Studies (IISS), vol. 8, no. 8 (October 2002), and "Dealing with the Axis of Evil," IISS vol. 8, no. 5 (June 2002).
2 Melvyn P. Leffler, "9/11 and the Past and Future of American Foreign Policy," *International Affairs* no.79, 5 (2003): 1045–1063. Also Strobe Talbott, "War in Iraq, Revolution in America," *International Affairs* no.79, 5 (2003): 1037–1043.
3 "The National Security Strategy of the United States of America," September 17, 2002. (www.whitehouse.gove/nsc/nssall.html).
4 Leffler, op. cit., 1046.
5 Ibid., 1047.
6 Even East Timor qualifies for this type of intervention. On a discussion of the significance of these experiences for the US, see Strobe Talbott, op. cit., 1037–1034.
7 Toby Dodge, "Consequences and Implications of US Military Intervention and Regime Change in Iraq," IISS Workshop "Intervention in the Gulf" at Gulf Research Center, Dubai, February 15–17, 2003.

8 Glenn Kessler, "US Decision on Iraq has Puzzling Past," *Washington Post*, January 12, 2002, A10.
9 See Faleh A. Jabar, "Iraq: Difficulties and Dangers of Regime Removal," Middle East Research and Information Project, *MERIP* no. 225 (Winter 2002): 18–19.
10 See, among other authorities, Mark Juergensmeyer, *Terror in the Mind of God* (Berkeley and Los Angeles, CA: University of California Press, 2000), 6–7; John Esposito, *Unholy War, Terror in the Name of Islam* (Oxford: Oxford University Press, 2002).
11 On readings into this tragic event see, among others, Fred Halliday, *Two Hours That Shook the World* (London: Saqi Books, 2002); and for an extensive historical background, see Bernard Lewis, *What Went Wrong? Western Impact and Middle Eastern Response* (Oxford, New York, NY: Oxford University Press, 2002).
12 For a discussion of this important point, see Anthony H. Cordesman, *Saudi Arabia Enters the Twenty-First Century, The Political, Foreign Policy, Economic and Energy Dimensions* (Westport, CT and London: Praeger, 2003), 164 and passim. On the Bush Doctrine and its impact on Saudi Arabia, see Geoff Simons, *Future Iraq, US Policy in Reshaping the Middle East* (London: Saqi Books, 2003),189–192 and 272–3, 275.
13 Cordesman, op. cit., 118–120.
14 "US Policy Towards the Middle East," *IISS* vol. 9, no. 1 (January 2003), 2.
15 Toby Dodge and Steven Simon (eds) *Iraq at the Cross-Roads* (London: IISS, January, 2003), 87.
16 These workshops included scores of Iraqi scholars, politicians and social activists who spent more than a year working in teams to produce documents on post-Saddam Iraq, among which is the important document, "Transition to Democracy."
17 George Packer, "War After the War: What Washington Does Not See in Iraq," *The New Yorker*, November 29, 2003.
18 Kenneth Katzman, "Iraq: Regime Change Efforts and Post-War Governance," Congressional Research Service, Reports for Congress, October 28, 2003, 21. On the past of Iraq, see also Michael Eisenstadt and Eric Mathewson (eds) *US Policy in Post Saddam Iraq: Lessons from the British Experience* (Washington DC: The Washington Institute for Near Eastern Policy, 2003).
19 Ray Salvatore Jennings, "The Road Ahead: Lessons in Nation Building from Japan, Germany, and Afghanistan for Postwar Iraq," US Institute for Peace (USIP), *Peaceworks* no. 49 (May 2003).

20 On the possibilities of a Vietnamese-type urban warfare in Iraq, see Toby Dodge and Steven Simon (eds), op. cit., 69–71.

21 Estimates on the number of technocrats are derived from: Republic of Iraq, Ministry of Planning, *Annual Abstract of Statistics* (Baghdad, 1992), 60-68. Our calculation included all holders of Ph.D., M.A. Higher Diploma, and B.A. academic degrees who are engaged in government employment, of both "rural" and "urban" origin.

22 On the size of the Iraqi armed forces and the order of battle, see Michael Eisenstadt, "Like a Phoenix from the Ashes? The Future of Iraqi Military Power," The Washington Institute for New East Policy, Policy Papers no. 36 (1993), 47 and 64; and David Ochmaneck, "A Possible US-led Campaign against Iraq: Key Factors and an Assessment," in Toby Dodge and Steven Simon (eds), op. cit., 55–56.

23 On the ratio of army and security forces to the civilian population during 1930–1995, see Faleh A. Jabar *Al-Dawla wal-Mujtama' al-Madani, wal Tatawur al-Demoqratif fil Iraq*, translated as State, Civil Society and Democratic Perspectives in Iraq (Cairo: Ibn Khaldoun Centre, 1995), Chapter 3, Tables 3-3, 3-4, 3-5 and 3-6, based on data derived from US Arms Control and Disarmament Agency, Washington DC, 1987/1988.

24 The Kurdistan Democratic Party's daily *Al-Ta'akhi*, Baghdad, June 16, 2003.

25 On the new Iraqi army see among others, the International Crisis Group (ICG) Middle East Report no. 20, "Iraq: Building a New Security Structure," December 23, 2003, 15–18.

26 Katzman, op. cit., 31.

27 ICG Middle East Report no. 20, op. cit.

28 During my Iraq research tour in June, July and August 2003, I had interviews and conversations with most of the Governing Council members, who, together with their advisors and assistants, expressed deep concern over the prospects of "communalization of politics." This view was equally voiced by Sunni and Shi'a liberals and secular-minded individuals. The International Crisis Group, for example, criticized the sectarian-ethnic quotas adopted before invasion (at the December 2002 London Conference of the Iraqi opposition) and after it, in the selection of the Governing Council. See ICG Report no.11, "War in Iraq: Political Challenges after Conflict," March 25, 2003, and ICG Report no.17, "Governing Iraq," August 2003.

29 On the intricacies of the constitutional process, the best source is the Preparatory Constitutional Committee's confidential report conveyed to the Governing Council in September 2003. The document was not

made available to the public, but parts of it were leaked to the press and to interested scholars. The best leakage available can be found in the *Washington Post*, November 13, 15 and 26, 2003; see ICG Middle East Report no.19, "Iraq's Constitutional Challenge," November 13, 2003.

30 Robert Looney, "The Neo-Liberal Model's Planner Role in Iraq's Economic Transition," *The Middle East Journal* vol. 57, no. 4 (Autumn 2003): 568–586.

31 Associated Press, December 6, 2003.

32 These figures are estimated on the basis of the 1992 membership of the Iraqi Industrial League, Union of Houses of Commerce, Contractors Union and Banker Union. These figures may be partly inflated by the multiple membership of some individuals in these unions. The figure also does not include an estimated 10% of the middle strata, who are people of property and capital.

33 On problems of market and liberalization models, see Kiren Aziz Chaudry, "Economic Liberalization and the Lineages of the Rentier State," *Comparative Politics* (October 1994): 1–24; Keiko Sakai, "Economic Liberalization in Iraq: Its Political Application," Paper submitted to the Middle East Workshop, *Institute of Developing Economies*, Tokyo, January 9, 1996.

34 On these and other debt-related issues see Justine Alexander and Colin Rowat, "A Clean Slate in Iraq, From Debt to Development," *MERIP* no. 228 (Fall 2003): 32–36; Sinan Al-Shabibi, "Prospects of Iraq's Economy," in *The Future of Iraq* (Middle East Institute, Washington DC, 1997); and Ahmed Jiyad, "An Economy in Debt Trap, 1980–2020," *Arab Studies Quarterly*, vol. 23, no. 4 (Winter 2001). For general prospects for the Iraq economy, see A. Nassrawi, *The Economy of Iraq: Oil, Wars, Destruction of Development and Prospects, 1950-2010* (Greenwood Publishing Group, 1994).

35 On the history of the formation of the Kingdom of Saudi Arabia, see Al-Rasheed, *A History of Saudi Arabia* (Cambridge: Cambridge University Press, 2000); A. Vassiliev, *The History of Saudi Arabia* (London: Saqi Books, 2000); and Katzman, op. cit. On the political, social and cultural history of the Gulf emirates, see (among others) D. Hawley, *The Trucial States* (London: Allen and Unwin, 1970); John D. Anthony, *Arab States of the Lower Gulf: People, Politics, Petroleum* (Middle East Institute, Washington DC, 1975); P. Khaldoun Hasan Al-Naqeeb, *Society and State in the Gulf and Arab Peninsula* (London: Routledge, 1990). Peter Lienhardt, *Shaikhdoms of Eastern Arabia*, edited by Ahmed Al-Shahi (Oxford: St. Anthony's College, 2001).

36 On overlapping histories in the 19th and 20th centuries, see Ibrahim Faseeh Al-Haidari, *'Unwan al-Majd fi bayan ahwal Baghdad wal Basra wa Najd,* translated as *Aspects of Glory in the History of Baghdad, Basra and Najd* (Baghdad, n.d., reprinted in London, 1999); also A. Vassiliev, *The History of Saudi Arabia* (London: Saqi Books, 2000).

37 On influences of Nassirism and other ideologies see A. Vassiliev, op. cit., 350–355.

38 On these episodes numerous authorities are available. On the early Saudi–Iraqi and Iraqi–Kuwaiti borders disputes, see, for example, Stephen Hemsley Longrigg, *Iraq 1900-1950, A Political, Social and Economic History* (Oxford: Oxford University Press, 1953), 217–220. An analysis of such disputes during the 1991 Gulf War is also illustrated in Majeed Khadduri and Edmund Gharib, *War in the Gulf, 1990-1991, The Iraq-Kuwait Conflict and its Implications* (Oxford: Oxford University Press, 1997), 6–20.

39 The Hanafite Ottomans instructed Muhammad Ali of Egypt to destroy the first Saudi dynasty; their pressures on the Baghdad Sunni Ulama to discredit Wahhabism are well-documented. See A. Vasiliev, op. cit., 140 and passim.

40 See Y. Nakash, *The Shi'is of Iraq* (Princeton, NJ: Princeton University Press, 1995), 25–30.

41 On the foreign policy of Iraq during this period, see Marion Farouk-Sluglett and Peter Sluglett, *Iraq Since 1958* (London: I.B. Tauris, 1987, 1990, revised edition 2001), 200–206. See also Phebe Marr, *The Modern History of Iraq* (Boulder, CO: Westview Press, 2004), 193–195.

42 See M. Al-Rasheed, op. cit., 128–134.

43 Farouk-Sluglett and Sluglett (London: IB Tauris, 1990) 123–126.

44 See Faleh A. Jabar, "The Gulf War and Ideologies, The Double-Edged Sword of Islam," in H. Bresheeth and Y. Y-Davis (eds) *The Gulf War and the New World Order* (London: ZED publishers, 1991), 211–217, and F.A. Jabar, "The Roots of an Adventure," in V. Britain (ed.) *The Gulf Between Us* (London: Verago, 1991), 27–41.

45 *Al-Sharq Al-Awsat* newspaper, London, January 6, 2004. Similar themes were reiterated by H.R.H. Prince Turki Al-Faisal Bin Abdul Aziz, the Saudi Ambassador in London, during the ECSSR Ninth Annual Conference on *The Gulf: Challenges of the Future*, Abu Dhabi, January 11-13, 2004. See chapter 13.

46 This loose estimate has to do with lack of reliable census. According to a secret study conducted by the security directorate in 1977, the

Shi'as constituted some 54%, including Shi'a Kurds and Shi'a Turkomen. If this ratio is still valid, Shi'a Arabs may range between 49%–50% of the population. See study by the General Directorate of Security, Baghdad.

47 On the divisions among the Shi'ites see F.A. Jabar, *The Shi'ite Movement in Iraq* (London: Saqi Books, 2003), chapter 1.

48 Conversation with Hussein Khomeini, the grandson of late Ayatollah Khomeini (Baghdad, July 2003). The new anti-Khomeini trends developing in religious schools in Qum are numerous and lively. One manifestation is the new theology developed in Qum. See, among others, Abdul-Jabar Al-Rifa'i, *The Neo-Theology and Philosophy of Religion, The Intellectual Landscape in Iran*, translated *as Ilm al-Kalam al-Jadid wa falsafat al-deen, al-Mash-had al-thaqafi fi Iran* (Beirut: Dar al-Hadi, 2002). See also *Qadhaya Islamiya Mu'asira* Quarterly (*Modern Islamic Problems*), Year 7, no. 22 (Winter 2003). This issue contains many essays defending "religious pluralism" and "religious difference," a theme that contradicts the monist centrality of the "governance of jurisprudence" principle.

49 In Saudi Arabia, Shi'ites conveyed to Crown Prince Abdullah, a petition of new demands. See Michael Scott Doran, "The Saudi Paradox," *Foreign Affairs* vol. 83, no.1 (January/February 2004): 35–51, see especially p. 46. See also Joshua Teitelbaum, "Holier Than Thou, Saudi Arabia's Islamic Opposition," Policy Papers no. 52 (Washington, DC: The Washington Institute for Near Eastern Policy, 2000); on Shi'a opposition see p. 83 and passim; M. Al-Rasheed, op. cit., on reforms p. 172 and passim; Richard F. Nyrop (ed.) *Saudi Arabia, A Country Study* (Washington DA, 1984); Gawdat Bahgat, "The Gulf Monarchies, New Economic and Political Realities," The Research Institute for the Study of Conflict and Terrorism (RISCT) *Conflict Studies* no. 296, February 1997, 6, 9–10, and 14-16.

50 See Michael Scott Doran, "The Saudi Paradox," *Foreign Affairs*, vol. 83, no.1 (January/February 2004): 35–51.

Chapter 16

1 Most famously, US Senator George Aiken, an independent-minded Republican from Vermont, ruefully suggested that the United States should get out of Vietnam and dedicate a fraction of its war spending to a global public relations campaign that declared America's victory. On October 19, 1966 he said, "the United States could well declare

unilaterally that this stage of the Vietnam war is over—that we have 'won' in the sense that our Armed Forces are in control of most of the field and no potential enemy is in a position to establish its authority over South Vietnam....Declare victory and get out."

2 L. Paul Bremer III, as quoted in *Washington Post*, November 2, 2003.

3 Douglas Feith's testimony as quoted in *Fortune* magazine, March 31, 2003.

4 British Ambassador to the United Nations, Jeremy Greenstock, speaking at a panel in Washington on May 8, 2003.

5 See Anthony H. Cordesman, CSIS, "Developments in Iraq at the End of 2003: Adapting US Policy to Stay the Course," December 29, 2003.

6 *A Wiser Peace*, Post-Conflict Reconstruction Project Report, CSIS, January 2003, 8-9 (http://www.csis.org/isp/wiserpeace.pdf).

7 "Iraqis Can do More," Jessica Mathews, *Washington Post*, September 29, 2003.

8 See "Attacks and Security put Karbala Residents on Edge," Edward Wong, *New York Times*, December 29, 2003.

9 Anthony H. Cordesman of CSIS in e-mail, "First Real National Poll in Iraq," December 1, 2003.

10 See Neil MacFarquhar, "Open War Over, Iraqis Focus on Crime and a Hunt for Jobs,"*New York Times*, September 16, 2003.

11 Anthony H. Cordesman, CSIS, e-mail, December 1, 2003.

12 Ibid.

13 See "For Many Iraqis, US-backed TV Echoes the Voice of its Sponsor; Station Staffers Acknowledge Their Reluctance to Criticize," Alan Sipress, *Washington Post*, January 8, 2004.

14 Anthony H. Cordesman, CSIS, e-mail, December 1, 2003.

15 See Thomas E. Ricks, "Marines to Offer New Tactics in Iraq: Reduced Use of Force Planned After Takeover from Army," *Washington Post*, January 7, 2004.

16 See Senator John McCain, "How to Win in Iraq," *Washington Post*, November 9, 2003.

17 See Marine Lt. Col. Carl E. Mundy III, "Spare the Rod, Save the Nation," *New York Times*, December 30, 2003.

18 See Senator John McCain, op. cit., *Washington Post*, November 9, 2003.

19 See Robin Wright and Rajiv Chandrasekaran, "Power Transfer in Iraq Starts This Week," *Washington Post*, January 4, 2004,

20 See Diane Orentlicher, "International Justice Can Indeed Be Local," *Washington Post*, December 21, 2003.

21 See James Glanz, "Rebuilding Iraq Takes Courage, Cash and Improvisation," *New York Times*, November 30, 2003.

22 See Milt Bearden, "Iraqi Insurgents Take a Page from the Afghan 'Freedom Fighters,'" *New York Times*, November 9, 2003.

Chapter 17

1 Quoted from the "Welcoming Remarks" by Dr. Jamal S. Al-Suwaidi, Director General of ECSSR, in the brochure of the Ninth Annual Conference of The Emirates Center for Strategic Studies and Research, entitled "The Gulf: Challenges of the Future," Abu Dhabi, January 11-13, 2004.

2 See Barrie Axford, *The Global System: Economic, Politics and Culture* (New York, NY: St. Martin's Press, 1995). See also Roland Robertson, *Globalization: Social Theory and Global Culture* (London: Sage, 1992, reprinted 1998).

3 On the Middle East as a subsystem composed of three regional parts, the most important of which is the Arabian Gulf, see Bassam Tibi, *Conflict and War in the Middle East: From Interstate War to New Security* (London: Macmillan Press, second edition 1998), 43-60.

4 Hedley Bull, *The Anarchical Society: Order in World Politics* (New York, NY: Columbia University Press, 1977 edition and reprints) 273.

5 Bassam Tibi, *The Challenge of Fundamentalism: Political Islam and the New World Disorder* (Berkeley, CA: University of California Press, 1998, updated 2002), Chapter V on the "Simultaneity of Structural Globalization and Cultural Fragmentation."

6 On the significance of the Arabian Gulf see Geoffrey Kemp (ed.) *Powder Keg in the Middle East: The Struggle for Gulf Security* (London: Rowman and Littlefield, 1995).

7 See Tibi, *Conflict and War in the Middle East: From Interstate War to New Security,* op. cit.

8 See Tibi, *The Challenge of Fundamentalism: Political Islam and the New World Disorder,* op. cit.

9 For a global perspective on democracy see David Held, *Democracy and Global Order: From the Modern State to Cosmopolitan Governance* (Stanford, CA: Stanford University Press, 1995), particularly the three chapters in Part IV.

10 For this particular perspective on democracy in Islam, see Bassam Tibi, "Democracy and Democratization in Islam," in Michéle

Schmiegelow (ed.) *Democracy in Asia* (New York, NY: St. Martin's Press, 1997), 127-146.

11 On Hellenization in Islam, refer to Herbert A. Davidson, *Alfarabi, Avicenna and Averroes on Intellect: Their Cosmologies, Theories of the Active Intellect, and Theories of Human Intellect* (New York, NY: Oxford University Press, 1992), and William Montgomery Watt, *Islamic Philosophy and Theology: An Extended Survey* (Edinburgh: University Press, 1962, second edition 1985), Parts 2 and 3.

12 On the contemporary Arab Turath debate see the contributions in *al-Turath wa Tahadiyyat al-asr fi al-Watan al-Arabi al-Asalah wa al-Mu'assarah* (Beirut: Center for Arab Unity Studies, 1985), and more recently, Mohammed Abed Al-Jabiri, in *al-Turath wa Hadatha* (Beirut: al-Markaz al Thaqafi, 1991).

13 Bruce Russett, *Grasping Democratic Peace: Principles for a Post-Cold War World* (Princeton, NJ: Princeton University Press, 1993).

14 Immanuel Kant, "Zum ewigen Frieden" in Zwi Batscha and Richard Saage (eds) *Friedensutopien: Kant, Fichte, Schlegel, Görres* (Frankfurt: Suhrkamp, 1979), 37-82.

15 Michael Walzer (ed.) and (trans), *Al-Farabi on the Perfect State* (Oxford: Clarendon Press, 1985). A translation of Abu Nasr Al-Farabi, *Mabadi ara' ahl al-Madina al-Fadila*.

16 These wars are dealt with in the following books: Anthony H. Cordesman and Abraham R. Wagner, *The Lessons of Modern War* in 4 volumes (Boulder, CO: Westview Press, 1990), Vol. II on the Iran-Iraq War. On the 2003 War in Iraq, see Bassam Tibi, *Der neue Totalitarismus: "Heiliger Krieg" und Westliche Sicherheit* (Darmstadt: Primus, 2004), Chapter V and on the Gulf War of 1991 see Stanley A. Renshon (ed.) *The Political Psychology of the Gulf War: Leaders, Publics, and the Process of Conflict* (Pittsburgh, PA: University of Pittsburgh Press, 1993), Part I: Origins.

17 On Iran's model for the Middle East see Rouhollah K. Ramazani, *Revolutionary Iran: Challenge and Response in the Middle East* (Baltimore, MD: John Hopkins University Press, 1986), Part 2: Gulf. See also Bassam Tibi, "The Failed Export of the Islamic Revolution into the Arab World," in Frédéric Grare (ed.), *Islamism and Security: Political Islam and the Western World* (Geneva: Program for Strategic and International Security Studies, 1999).

18 On Saddam and Pan-Arabism, see Samir Al-Khalil, *The Republic of Fear: The Politics of Modern Iraq* (Berkeley, CA: University of California Press, 1989), chapters of Part II.

19 Lawrence E. Harrison and Samuel P. Huntington (eds.) *Culture Matters: How Values Shape Human Progress* (New York: Basic Books, 2000).
20 Bassam Tibi, "The Interplay between Social and Cultural Change: The Case of Germany and the Middle East," in Ibrahim M. Oweiss and George N. Atiyeh (eds) *Arab Civilization: Challenges and Responses—Studies in Honor of Constantine K. Zurayk* (Albany, NY: State University of New York Press, 1988), 166-182.
21 Juergen Habermas, *The Philosophical Discourse of Modernity: Twelve Lectures* (Cambridge, MA: MIT Press, 1987).
22 Fazlur Rahman, *Islam and Modernity: Transformation of an Intellectual Tradition* (Chicago, IL: University of Chicago Press, 1982).
23 See Bassam Tibi, "International Morality and Cross-Cultural Bridging," Roman Herzog et al., *Preventing the Clash of Civilizations: A Peace Strategy for the Twenty-First Century* (ed.) by Henrik Schmiegelow (New York, NY: St. Martin's Press, 1999), 107-126.
24 On Gulf oil, see the chapter by Paul Stevens, in Gary G. Sick and Lawrence G. Potter (eds) *The Persian Gulf at the Millennium: Essays in Politics, Economy, Security, and Religion* (New York, NY: St. Martin's Press, 1997), 85-114.
25 On the Gulf during the Cold War, see Aryeh Yadfat and Mordechai Abir, *In the Direction of the Gulf: The Soviet Union and the Persian Gulf* (London: Frank Cass, 1977). On the changed Gulf, see F. Gregory Gause, *Oil Monarchies: Domestic and Security Challenges in the Arab Gulf States* (New York: Council on Foreign Relations Press, 1994).
26 On Saddam Hussein, see among several books, Said K. Aburish, *Saddam Hussein: The Politics of Revenge* (New York, NY: Bloomsbury, 2000). See also the reference in note 18 above.
27 On this issue, see chapter 8 on education in: Bassam Tibi, *Islam: Between Culture and Politics* (New York, NY: Palgrave, 2001), 167-185.
28 See Abdullahi An-Na'im, *Toward an Islamic Reformation* (Syracuse, NY: Syracuse University Press, 1990), and also Bassam Tibi, "The Human Tradition of Human Rights and the Culture of Islam," in Francis M. Deng (ed.) *Human Rights in Africa: Cross-Cultural Perspectives* (Washington, DC: The Brookings Institution, 1990).
29 See Harrison and Huntington (eds.) *Culture Matters: How Values Shape Human Progress*, op. cit.

30 The findings related to the "Culture Matters" research project as edited by Lawrence Harrison are due to be published by Routledge and Oxford University Press in 2005.
31 See Mohammed Shahrur, *al-Kitab wa al-Qur'an: Qira'a Mu'asira* (Beirut: Sharikat al-Matbu'at, sixth edition, 2000), in particular on the separation between religion and politics, 719-723.
32 John Waterbury, *The Commander of the Faithful: The Moroccan Political Elite—A Study in Segmented Politics* (New York, NY: Colombia University Press, 1970), 5.
33 See Surah Al-Nisa (4), ayah 79.
34 This recommendation is also found in Mohammed Abed Al-Jabiri, *Arab Islamic Philosophy: A Contemporary Critique* (Austin, TX: Center for Middle Eastern Studies, 1999).
35 On these counter-elites see Bassam Tibi, "The Fundamentalist Challenge in the Middle East," in *Fletcher Forum of World Affairs*, vol. 23, no.1 (Winter/Spring, 1999): 191-210.
36 See Shahrur, *al-Kitab wa al-Qur'an: Qira'a Mu'asira*, op. cit.
37 See Barbara F. Stowasser, *Women in the Qur'an: Traditions and Interpretations* (New York, NY: Oxford University Press, 1994).
38 On this concept in Indonesia see Robert W. Hefner, *Civil Islam: Muslims and Democratization in Indonesia* (Princeton, NJ: Princeton University Press, 2000).
39 See Bassam Tibi."International Morality and Cross-Cultural Bridging," and also the chapter on civilization and dialogue in Bassam Tibi, *Islam: Between Culture and Politics*, op. cit., 210-230. See also the chapter on education in the same work.

Chapter 18

1 See for example, Richard N. Rosecrance, *The Rise of the Trading State: Commerce and Conquest in the Modern World* (New York, NY: Basic Books, 1986).
2 Rosecrance, op.cit., 27.
3 According to World Bank statistics, GCC official unemployment statistics are as follows: Bahrain: 12.6%, Qatar: 11.6%, Oman: 10%, Saudi Arabia: 9.66%, Kuwait: 2.8%, UAE: 2.3%. See *Al-Watan* newspaper, Saudi Arabia, December 13, 2003.

Chapter 19

1 Abdallah Hadiya, *Political Middle Eastern Issues* [in Arabic] (Cairo, Al-Tobji Commercial Press Co., 1998), 76-119.
2 On concepts of peaceful coexistence, détente, common security and peace-building, see M. Griffiths and T. O'Callaghan, *International Relations: The Key Concepts* (London and New York: Routledge, 2002), 22-25.
3 To review the literature on the transformations in the strategic environment in the Middle East and the Gulf region see Nora Ben Sahel and Daniel Byman (eds.), *The Future Security Environment in the Middle East: Conflict, Stability and Political Change* (Santa Monica, CA: RAND, 2004), 10-35. See also *The Gulf in 2003* (Dubai: Gulf Research Center, 2004), 191-203. See also Amani Gindeal, "Activating the Role of Civil Public Institutions in the GCC States," *Debate on the Reality and Future of the Organizations of Civil Society in the GCC States* (Kuwait: Center for Gulf and Arabian Peninsula Studies, University of Kuwait, April 3-4 2000) 43-64.
4 For a review of the dimensions of this strategy and topics relating to weapons of mass destruction, see, Shireen Mazari, "The New US Doctrine: Implications for the South Asian Region," *Strategic Studies* (Islamabad: Spring 2003), 18-23.
5 See *Arab News*, Kuwait, September 15, 2003.
6 See Mazari, op. cit, 8-31.
7 See *Al Siyasa,* Kuwait, April 19, 2004.
8 See Najam Rafique, "9/11, A Year After: Appraising US Regional Strategies," *Strategic Studies* (Islamabad: Autumn 2002), 220-23.
9 See *Arab Strategic Report for 2002-2003* [in Arabic] (Cairo: Al Ahram Center for Political and Strategic Studies, 2003) 275-280.
10 Ibid. 299.
11 Vahan Zanoyan, *Time for Making Historic Decisions in the Middle East* (Kuwait Center for Strategic and Future Studies, Kuwait University, 2002).
12 Paul Holt (ed.), *The Democratic Peace and International Conflict in the Twentieth Century* (Cambridge: Cambridge University Press, 2002). See also Khalil Ismail Al-Hadithi, *Functionalism and Functional Approach in the Sphere of the Arab League* [in Arabic], *Strategic Studies Series*, no. 62 (Abu Dhabi: ECSSR, 2001), 7-31.
13 See *Debate on the Reality and Future of the Organizations of Civil Society in the GCC States* [in Arabic], April 3-4, 2000, op. cit; Also Ahmed Al-Yen, "Rights of Political Reform in the GCC States" in

debate on *The GCC and Regional and International Variables* [in Arabic], (Kuwait: Center for Strategic and Futuristic Studies, University of Kuwait, 2004).

14 Saleh Al-Mani, "Capabilities and Mechanisms of Enhancing Confidence in the Gulf Region on the Security and Military Levels," *Record of Current Events in the Gulf Region and the Arabian Peninsula and Their Geographical Neighborhood* [in Arabic], no. 21 (Kuwait: Center for Gulf and Arabian Peninsula Studies, January-March 2002), 43-53.

15 Mustafa Abdul Aziz Morsi, "Economic Integration and Neo-Functionalism: A Theoretical Approach with Reference to the Gulf Experience" [in Arabic], *Annals of Arts and Social Sciences*, Thesis no. 201, Annals 27 (Kuwait: University of Kuwait, 2003), 53-58.

16 *The GCC in 2002* [in Arabic] (Kuwait: Gulf Center for Strategic Studies, May 2002), 90-95.

17 For an evaluation of regional security in the Gulf region, see Andrew Rathmell et al., *A New Persian Gulf Security System* (Santa Monica, CA: RAND, 2003), 1-11.

Chapter 21

1 Muhammad Ghanim Al-Rumaihi, *Al Bahrain: Mushkilat al Taghyeer al Siyassi wa Alijtima'i* [Bahrain: Problems of Political and Social Change] (Kuwait: Kadhima Co. for Publication, Translation and Distribution, 1984), 163.

2 Thuraya Al-Turki and Huda Zraiq, *Taghayor al qiyam fi al a'aila al a'arabiyya* [Changing Values of Arab Family], serial studies on Arab Women and Development 21, Economic and Social Commission for Western Asia, ESCWA (New York, NY: United Nations, December 27, 1995), 9.

3 See William F. Halloran: "Zayed University: A New Model for Higher Education in the United Arab Emirates," in *Education and the Arab World: Challenges of the Next Millennium* (Abu Dhabi, UAE: The Emirates Center for Strategic Studies and Research, 1999), 324.

4 Kaltham 'Ali Al-Ghanim, *Al Mara'a wa al Tanmiya fi al Mujtama' al Qatari* [Women and Development in Qatari Society], an analytical study on human development opportunities allowed to Qatari women, Symposium on "Women and Politics and Their Role in Development," April 21-23, 2002, the Supreme Council for Family Affairs, Doha, 38, 164.

5 Baqir Al Najjar, *Sira'a al Ta'aleem wa al Mujtama'a fi al Khaleej al 'Arabi* [The Conflict between Education and Society in the Arab Gulf] (Beirut: Dar Al Saqi, 2003), 62.
6 Yasameen bint Ahmed Ja'afar, "Health Care for Women in the Sultanate of Oman," Advisory Symposium on Preparing a National Strategy for Omani Women, March 3-5, 2002, Committee on Coordination of Female Voluntary Works, in cooperation with the Ministry of Social Development, Sultanate of Oman, 6.
7 ESCWA, Report by the *Arab Women's Center 2003*, (New York, NY: United Nations, 2003), 331.
8 Ibid, Table 12, 25.
9 Ibid, 334.
10 Ibid, 334.
11 Ibid, 335.
12 See Baqir Al-Najjar, *Al Mara'a fi al Khaleej al 'Arabi wa Tahwoulat al Hadatha al 'Aseera* [Women in the Arab Gulf and the Difficult Changes of Modernization] (Beirut: the Arab Cultural Center, 2000), 48-49.
13 Kaltham Al-Ghanim, op. cit., 14.
14 Al Khaleej Center for Strategic Studies, "Examining the Present and Future of Gulf Women" *Akhbar al Khaleej,* no. 8682 (Manama, Beirut: December 30, 2001), 11.
15 Yasameen bint Ahmed Ja'afar, op. cit., 3.
16 Maytha' Al-Shamsi, "Gulf Women...Where to?" *Al Mustaqbal al 'Arabi*, Year 24, no. 273 (Beirut: November 2001), 79-130.
17 "Partnership in the Arab Family," serial studies on Arab Women and Development 31, Economic and Social Commission for Western Asia, ESCWA (New York, NY: United Nations, 2001), 12.
18 Ibid. 18.
19 Baqir Al-Najjar, "Women in the Arab Gulf," op. cit., 117.
20 Ibid. 120.
21 Hassan Krayim, *Ta'atheer al Siyasat al Ijtima'iya ala al Ussra* [The Impact of Social Policies on the Family], the Arab meeting for the tenth evaluation of the International Year of Family, ESCWA, Beirut, October 7-9, 2003, 5.
22 Report by the Arab Women's Center, op. cit., 6.
23 Ibid. 337-343.
24 Ibid. 345-346.
25 Sa'ad Al-Hajji, "Women's Social Societies in the GCC States," documentary study (Kuwait, 2000), 709-761.
26 Ibid. 23-170.

27 Abdul Khaliq Abdullah et al., *Al Mujtama'* al *Madani wa al Tahawol al Demoqrati fi al Emarat al Arabiya al Mutahida* [The Civil Society and Democratic Transformation in the UAE] (Cairo, Ibn Khaldoon Center for Developmental Studies, jointly with Dar al Ameen for Publication and Distribution, 1995), 32.

28 Ibid. 33.

29 Sa'ad Al-Hajji, op. cit., 555-703.

30 Al-Mustafa Al-Assri, "Morocco Debates the Law of Family Code [*Muddawana*]," *Al Wasat* daily, no. 401 (Manama: October 21, 2003), 14.

Chapter 22

1 See: "Globalization: Threat or Opportunity?" *IMF Issues Brief* (2000-2003), 1-9.

2 Ibid., 7-9.

3 The Globalization Website Glossary (www.emory.edu/soc/globalization/glossary). The three quotations are respectively from T.L. Friedman, *The Lexus and the Olive Tree*, 1999, 7-8; M. Waters, *Globalization*, 1995, 3; and J.H. Mittelman, *The Globalization Syndrome*, 2000.

4 Dr. Yousuf Al-Qardhawi, in an interview by Al Jazeera channel. See Al Jazeera website, November18, 2001, 1-19

5 Sawsan Al-Abtah, "For Women Only" program, Al Jazeera channel, June 19, 2002. See Al Jazeera website,1-18

6 Dr. Yousuf Al-Qardhawi, op.cit.

7 Dr. Yousuf Al-Qardhawi, at the opening of *"Islam Online"* website, 1999.

8 See the conclusions of the *First Arab Human Development Report*, page 19.

9 See *The Empowerment of Women and Integration of Gender Perspectives in the Promotion of Economic Growth, Poverty Eradication and Sustainable Development,* Report of the Secretary-General, UN General Assembly, July 2003.

10 In Kuwait, three women, Dr. Latifa Al-Rujaib, Mrs. Siham Al-Rizouqi, and Mrs. Fidha Al-Khalid, were appointed in 1979 to the posts of Assistant Deputy Ministers at the ministries of Social Affairs and Labor, Finance, and Education, respectively. In 1992, a fourth woman was appointed as Deputy-Minister at the Ministry of Education. In 1995, a woman was appointed as Director of Kuwait University, and in 1996 a Kuwaiti woman was appointed as Kuwait's ambassador to South Africa. She is now Kuwait's Permanent Representative at the United Nations in New York. In Oman, a

woman was appointed as Chairperson of the General Commission for Handicraft Industries (with the rank of Minister). In 2003, a woman was appointed as the Minister of Education.

11 See Badria Al-Awadhi, "The Role of Gulf Women in Parliamentary and Political Life," in a seminar on "The Role of Women in Parliamentary and Political Life" (State of Qatar, March 2003).

12 This situation is identical to the Kuwaiti women's experience in confronting the hostile internal stances against women's right to political participation during the period from early 1970s till end of 1990s, before the Emiri declaration endorsing women's political rights was declared on May 16, 1999.

13 Until November 2003, Qatar, the UAE and Oman had not yet endorsed the International Convention on the Elimination of all Forms of Discrimination against Women, 1979. On January 13, 2004 the UAE's *Al Ittihad* daily newspaper reported that the UAE cabinet had agreed to join the Convention. See *Al Ittihad* daily No. 10418, January 13, 2004, 1.

14 See Badria Al-Awadhi, op. cit.

15 In July 2002, Bahrain ratified the International Convention on the Elimination of all Forms of Discrimination against Women, 1979.

16 See Badria Al-Awadhi, "Arab States and Adherence to the International Convention on the Elimination of all Forms of Discrimination Against Women (1979)," at the Second World Conference on the Role of Women in Cultural, Social and Economic Development (Kuwait, Women's Cultural Association, 2001).

17 Approximately 77 states are applying the "quota system" under their respective constitutions or election laws, including several Arab states like Sudan, Yemen, Morocco, Jordan, Egypt, Syria and Lebanon.

18 See Carolina Rodriguez Bello, *Women and Political Participation,* Women's Human Rights Net (November, 2003). The percentage of females in Parliaments of the United States, France, Japan, Uganda, Rwanda, Sweden, Denmark, Finland and The Netherlands are 14%, 11.8%, 10%, 24.7%, 48.8%, 45.3%, 38%, 37.5% and 36.7%, respectively.

19 Concerns about applying the principles of human rights instead of the Islamic Law provisions were among the findings of the questionnaire conducted by the Kuwaiti *Al Ra'y Al A'am* daily on June 22, 1999 after the issuing of the Emiri Decree which recognizes the political rights of Kuwaiti women. The questionnaire covered a random sample of 400 citizens of both sexes, of whom 31% were in favor and 17% responded "don't know," while 58% stated that Kuwaiti women are

not qualified to exercise their political rights as against 28% who thought that they were entitled to do so.

20 See paragraph (f) of Article (2) of the Convention on the Elimination of all Forms of Discrimination against Women (1979). Paragraph (g) is also calling member-states "to repeal all national penal provisions which constitute discrimination against women." Under paragraph (e) of the same Article, parties to the Convention are committed "to take all appropriate measures to eliminate discrimination against women by any person, organization or enterprise."

21 Pursuant to the Emiri declaration, the Emiri Decree No. 9 (1999) was issued on May 25, 1999, endorsing the full political rights of Kuwaiti women by amending Article (1) of Elections Law No. 35 (1962).

22 See Badria Al-Awadhi, "Women's Political Rights and the Principle of Equity," Post-Graduates Association seminar (Kuwait, 1996); and "Women's Political Right from Constitutional Perspective" (Kuwait, The National Council for Culture, Arts and Literature, 2003).

23 For further details on the Forum's final communiqué and recommendations, see *Al-Sharq Al-Awsat* daily No. 9168 (January 4, 2004), 19.

24 See *Al-Sharq Al-Awsat* daily of January 7, 2004, 4, and No. 9168, January 4, 2004, 19.

25 See Badria Al-Awadhi, "Arab States and Adherence to the International Convention on the Elimination of all Forms of Discrimination against Women (1979)," op. cit.

26 See Article (5) of Kuwait's Nationality Law, 1959, as amended in 1968.

27 See the study by Badria Al-Awadhi on "Women's Right to Terminate Contract of Marriage in Arab Personal Status Laws," Legal Forum on "Women and Law: Challenges of the Present and Prospects of Future" (Bahrain, 2001).

28 By 2003, only two states have passed personal status laws despite the importance of such a law in protecting women's rights during marriage and at the time of its dissolution.

29 See articles 153, 163-164, 174-177, and 195-197 of the Kuwaiti Penal Code.

30 See a study by Badria Al-Awadhi on "The International Labor Standards related to Working Women and their Application in Gulf States: A Comparative Study" (Qatar: Family Development Center, 1997).

31 See the Kuwaiti *Al Ra'y Al A'am* daily of December 11, 2003, 31.

32 See "Kuwait Curricula at the Moment of Truth," by Dr. Fawzi Ayoub, *Al Qabas* daily, January 5, 2004, Part 8, 36.

33 Ibid.

34 See paragraph (c) of Article (10) of the International Convention on the Elimination of all Forms of Discrimination against Women, 1979, ratified by Kuwait in 1994.

35 See Baqir Salman Al-Najar, "The Development of Civil Society in the GCC States," Gulf Research Center, The 2003 First Annual Report, *Al Khaleej* daily No. 1424, January 13, 2004, 16.

36 See a study by Badria Al-Awadhi on "The Kuwaiti Women and Challenges of the Upcoming Stage" at the seminar on the same topic (Kuwait, Center for Community Services, Kuwait University, 1999).

BIBLIOGRAPHY

A Wiser Peace. Post-Conflict Reconstruction Project Report. CSIS, January 2003 (http://www.csis.org/isp/wiserpeace.pdf).

Abdullah, Abdul Khaliq et al. *Al Mujtama'* al *Madani wa al Tahawol al Demoqrati fi al Emarat al Arabiya al Mutahida* [The Civil Society and Democratic Transformation in the UAE] (Cairo, Ibn Khaldoon Center for Developmental Studies, jointly with Dar al Ameen for Publication and Distribution, 1995).

Aburish, Said K. *Saddam Hussein: The Politics of Revenge* (New York, NY: Bloomsbury, 2000).

Al-Assri, Al-Mustafa. "Morocco Debates the Law of Family Code [*Muddawana*]," *Al Wasat* daily, no. 401 (Manama: October 21, 2003).

Al-Dakhil, Abdul Aziz. "Economic Globalization and the GCC States," in *Gulf States and Globalization* [in Arabic]. Development Forum, Twenty-First Annual Meeting, Dubai, February 3-4, 2000.

Al-Ghanim, Kaltham 'Ali. *Al Mara'a wa al Tanmiya fi al Mujtama' al Qatari* [Women and Development in Qatari Society], Symposium on "Women and Politics and their Role in Development," April 21-23, 2002, the Supreme Council for Family Affairs, Doha.

Al-Ghilani, Abdallah bin Muhammad. "Elements of Stability in the Gulf" [in Arabic] (http://www.alwatan.com/graphics/2001/mar/11.3/heads/ot6.htm).

Al-Hadidi, Mona. "Satellite Channels and the Arab World under Globalization," in *Globalization: Towards a Different Vision* [in Arabic] (Cairo: Department of Political Science, Cairo University, 2001).

Al-Hajji, Sa'ad. "Women's Social Societies in GCC States," documentary study (Kuwait, 2000).

Al-Hamad, Torki. "Globalization and the Search for a Definition" in *Gulf States and Globalization* [in Arabic]. Development Forum, Twenty-First Annual Meeting, Dubai, February 3-4, 2000.

Al-Ibrahim, Yusuf Hamad. "Reforming Production Flaws in the GCC States: An Economic-Political Approach" [in Arabic] in *Radical Reform: A Vision from Within*. Development Forum, Twenty Fifth Annual Meeting, Bahrain, January 4-6, 2004.

Al Ittihad, no. 10418, January 13, 2004.

Al-Khafaji, As'ad. "Islamic Globalization and American Change" [in Arabic], (http://www.elaph.com9090/elaph/arabic/frontendprocessArtica1). January 23, 2004.

Al-Khaleej Center for Strategic Studies. "Examining the Present and Future of Gulf Women." *Akhbar al Khaleej,* no. 8682 (Manama, Beirut: December 30, 2001).

Al-Mushaghiba, Amin and Shamlan Al-Issa. "Political Reform in the GCC States," in *Political Reform in the Arab World* [in Arabic] (Center for Developing Countries Studies and Research, Cairo University, May 3-4, 2004).

Al-Najjar, Baqir. *Al Mara'a fi al Khaleej al 'Arabi wa Tahwoulat al Hadatha al 'Aseera* [Women in the Arab Gulf and the Difficult Changes of Modernization] (Beirut: The Arab Cultural Center, 2000).

Al-Najjar, Baqir. *Sira'a al Ta'aleem wa al Mujtama'a fi al Khaleej al 'Arabi* [The Conflict between Education and Society in the Arab Gulf] (Beirut: Dar Al Saqi, 2003).

Al Ra'y Al A'am daily Kuwait, December 11, 2003.

Al Ra'y Al A'am daily, Kuwait, June 22, 1999.

Al-Rashidi, Ahmed. "The Phenomenon of Globalization and the Principle of National Sovereignty" [in Arabic] in Hassan Naf'i'a et. al., *Globalization: Issues and Concepts* [in Arabic] (Cairo: Department of Political Science, Cairo University, 2000).

Al-Rumaihi, Muhammad Ghanim. *Al Bahrain: Mushkilat al Taghyeer al Siyassi wa Alijtima'i* [Bahrain: Problems of Political and Social Change] (Kuwait: Kadhima Company for Publication, Translation and Distribution, 1984).

Al-Sabban, Rima. "Issues and Concerns of the Civil Society in the GCC States: Institutions, Regulations, Minorities" [in Arabic], in *Issues and Concerns of the Civil Society in the GCC States*. Development Forum, Nineteenth Annual Meeting, Dubai, February 19-20, 1998.

Al-Shamsi, Maytha.' "Gulf Women...Where to?" *Al Mustaqbal al 'Arabi,* Year 24, no. 273 (Beirut: November 2001).

Al-Sharq Al-Awsat daily no. 9168, January 4, 2004.

Al-Sharq Al-Awsat daily, January 7, 2004.

Al Siyasa. Kuwait, April 19, 2004.

Al-Tarah, Ali. "The Social Dimensions of Globalization and Their Impact on the Role of Gulf Women" [in Arabic]. *Journal of the Faculty of Arts*, vol. 60, no. 4 (Cairo: Faculty of Arts, Cairo University, October 2000).

Al-Turki, Thuraya and Huda Zraiq. *Taghayor al qiyam fi al a'aila al a'arabiyya* [Changing Values of Arab Family], serial studies on Arab

Women and Development 21, Economic and Social Commission for Western Asia (ESCWA) (New York, NY: United Nations, December 27, 1995).

Al-Awadhi, Badria. "The International Labor Standards related to Working Women and their Application in Gulf States: A Comparative Study" (Qatar: Family Development Center, 1997).

Al-Awadhi, Badria. "Arab States and Adherence to the International Convention on the Elimination of all Forms of Discrimination against Women, 1979." Second World Conference on the Role of Women in Cultural, Social and Economic Development. Women's Cultural Association, Kuwait, 2001.

Al-Awadhi, Badria. "The Kuwaiti Women and Challenges of the Upcoming Stage." Center for Community Services seminar, Kuwait University, Kuwait, 1999.

Al-Awadhi, Badria. "The Role of Gulf Women in Parliamentary and Political Life," in the seminar on "The Role of Women in Parliamentary and Political Life." State of Qatar, March 2003).

Al-Awadhi, Badria. "Women's Political Right from Constitutional Perspective." Kuwait, The National Council for Culture, Arts and Literature, 2003.

Al-Awadhi, Badria. "Women's Political Rights and the Principle of Equity." Post-Graduates Association seminar, Kuwait, 1996.

Al-Awadhi, Badria. "Women's Right to Terminate Contract of Marriage in Arab Personal Status Laws." Legal Forum on "Women and Law: Challenges of the Present and Prospects of Future." Bahrain, 2001.

Alexander, Justine and Colin Rowat. "A Clean Slate in Iraq, From Debt to Development." Middle East Research and Information and Research Project, *MERIP* no. 228 (Fall 2003): 32–36.

Al-Hadithi, Khalil Ismail. *Functionalism and Functional Approach in the Sphere of the Arab League* [in Arabic], *Strategic Studies Series*, no. 62 (Abu Dhabi: ECSSR, 2001).

Al-Haidari, Ibrahim Faseeh. *'Unwan al-Majd fi bayan ahwal Baghdad wal Basra wa Najd,* translated as *Aspects of Glory in the History of Baghdad, Basra and Najd* (Baghdad, n.d.; reprinted in London, 1999).

Al-Jabiri, Mohammed Abed in *al-Turath wa hadatha* (Beirut: al-Markaz al-Thaqafi, 1991).

Al-Jabiri, Mohammed Abed. *Arab Islamic Philosophy: A Contemporary Critique* (Austin, TX: Center for Middle Eastern Studies, 1999).

Al-Khalil, Samir. *The Republic of Fear: The Politics of Modern Iraq* (Berkeley, CA: University of California Press, 1989).

Al-Mani, Saleh. "Capabilities and Mechanisms of Enhancing Confidence in the Gulf Region on the Security and Military Levels." *Record of Current Events in the Gulf Region and the Arabian Peninsula and Their Geographical Neighborhood* [in Arabic], no. 21 (Kuwait: Center for Gulf and Arabian Peninsula Studies, January-March 2002).

Al-Najar, Baqir Salman. "The Development of Civil Society in the GCC States." Gulf Research Center, The 2003 First Annual Report, *Al Khaleej* daily No. 1424, January 13, 2004.

Al-Naqeeb, P. Khaldoun Hasan. *Society and State in the Gulf and Arab Peninsula* (London: Routledge, 1990).

Al-Rasheed, M. *A History of Saudi Arabia* (Cambridge: Cambridge University Press, 2000);.

Al-Rifa'i, Abdul-Jabar. *The Neo-Theology and Philosophy of Religion, The Intellectual Landscape in Iran*, translated *as Ilm al-Kalam al-Jadid wa falsafat al-deen, al-Mash-had al-thaqafi fi Iran* (Beirut: Dar al-Hadi, 2002).

Al-Shabibi, Sinan. "Prospects of Iraq's Economy," in *The Future of Iraq* (Middle East Institute, Washington DC, 1997).

Al-Sharq Al-Awsat, London, January 6, 2004.

Al-Turath wa Tahadiyyat al-asr fi al-Watan al-Arabi al-Asalah wa al-Mu'assarah (Beirut: Center for Arab Unity Studies, 1985),

Al-Yen, Ahmed. "Rights of Political Reform in the GCC States" in *Debate on the GCC and Regional and International Variables* [in Arabic] (Kuwait: Center for Strategic and Futuristic Studies, University of Kuwait, 2004).

Amuzegar, Jahangir. "Iran's Crumbling Revolution." *Foreign Affairs*, vol. 82, no.1 (January-February 2003): 44-57.

An-Na'im, Abdullahi. *Toward an Islamic Reformation* (Syracuse, NY: Syracuse University Press, 1990).

Ansari, Ali M. "Iran: Continuous Regime Change from Within." *Washington Quarterly*, vol. 24, no. 4 (Autumn 2003).

Anthony, John D. *Arab States of the Lower Gulf: People, Politics, Petroleum* (Middle East Institute, Washington DC, 1975).

Arab News, Kuwait, September 15, 2003.

Arab Strategic Report for 2002-2003 [in Arabic] (Cairo: Al Ahram Center for Political and Strategic Studies, 2003).

Arjomand, Said Amir. *From Nationalism to Revolutionary Iran* (New York, NY: State University of New York, 1984).

Axford, Barrie. *The Global System: Economics, Politics and Culture* (New York, NY: St. Martin's Press, 1995).

Ayoub, Fawzi. "Kuwait Curricula at the Moment of Truth." *Al Qabas* daily, January 5, 2004, Part 8, 36.

Bahgat, Gawdat. "The Gulf Monarchies, New Economic and Political Realities." The Research Institute for the Study of Conflict and Terrorism (RISCT) *Conflict Studies* no. 296, February 1997.

Barber, Benjamin R. *Fear's Empire: War, Terrorism and Democracy* (New York, NY: W.W. Norton and Company, 2003)

Bearden, Milt. "Iraqi Insurgents Take a Page from the Afghan 'Freedom Fighters.'" *New York Times*, November 9, 2003.

Bello, Carolina Rodriguez. *Women and Political Participation*. Women's Human Rights Net, November 2003.

Ben Sahel, Nora and Daniel Byman (eds.). *The Future Security Environment in the Middle East: Conflict, Stability, and Political Change* (Santa Monica, CA: RAND, 2004).

Berger, Samuel R. "Power and Authority: America's Path Ahead." Brookings Institution Leadership Forum, June 17, 2003.

Bishara, Abdullah. "Globalization and State Sovereignty: The Case of the Gulf," in *Globalization and Its Impact on Society and State* [in Arabic] (Abu Dhabi: ECSSR, 2002).

Bull, Hedley. *The Anarchical Society: Order in World Politics* (New York, NY: Columbia University Press, 1977).

Chaudry, Kiren Aziz. "Economic Liberalization and the Lineages of the Rentier State." *Comparative Politics* (October 1994).

Chubin, Shahram and Robert S. Litwack, "Debating Iran's Nuclear Aspirations." *Washington Quarterly*, vol. 24, no. 4 (Autumn 2003).

Cordesman, Anthony H. and Abraham R. Wagner. *The Lessons of Modern War* (Boulder, CO: Westview Press, 1990).

Cordesman, Anthony H. "Developments in Iraq at the End of 2003: Adapting US Policy to Stay the Course." CSIS, December 29, 2003.

Cordesman, Anthony H. *Saudi Arabia Enters the Twenty-First Century, The Political, Foreign Policy, Economic and Energy Dimensions* (Westport, CT and London: Praeger, 2003).

Davidson, Herbert A. *Alfarabi, Avicenna and Averroes on Intellect: Their Cosmologies, Theories of the Active Intellect, and Theories of Human Intellect* (New York, NY: Oxford University Press, 1992).

Dodge, Toby and Steven Simon (eds). *Iraq at the Cross-Roads* (London: IISS, January, 2003).

Dodge, Toby. "Consequences and Implications of US Military Intervention and Regime Change in Iraq." IISS Workshop "Intervention in the Gulf" at Gulf Research Center, Dubai, February 15–17, 2003.

Doran, Michael Scott. "The Saudi Paradox." *Foreign Affairs* vol. 83, no.1 (January/February 2004).

Ebtehaj, Abolhassan. *Memoirs* (Tehran: Elmi Publications, 1993, second volume), 525–526.

Economic and Social Commission for Western Asia (ESCWA). "Partnership in the Arab Family." Serial studies on Arab Women and Development 31 (New York, NY: United Nations, 2001).

Ehteshami, Anoushiravan. "Iran-Iraq Relations after Saddam." *Washington Quarterly*, vol. 26, no. 4 (Autumn 2003): 115-129.

Eisenstadt, Michael and Eric Mathewson (eds). *US Policy in Post Saddam Iraq, Lessons from the British Experience* (Washington DC: The Washington Institute for Near Eastern Policy, 2003).

Eisenstadt, Michael. "Like a Phoenix from the Ashes? The Future of Iraqi Military Power." The Washington Institute for Near East Policy, Policy Papers no. 36 (1993).

ESCWA. Report by the *Arab Women's Center 2003* (New York, NY: United Nations, 2003).

Esposito, John. *Unholy War, Terror in the Name of Islam* (Oxford: Oxford University Press, 2002).

Farouk-Sluglett, Marion and Peter Sluglett. *Iraq Since 1958* (London: I.B. Tauris, 1987, 1990, revised edition 2001).

Fuller, Graham. *The Center of the Universe* (Boulder, CO: Westview Press, 1991).

Gause, F. Gregory. *Oil Monarchies: Domestic and Security Challenges in the Arab Gulf States* (New York, NY: Council on Foreign Relations Press, 1994).

General Department of Legal Affairs, Ministry of Interior, Kuwait. "National Assembly Elections Results, 2003." July 2003.

Ghubash, Muhammad. "The Democratic March in the Arab Gulf and the Arab World and Future Horizons," debate on *Current International Changes and Their Impact on the Future of the Gulf* [in Arabic], The National Council for Culture, Art and Literature, Kuwait, January 5-7, 2003.

Ghubash, Muhammad. "The Gulf State: A More Than Absolute Authority and a Less Than Capable Society," in *Radical Reform: A Vision from Within.* Development Forum, Twenty Fifth Annual Meeting, Bahrain, January 4-6, 2004.

Gindeal, Amani. "Activating the Role of Civil Public Institutions in the GCC States." *Debate on the Reality and Future of the Organizations of Civil Society in the GCC States* (Kuwait: Center for Gulf and Arabian Peninsula Studies, University of Kuwait, April 3-4, 2000).

Glanz, James. "Rebuilding Iraq Takes Courage, Cash and Improvisation." *New York Times*, November 30, 2003.

Global Governance Initiative, Annual Report 2004, World Economic Forum. Available at http://www.weforum.org/

"Globalization: Threat or Opportunity?" *IMF Issues Brief,* 2000-2003.

Gompert, David C., Jerrold D. Green, F. Stephen Larrabee. "How an Atlantic Partnership could Stabilize the Middle East." *RAND Review* (Spring 1999).

Griffiths M. and T. O'Callaghan. *International Relations: The Key Concepts* (London and New York: Routledge, 2002).

Gulf Strategic Report 2003-2004 [in Arabic] (Sharjah: Studies Unit, Dar al Khaleej for Press, Printing and Publication, 2004).

Habermas, Juergen. *The Philosophical Discourse of Modernity: Twelve Lectures* (Cambridge, MA: MIT Press, 1987).

Habib, Kazim. "Globalization and Arab World Fears," Ibn-Rushd Forum for Free Thinking, no. 2, Summer 2001 (http://www.ibn-rushd.org/forum/GlobalA.html.)

Hadiya, Abdallah. *Political Middle Eastern Issues* [in Arabic] (Cairo, Al-Tobji Commercial Press Co., 1998).

Halliday, Fred. *Two Hours That Shook the World* (London: Saqi Books, 2002).

Harrison, Lawrence (ed.) "Culture Matters." Findings of research project (Forthcoming, Routledge and Oxford University Press, 2005).

Harrison, Lawrence E. and Samuel P. Huntington (eds.) *Culture Matters: How Values Shape Human Progress* (New York, NY: Basic Books, 2000).

Hawley, D. *The Trucial States* (London: Allen and Unwin, 1970).

Hefner, Robert W. *Civil Islam: Muslims and Democratization in Indonesia* (Princeton, NJ: Princeton University Press, 2000).

Held, David. *Democracy and Global Order: From the Modern State to Cosmopolitan Governance* (Stanford, CA: Stanford University Press, 1995).

Higgott, Richard. "Globalization and Regionalization: Two New Trends in World Politics." *Emirates Lecture Series*, no. 25 [in Arabic] (Abu Dhabi: ECSSR, 1995).

Holt, Paul (ed.) *The Democratic Peace and International Conflict in the Twentieth Century* (Cambridge: Cambridge University Press, 2002).

Human Security Report. Edited by Human Security Centre (Oxford University Press, 2004).

Hunter, Shireen. "Iran's Pragmatic Regional Policy." *Journal of International Affairs*, vol. 56, no. 2 (Spring 2003):133-147.

Ibrahim, Fouad. "Civil Society in the Gulf: Hopes and Roles" [in Arabic], unpublished research paper.

ICG Middle East Briefing. *The Challenge of Political Reform in Egypt: After the Iraq War*, September 30, 2003.

ICG Middle East Briefing. *The Challenge of Political Reform: Jordanian Democratisation and Regional Instability*, October 8, 2003.

ICG Middle East Report No 2. "Middle East Endgame I: Getting To A Comprehensive Arab-Israeli Peace Settlement." July 16, 2002.

ICG Middle East Report No 3. "Middle East Endgame II: How a Comprehensive Israeli-Palestinian Peace Settlement Would Look." July 16, 2002.

ICG Middle East Report No 4. "Middle East Endgame III: Israel, Syria and Lebanon – How Comprehensive Peace Settlements Would Look." July 16, 2002.

ICG Middle East Report No. 11, "War in Iraq: Political Challenges after Conflict." March 25, 2003.

ICG Middle East Report No.17. "Governing Iraq," August 25, 2003, 12.

ICG Middle East Report No. 19. "Iraq's Constitutional Challenge." November 13, 2003.

ICG Middle East Report No. 20. "Iraq: Building a New Security Structure." December 23, 2003.

Ignatieff, Michael. "Why are we in Iraq? (And Liberia? And Afghanistan?)." *New York Times,* September 7, 2003.

Implementation of the United Nations Millennium Declaration, Report of the Secretary-General, September 8, 2003.

International Atomic Energy Agency. Report by the Director General of the IAEA Board of Governors, "Implementation of NPT Safeguards Agreement in the Islamic Republic of Iran." GOV/2003/75, November 10, 2003.

International Crisis Group (ICG). website www.crisisweb.org .

International Institute for Strategic Studies. "Dealing with the Axis of Evil." IISS vol. 8, no. 5 (June 2002).

International Institute for Strategic Studies. "The Bush National Security Strategy." IISS, vol. 8, no. 8 (October 2002).

International Institute for Strategic Studies. "US Policy Towards the Middle East." IISS vol. 9, no. 1 (January 2003).

"Iran's Nuclear Calculations." *World Policy Journal*, vol 20, no. 2 (Summer 2003/Special edition).

Ja'afar, Yasameen bint Ahmed. "Health Care for Women in the Sultanate of Oman." Advisory Symposium on Preparing a National Strategy for Omani Women, March 3-5, 2002, Committee on Coordination of Women's Voluntary Works, in cooperation with the Ministry of Social Development, Sultanate of Oman.

Jabar, Faleh A. "Iraq: Difficulties and Dangers of Regime Removal." Middle East Research and Information Project, *MERIP* no. 225 (Winter 2002): 18–19.

Jabar, Faleh A. "The Gulf War and Ideologies, The Double-Edged Sword of Islam," in H. Bresheeth and Y. Y-Davis (eds) *The Gulf War and the New World Order* (London: ZED publishers, 1991).

Jabar, Faleh A. *Al-Dawla wal-Mujtama' al-Madani, wal Tatawur al-Demoqratif fil Iraq*, translated as "State, Civil Society and Democratic Perspectives in Iraq" (Cairo: Ibn Khaldoun Centre, 1995).

Jabar, Faleh A. *The Shi'ite Movement in Iraq* (London: Saqi Books, 2003).

Jabar, Faleh.A. "The Roots of an Adventure," in V. Britain (ed.) *The Gulf Between Us* (London: Verago, 1991).

Jacques, Isabelle. "Political, Economic and Social Reform in the Arab World," in *Post-Iraq: What Implications for Middle East Arab Security?* Conference Report on Wilton Park Conferences, March 31-April 4, 2003.

Jennings, Ray Salvatore. "The Road Ahead, Lessons in Nation Building from Japan, Germany, and Afghanistan for Postwar Iraq." US Institute for Peace, Peaceworks no. 49 (May 2003).

Jiyad, Ahmed. "An Economy in Debt Trap, 1980–2020." *Arab Studies Quarterly*, vol. 23, no. 4 (Winter 2001).

Juergensmeyer, Mark. *Terror in the Mind of God* (Berkeley and Los Angeles, CA: University of California Press, 2000), 6–7.

Kant, Immanuel. "*Zum ewigen Frieden,*" in Zwi Batscha and Richard Saage (eds) *Friedensutopien: Kant, Fichte, Schlegel, Görres* (Frankfurt: Suhrkamp, 1979).

Katouzian, Homa. *A Political Economy of Modern Iran* (London: I. B. Tauris, 1985) 24–65.

Katzenstein, Peter. *Between Power and Plenty* (Wisconsin, WI: Wisconsin University Press, 1978).

Katzman, Kenneth. "Iraq: Regime Change Efforts and Post-War Governance." Congressional Research Service (CRS) Reports for Congress, October 28, 2003.

Keddi, Nikki. *Iran and the Muslim World: Resistance and Revolution* (New York, NY: New York University Press, 1995).

Kemp, Geoffrey (ed.). *Powder Keg in the Middle East: The Struggle for Gulf Security* (London: Rowman & Littlefield, 1995).

Kessler, Glenn. "US Decision on Iraq has Puzzling Past." *Washington Post*, January 12, 2002, A10.

Khadduri, Majeed and Edmund Gharib. *War in the Gulf, 1990-1991, The Iraq-Kuwait Conflict and its Implications* (Oxford: Oxford University Press, 1997), 6–20.

Khaliq, Goda Abdul. "Globalization and Political Economy for the Nation-State," in Hassan Naf'i'a et. al., *Globalization: Issues and Concepts* [in Arabic] (Cairo: Department of Political Science, Cairo University, 2000).

Khosrokhavar, Farhad. "Iran: The Fear of Encirclement" (unpublished, 2003).

Khosrokhavar, Farhad. "La politique étrangere en Iran: de la révolution à l'"axe du Mal," *Politique Etrangere* (1/2003): 77-91.

Kinzer, Stephen. *All the Shah's Men: An American Coup and the Roots of Middle East Terror* (Hoboken, NJ: John Wiley & Sons, 2003).

Krayim, Hassan. *Ta'atheer al Siyasat al Ijtima'iya ala al Ussra* [The Impact of Social Policies on the Family], the Arab meeting for the tenth evaluation of the International Year of Family, ESCWA, Beirut, October 7-9, 2003.

Leffler, Melvyn P. "9/11 and the Past and Future of American Foreign Policy," *International Affairs* no.79, 5 (2003): 1045–1063.

Lewis, Bernard. *What Went Wrong? Western Impact and Middle Eastern Response* (Oxford, New York, NY: Oxford University Press, 2002).

Lienhardt, Peter. *Shaikhdoms of Eastern Arabia*, edited by Ahmed Al-Shahi (Oxford: St. Anthony's College, 2001).

Longrigg, Stephen Hemsley. *Iraq 1900-1950: A Political, Social and Economic History* (Oxford: Oxford University Press, 1953).

Looney, Robert. "The Neo-Liberal Model's Planner Role in Iraq's Economic Transition." *The Middle East Journal* vol. 57, no. 4 (Autumn 2003): 568–586.

MacFarquhar, Neil. "Open War Over, Iraqis Focus on Crime and a Hunt for Jobs." *New York Times*, September 16, 2003.

Marr, Phebe. *The Modern History of Iraq* (Boulder, CO: Westview Press, 2004).

Mathews, Jessica. "Iraqis Can do More." *Washington Post*, September 29, 2003.

Mazari, Shireen. "The New US Doctrine: Implications for the South Asian Region." *Strategic Studies* (Islamabad: Spring 2003).

McCain, John. "How to Win in Iraq." *Washington Post*, November 9, 2003.

Ministry of Planning, Kuwait. "The Status of Population and Labor Force in Kuwait in the Years 1994-2002" [in Arabic], May 2003.

Morsi, Mustafa Abdul Aziz. "Economic Integration and Neo-Functionalism: A Theoretical Approach with Reference to the Gulf Experience" [in Arabic] *Annals of Arts and Social Sciences*, Thesis no. 201, Annals 27 (Kuwait: University of Kuwait, 2003).

Muhyo, Saad. "Greater Middle East or a Grand Maneuver? *Al Wasat*, no. 630, February 23, 2004.

Mundy III, Marine Lt. Col. Carl E. "Spare the Rod, Save the Nation." *New York Times*, December 30, 2003.

Naim, Moses. YaleGlobal Online, December 29, 2004 (http://yaleglobal.yale.edu.).

Nakash, Y. *The Shi'is of Iraq* (Princeton, NJ: Princeton University Press, 1995), 25–30.

Nassrawi, A. *The Economy of Iraq: Oil, Wars, Destruction of Development and Prospects, 1950-2010* (Greenwood Publishing Group, 1994).

Nili, Massod et al. *Iran's Industrial Strategy* (Tehran: Center for Advanced Research and Planning of Development, Farsi edition, 2003), 65–121.

Nimir, Suleiman. "Saudi Arabia and Egypt Refuse." *Al Hayat*, no. 14943, February 25, 2004.

Nyrop, Richard F. (ed.) *Saudi Arabia: A Country Study* (Washington DA, 1984).

Ochmaneck, David. "A Possible US-led Campaign against Iraq: Key Factors and an Assessment," in Toby Dodge and Steven Simon (eds), *Iraq at the Cross-Roads* (London: IISS, January, 2003).

Orentlicher, Diane. "International Justice Can Indeed Be Local." *Washington Post*, December 21, 2003.

Packer, George. "War after the War: What Washington Does Not See in Iraq." *The New Yorker*, November 29, 2003.

"Pakistan Admits Possible Nuke Ties with Iran." *Los Angeles Times*, December 24, 2003, A2.

Preparatory Constitutional Committee's report to the Governing Council in September 2003, report in *Washington Post*, November 13, 15 and 26, 2003.

Qadhaya Islamiya Mu'asira Quarterly (*Modern Islamic Problems*). Year 7, no. 22 (Winter 2003).

Rafique, Najam. "9/11, A Year After: Appraising US Regional Strategies." *Strategic Studies* (Islamabad: Autumn 2002).

Rahman, Fazlur. *Islam and Modernity: Transformation of an Intellectual Tradition* (Chicago, IL: University of Chicago Press, 1982).

Ramazani, Rouhollah K. *Revolutionary Iran: Challenge and Response in the Middle East* (Baltimore, MD: John Hopkins University Press, 1986).

Rathmell Andrew et al. *A New Persian Gulf Security System* (Santa Monica, CA: Rand, 2003).

Renshon, Stanley A. (ed.) *The Political Psychology of the Gulf War: Leaders, Publics, and the Process of Conflict* (Pittsburgh, PA: University of Pittsburgh Press, 1993).

Report of the UN Secretary-General. *The Empowerment of Women and Integration of Gender Perspectives in the Promotion of Economic Growth, Poverty Eradication and Sustainable Development.* UN General Assembly, July 2003.

Republic of Iraq, Ministry of Planning, *Annual Abstract of Statistics* (Baghdad, 1992).

Robertson, Roland. *Globalization: Social Theory and Global Culture* (London: Sage, 1992 and 1998).

Rohwer, Jim. *Asia Rising: Why America Will Prosper as Asia's Economies Prosper* (New York, NY: Simon and Schuster, 1995), 207–278.

Rosecrance, Richard N. *The Rise of the Trading State: Commerce and Conquest in the Modern World* (New York, NY: Basic Books, 1986).

Russett, Bruce. *Grasping Democratic Peace: Principles for a Post-Cold War World* (Princeton, NJ: Princeton University Press, 1993).

Sakai, Keiko. "Economic Liberalization in Iraq: Its Political Application." Paper submitted to the Middle East Workshop, *Institute of Developing Economies*, Tokyo, January 9, 1996.

Sariolghalam, Mahmood. "Arab-Iranian Rapprochement: Regional and International Impediments," in Khair-din Haseeb (ed.) *Arab-Iranian Relations*, (Beirut: Centre for Arab Unity Studies, 1998), 425–427.

Sariolghalam, Mahmood. "The Future of the Middle East: The Impact of the Northern Tier." *Security Dialogue* vol. 27, no. 3 (September 1996): 312–315.

Sariolghalam, Mahmood. "Understanding Iran: Getting Past Stereotypes and Mythology." *Washington Quarterly* vol. 26, no. 4 (Autumn 2003): 73–78.

Sariolghalam, Mahmood. *Globalization and the Issue of Sovereignty of the Islamic Republic of Iran* (Tehran: Center for Strategic Studies, 2004), 63–75.

Sariolghalam, Mahmood. *Rationality and the Future of Iran's Development* (Tehran: Center for Scientific Research and Middle East Strategic Studies, 2004, third edition, in Farsi), 9–29.

See Halloran, William F. "Zayed University: A New Model for Higher Education in the United Arab Emirates," in *Education and the Arab World: Challenges of the Next Millennium* (Abu Dhabi: The Emirates Center for Strategic Studies and Research, 1999).

Shahrur, Mohammed. *al-Kitab wa al-Qur'an: Qira'a Mu'asira* (Beirut: Sharikat al-Matbu'at, sixth edition, 2000).

Shihab, Ahmed. "Openness to Build a Contemporary State" [in Arabic] (http://www.alhalem.net/magalat/almosaraha.htm).

Shokri, Yadollah. *A History of Safavids* (Tehran: Ettellat Publications, second edition, in Farsi, 1985), 3–17.

Sick, Gary G. and Lawrence G. Potter (eds). *The Persian Gulf at the Millennium: Essays in Politics, Economy, Security, and Religion* (New York, NY: St. Martin's Press, 1997).

Sick, Gary. "Iran: Confronting Terrorism." *Washington Quarterly*, vol. 26, no. 4 (Autumn 2003): 83-98.

Simons, Geoff. *Future Iraq, US Policy in Reshaping the Middle East* (London: Saqi Books, 2003).

Sipress, Alan. "For Many Iraqis, US-backed TV Echoes the Voice of its Sponsor; Station Staffers Acknowledge Their Reluctance to Criticize." *Washington Post*, January 8, 2004.

Stowasser, Barbara F. *Women in the Qur'an: Traditions and Interpretations* (New York, NY: Oxford University Press, 1994).

Tabatabaee, Seyed Javad. *An Introduction to a History of Iran's Deterioration* (Tehran: Zaman Moaser Publications, 2002), 112–146.

Takeyh, Ray. "Iranian Options: Pragmatic Mullahs and America's Interests." *The National Interest* (Fall 2003): 49-56.

Talbott, Strobe. "War in Iraq, Revolution in America." *International Affairs* no.79, 5 (2003): 1037–1043.

Teitelbaum, Joshua. "Holier Than Thou, Saudi Arabia's Islamic Opposition." Policy Papers no. 52 (Washington, DC: The Washington Institute for Near Eastern Policy, 2000).

The GCC in 2002 [in Arabic] (Kuwait: Gulf Center for Strategic Studies, May 2002).

The Globalization Website Glossary (www.emory.edu/soc/globalization/glossary)

The Gulf in 2003 (Dubai: Gulf Research Center, 2004).

The Kurdistan Democratic Party's daily *Al-Ta'akhi*, Baghdad, June 16, 2003.

The Military Balance, 2003-2004. International Institute for Strategic Studies (Oxford: Oxford University Press, 2003), 121.

"The National Security Strategy of the United States of America." September 17, 2002. (www.whitehouse.gove/nsc/nssall.html).

The Responsibility to Protect, Report of the International Commission on Intervention and State Sovereignty. International Development Research Centre, Ottawa, Canada, December 2001. Available at http://dfait-maeci.gc.iciss-ciise/

Tibi, Bassam. "Democracy and Democratization in Islam," in Michéle Schmiegelow (ed.) *Democracy in Asia* (New York, NY: St. Martin's Press, 1997).

Tibi, Bassam. "International Morality and Cross-Cultural Bridging," in Roman Herzog et al., *Preventing the Clash of Civilizations: A Peace Strategy for the Twenty-First Century* edited by Henrik Schmiegelow (New York, NY: St. Martin's Press, 1999).

Tibi, Bassam. "The Failed Export of the Islamic Revolution into the Arab World," in Frédéric Grare (ed.), *Islamism and Security: Political Islam and the Western World* (Geneva: Program for Strategic and International Security Studies, 1999).

Tibi, Bassam. "The Fundamentalist Challenge in the Middle East." *Fletcher Forum of World Affairs*, vol. 23, no.1 (Winter/Spring, 1999).

Tibi, Bassam. "The Human Tradition of Human Rights and the Culture of Islam," in Francis M. Deng (ed.) *Human Rights in Africa: Cross-Cultural Perspectives* (Washington, DC: The Brookings Institution, 1990).

Tibi, Bassam. "The Interplay between Social and Cultural Change: The Case of Germany and the Middle East," in Ibrahim M. Oweiss and George N. Atiyeh (eds) *Arab Civilization: Challenges and Responses: Studies in Honor of Constantine K. Zurayk* (Albany, NY: State University of New York Press, 1988).

Tibi, Bassam. *Conflict and War in the Middle East: From Interstate War to New Security* (London: Macmillan Press, second edition 1998).

Tibi, Bassam. *Der neue Totalitarismus: "Heiliger Krieg" und Westliche Sicherheit* (Darmstadt: Primus, 2004).

Tibi, Bassam. *Islam: Between Culture and Politics* (New York, NY: Palgrave, 2001),

Tibi, Bassam. *The Challenge of Fundamentalism: Political Islam and the New World Disorder* (Berkeley, CA: University of California Press, 1998 and 2002).

Vassiliev, A. *The History of Saudi Arabia* (London: Saqi Books, 2000).

Walzer Michael (ed.) and (trans.) *Al-Farabi on the Perfect State.* Translation of Abu Nasr al-Farabi, *Mabadi ara' ahl al-Madina al-Fadila* (Oxford: Oxford University Press, 1985).

Washington Post, January 4, 2004.

Waterbury, John. *The Commander of the Faithful: The Moroccan Political Elite—A Study in Segmented Politics* (New York, NY: Colombia University Press, 1970), 5.

Watt, William Montgomery. *Islamic Philosophy and Theology: An Extended Survey* (Edinburgh: University Press, 1962, second edition 1985).

Wong, Edward. “Attacks and Security put Karbala Residents on Edge.” *New York Times*, December 29, 2003.

Wright, Lawrence. “The Kingdom of Silence.” *The New Yorker*, January 5, 2004.

Wright, Robin and Rajiv Chandrasekaran. “Power Transfer in Iraq Starts This Week.”

Yadfat, Aryeh and Mordechai Abir. *In the Direction of the Gulf: The Soviet Union and the Persian Gulf* (London: Frank Cass, 1977).

Yousuf, Ahmed. “Globalization and the Arab Regional Order,” in Hassan Naf'i’a et. al., *Globalization: Issues and Concepts* [in Arabic] (Cairo: Department of Political Science, Cairo University, 2000).

Zanoyan, Vahan. *Time for Making Historic Decisions in the Middle East* (Kuwait Center for Strategic and Future Studies, Kuwait University, 2002).

Index